2025 ANNUAL
INTERIOR DETAIL

foreword

〈2025 애뉴얼 인테리어 디테일〉이 발행되었습니다. 사진 위주의 작품집에 상세 도면을 함께 수록하여 실무에 접근성을 더하였습니다. 매년 1회씩 발행되는 본 연감은 54개의 작품을 레스토랑 / 카페 / 병원 / 종교 / 플래그십 스토어 / 상업 / 오피스 / 교육 / 전시 / 쇼룸 / 뷰티로 분류하여 설계시 초기 상품 구상에 편리하도록 최선을 다하였습니다. 인테리어 트랜드는 매년 빠른 주기로 변화하므로 당 해년도의 작품을 도면과 함께 정리한 본 연감은 최근 작품을 한 눈에 비교할 수 있는 좋은 자료가 되리라 생각됩니다. 이러한 기획하에 발행되는 연감이 실무에 많은 참고가 되기를 바라며, 끝으로 자료 협조에 노력을 아끼지 않으신 작가 및 실무진 여러분들께 감사드립니다.

Contents _2025 ANNUAL INTERIOR DETAIL 41

2025 ANNUAL INTERIOR DETAIL 41

발행	아키랩
등록	제2014-000167호
발행인	조배연
편집	BOB 매거진_ 이윤형 기자 (2021annual@gmail.com)
디자인	BOB 매거진_ 신민기 디렉터 (dbenfdus@naver.com)
주소	서울특별시 서초구 양재천로 13길 18(양재동)
전화	02-579-7747
이메일	1979anc@naver.com

ⓒ 아키랩

한국간행물 윤리위원회의 윤리강령 및 실천요강을 준수합니다.
본지에 게재된 내용을 사전허가 없이 무단 복제 및 전제를 금합니다.

정가 92,000원

RESTAURANT 레스토랑

- 010 **TEST KITCHEN** 테스트 키친
- 020 **IMOK SMOKE DINING** 이목 스모크 다이닝
- 028 **OBALTAN SAMSUNG BRANCH** 오발탄 삼성점
- 036 **CHEESEWAVE** 치즈웨이브
- 042 **STINKY BACON TRUCK** 스팅키 베이컨 트럭
- 050 **DEOKBOON KOREAN DINING & GRILL** 덕분 한식당
- 066 **RAINBOW BRIDGE GWANGGYO BRANCH** 레인보우 브릿지 광교점
- 078 **HAENAM CHEONILGWAN** 해남천일관
- 088 **VIKEN HUS** 비킹후스
- 096 **VARIEGATA GROTTA** 바리에가타 그로타

CAFE 카페

- 104 **SETT** 세트
- 118 **KNOTTED GIMPO** 노티드 김포
- 126 **INC COFFEE** 인크커피
- 134 **SECOND ONE LAKE ASAN** 세컨드원 레이크 아산
- 148 **OHVENU HANNAM** 오베뉴 한남
- 154 **LES MAINS DORÉES** 레망도레
- 164 **AISO SOUND** 아이소 사운드
- 180 **HORONG** 호롱
- 188 **FILLMATE** 필메이트
- 206 **SOOSOO COFFEE** 수수커피

CLINIC · RELIGION 병원 · 종교

- 216 **OGANACELL DERMATOLOGY CLINIC** 오가나셀 피부과 의원 잠실점
- 224 **CELLIN CLINIC** 셀린의원 홍대점
- 236 **JEILSOMANG CHURCH** 제일소망교회

FLAGSHIP STORE 플래그십 스토어

- 254 **INSILENCE SEONGSU** 인사일런스 성수
- 260 **PYUNKANG YUL FLAGSHIP STORE** 편강 율 플래그십 스토어
- 268 **MEDICUBE FLAGSHIP STORE HONGDAE** 메디큐브 플래그십 스토어 홍대
- 274 **JAVIN DE SEOUL** 자빈드서울

Contents _2025 ANNUAL INTERIOR DETAIL 42

2025 ANNUAL INTERIOR DETAIL 42

Publication	ARCHI-LAB CO.
Registration	2014-000167
Publisher	Baeyeon Cho
Edit	BOB Magazine_ Yooonhyung Lee (2021annual@gmail.com)
Design	BOB Magazine_ Minki Shin (dbenfdus@naver.com)
Address	18, Yangjaecheon-ro 13-gil, Seocho-gu, Seoul, Republic of Korea
Tel	+82-2-579-7747
E-mail	1979anc@naver.com

Copyright © 2024 by ARCHI-LAB Co. and may not be reproduced in any manner or from without permission

The exclusive distributorship in Taiwan is offered to ArchiHeart Corporation. Any infringement shall be subject to penalties.

Price $92

COMMERCE 상업

- 010 **EQL GROVE** EQL 숲
- 024 **LENSME** 렌즈미
- 034 **WOOALONG STARFIELD SUWON** 우알롱 스타필드 수원
- 042 **WILSON** 윌슨
- 050 **MONAMI 153 MANSION** 모나미 153 맨션
- 058 **LOTTE CINEMA SUWON HALL RENEWAL** 롯데시네마 수원관 홀 리뉴얼
- 068 **GOURMET STREET STARFIELD SUWON** 고메스트리트 스타필드 수원
- 076 **EATOPIA STARFIELD SUWON** 잇토피아 스타필드 수원
- 092 **SMART SCORE STORE DAEGU** 스마트스코어 스토어 대구점
- 104 **THE LU'PIUM HOUSE** 루피움 하우스
- 112 **TAP SHOP BAR DOSAN BRANCH** 탭샵바 도산대로점
- 122 **SAPPUN** 사뿐
- 130 **INSTANTFUNK SHINSEGAE GANGNAM** 인스턴트펑크 신세계 강남점

OFFICE 오피스

- 138 **BRIGHTEN LOUNGE** 브라이튼 라운지
- 148 **JOBIS & VILLAINS3.3** 자비스앤빌리언즈 삼쩜삼
- 158 **SUPERBIN OFFICE** 수퍼빈 오피스
- 168 **DESIGN TOKEN OFFICE** 디자인토큰 오피스
- 176 **STUDIO CURIOSITY** 호기심 스튜디오

EDUCATION · EXHIBITION 교육 · 전시

- 186 **SeSAC : SEOUL SOFTWARE ACADEMY** 청년취업사관학교
- 198 **SEOUL WOMEN'S COLLEGE OF NURSING_ A NURSING PRACTICE ROOM** 서울여자간호대학교 간호실습실
- 204 **SEOUL WOMEN'S COLLEGE OF NURSING_ GAONNURI LOUNGE** 서울여자간호대학교 가온누리
- 210 **NEW:SPACE** 뉴스페이스

SHOWROOM · BEAUTY 쇼룸 · 뷰티

- 226 **HANSSEM STARFIELD SUWON** 한샘 스타필드 수원
- 242 **SIDIZ THE PROGRESSIVE HAPJEONG** 시디즈 더 프로그레시브 합정
- 256 **STUDIO TLE** 스튜디오 틀
- 266 **IMTYPE SEOUL** 아임타입 서울
- 276 **GUNGSEOCHAE HAIR PANGYO** 궁서채 헤어 판교

2025 ANNUAL

RESTAURANT

INTERI

DETAI

010	**TEST KITCHEN** 테스트 키친
020	**IMOK SMOKE DINING** 이목 스모크 다이닝
028	**OBALTAN SAMSUNG BRANCH** 오발탄 삼성점
036	**CHEESEWAVE** 치즈웨이브
042	**STINKY BACON TRUCK** 스팅키 베이컨 트럭
050	**DEOKBOON KOREAN DINING & GRILL** 덕분 한식당
066	**RAINBOW BRIDGE GWANGGYO BRANCH** 레인보우 브릿지 광교점
078	**HAENAM CHEONILGWAN** 해남천일관
088	**VIKEN HUS** 비킹후스
096	**VARIEGATA GROTTA** 바리에가타 그로타

TEST KITCHEN

HEDURBAN STUDIO | Bumgyu Kim

테스트 키친은 외식기업에서 메뉴 개발, 메뉴 퀄리티 컨트롤, 신규 브랜드 인큐베이팅과 같은 목적을 가지고 운영되는 셰프들의 실험실과 같은 곳을 말한다. 헤드어반은 외식기업의 대가 SG 다인힐의 테스트 키친 설계 및 시공을 맡아 전에 없던 새로운 콘셉트의 공간을 디자인했다. 테스트 키친의 공간 키워드는 'Scent Cruise'로 'Scent'(풍기다, 향을 맡다)와 'Cruise'(항해하다)의 결합어로 음식으로 온전한 휴식을 취할 수 있는 여행을 만든다는 의미를 내포하고 있다. 감각을 깨우는 여행의 공간, 바다를 향해 하는 크루즈 같은 공간. 이것이 헤드어반이 만들어내고 싶었던 테스트 키친의 모습이다. 마치 유럽의 크루즈에 탑승한 것만 같은 이국적인 무드와 동시에 무게감이 느껴지는 주출입구, 그리고 주방과 연결된 짧지만 강렬한 복도를 지나면 메인 다이닝 공간이 펼쳐진다. 크루즈라는 주제에 맞게 클래식한 샹들리에와 빈티지한 콘셉트의 가구, 답한 우드톤 마감재를 활용해 공간을 세련되게 풀어 냈다. 또한 '테스트 키친'이라는 공간의 목적성에 맞게 다이닝과 주방 간 포켓도어를 설치해 소통의 용이성을 높였다.

A test kitchen refers to a place such as a chef's laboratory operated by a food service company for the purpose of developing menus, controlling menu quality, and incubating new brands. HED URBAN is in charge of designing and constructing a test kitchen for SGDineHill, a master of food service enterprises, and has designed a new concept space that has never been before. The test kitchen's space keyword is 'Scent Cruise', combination of 'Scent' (to smell) and 'Cruise' (to sail), which means creating a trip where you can relax completely with food. A space for sensational travel, a space like a cruise in the sea, is what HED URBAN wanted to create. After passing the main entrance, which feels exotic and weighty like the boarding gate of a European cruise, and the short but intense corridor connected to the kitchen, the main dining space appears. In line with the theme of cruise, classical chandeliers, vintage concept furniture, and deep wood-toned finishing materials are used to stylishly fill the space. In addition, the pocket door between the kitchen and a dining have been installed for the ease of communication according to the purpose of the 'test kitchen'.

디자인 김범규 / 헤드어반스튜디오
위치 서울특별시 강남구 언주로 835, 3층
용도 레스토랑
면적 132m²
마감 바닥 – 우드플로링 / 벽 – 도장, 우드 패널, 몰딩 / 천정 – 도장, 몰딩
디스플레이 변아라
클라이언트 SG다인힐
디자인팀 백수연, 김혜은
사진 이요셉

Location 3F, 835, Eonju-ro, Gangnam-gu, Seoul
Use Restaurant
Area 132m²
Finishing Floor - Wood flooring / Wall – Painting, Wood panel, Moulding / Ceiling - Painting, Moulding
Display Ara Byun
Client SG DINEHILL
Design team Sooyeon Baek, Hyeeun Kim
Photographs Joseph Lee

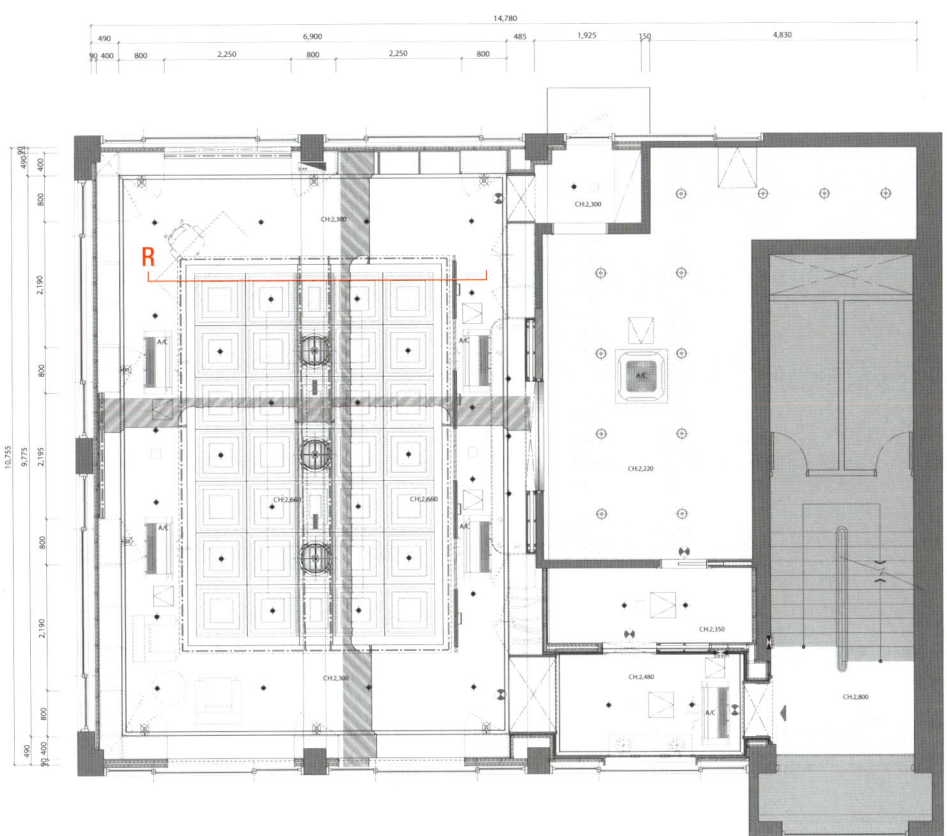

천장도 / ceiling plan

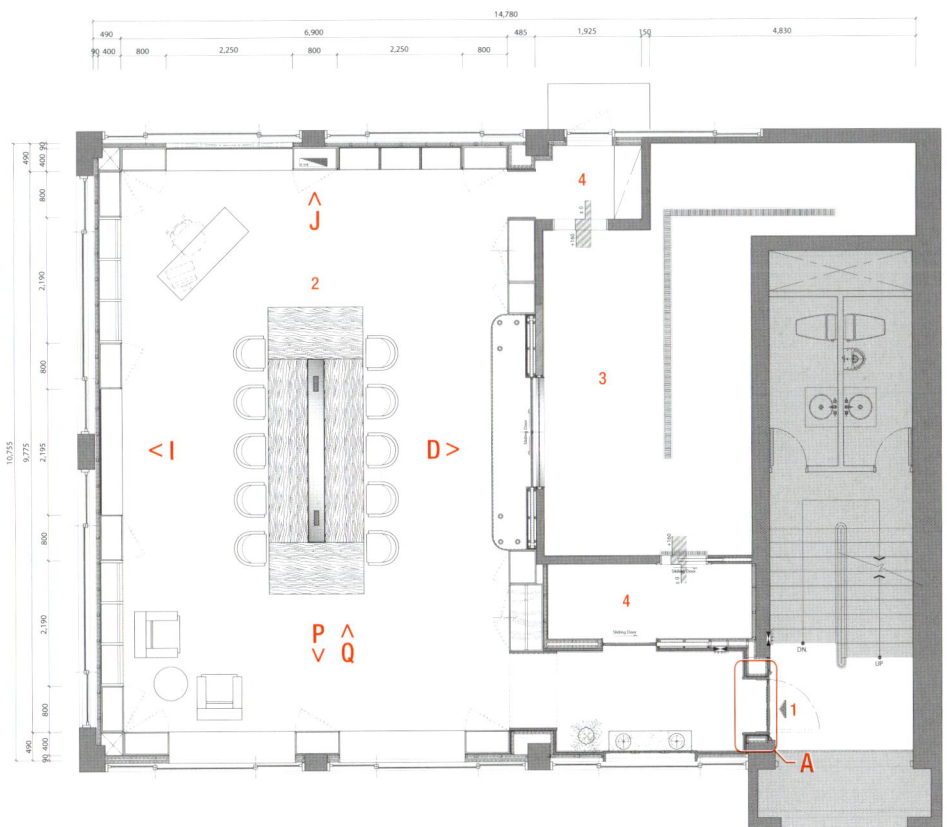

평면도 / floor plan

1 입구 2 홀 3 주방 4 식품 저장고

1 Entrance 2 Hall 3 Kitchen 4 Pantry

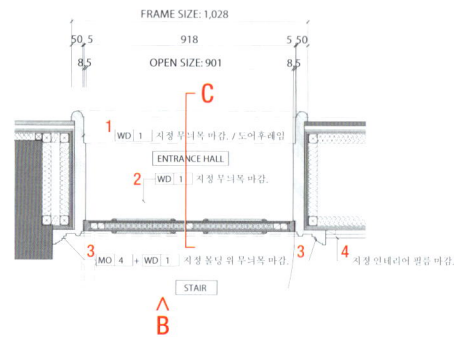

주출입문 평단면 A / entrance door top section A

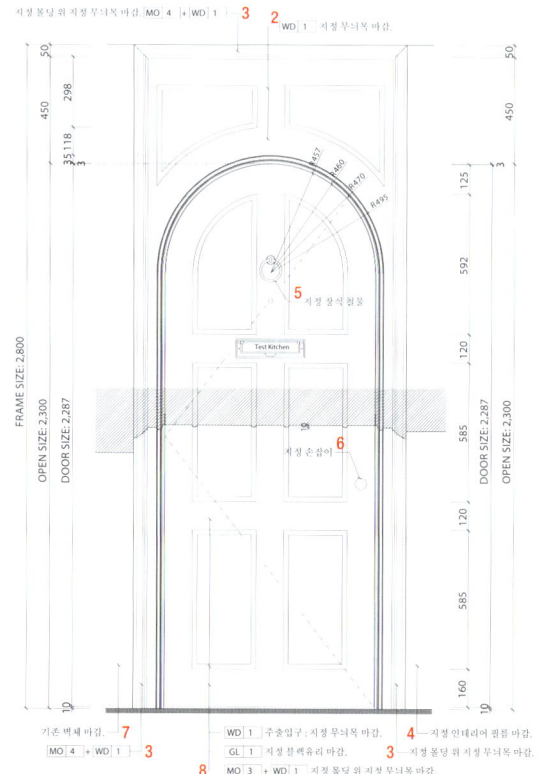

주출입문 정면 B / entrance door front view B

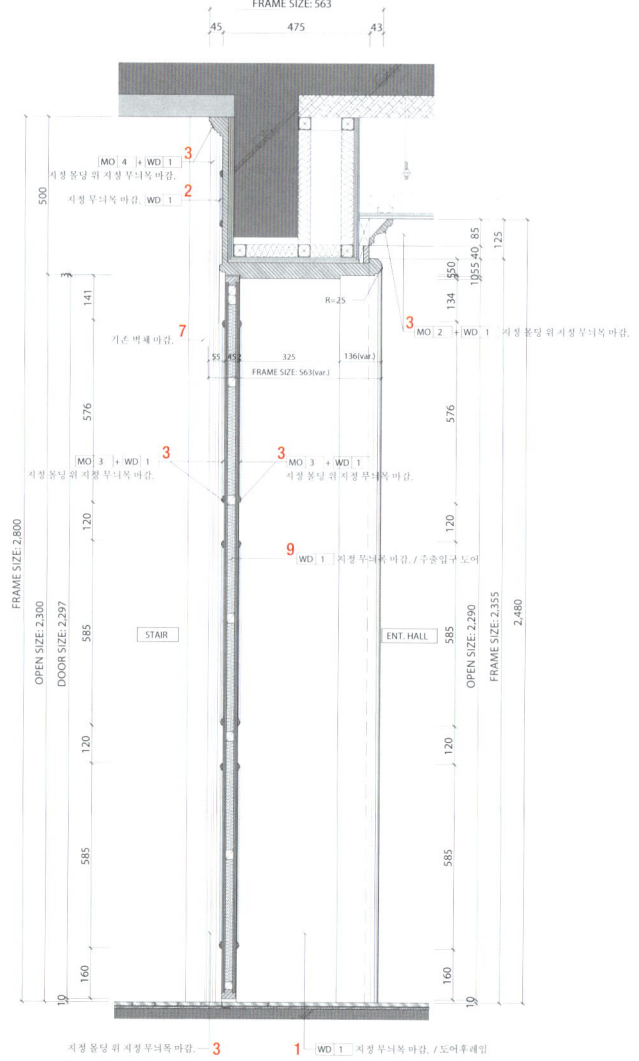

주출입문 단면 C / entrance door section C

1 지정 무늬목 / 도어 프레임 2 지정 무늬목 3 지정 몰딩 위 무늬목 4 지정 인테리어 필름 5 지정 장식 철물 6 지정 손잡이 7 기존 벽체 8 주출입구 : 지정 무늬목 / 지정 블랙 유리 / 지정 몰딩 위 지정 무늬목 9 주출입구 문 : 지정 무늬목 10 지정 무늬목 / 지정 패브릭 11 지정 무늬목 / Ø20 파이프 위 지정 도장 12 지정 흑경 13 지정 몰딩 위 지정 무늬목 / 오픈프레임 : 지정 무늬목 14 오픈프레임 : 지정 무늬목 15 지정 패브릭 / 지정 무늬목 16 지정 무늬목 / 상판 : 지정 인조대리석 17 지정 블라인드

1 Door frame : App. wood veneer 2 App. wood veneer 3 App. wood veneer on moulding 4 App. interior film 5 App. decorative hardware 6 App. door grip 7 Existing wall finish 8 Main entrance : App. wood veneer / App. black mirror / App. wood veneer on moulding 9 Main entrance door : App. wood veneer 10 App. wood veneer / App. fabric 11 App. wood veneer / App. painting on Ø20 pipe 12 App. black mirror 13 App. wood veneer on moulding / Opened frame : App. wood veneer 14 Opened frame : App. wood veneer 15 App. fabric / App. wood veneer 16 App. wood veneer / Top : App. composite marble 17 App. blinds

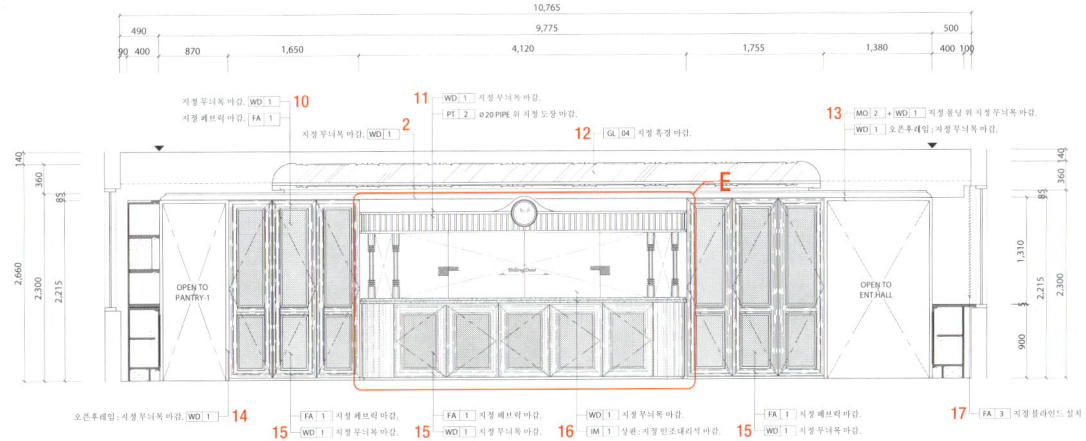

홀 입면 D / hall elevation D

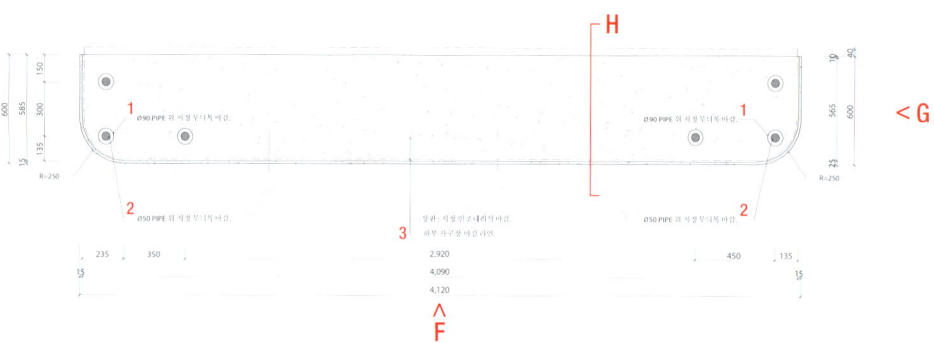

바 카운터 평단면 E / bar counter top section E

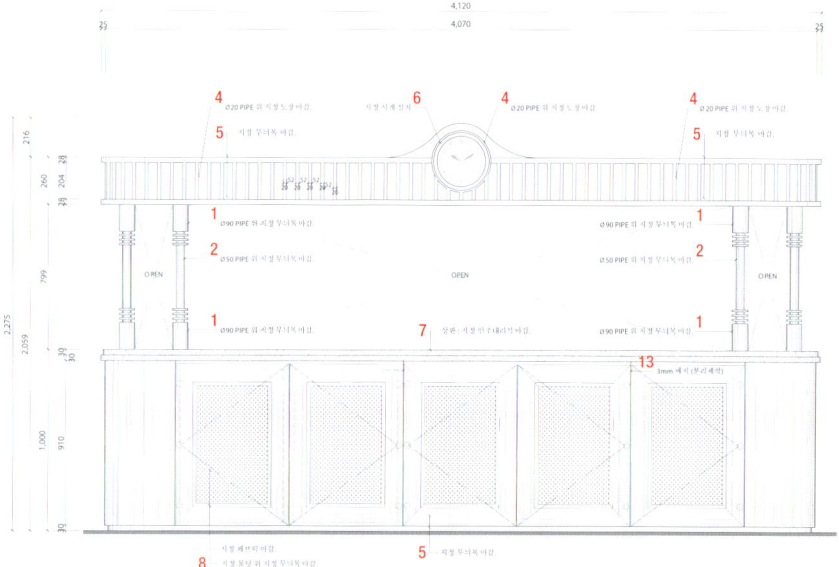

바 카운터 정면 F / bar counter front view F

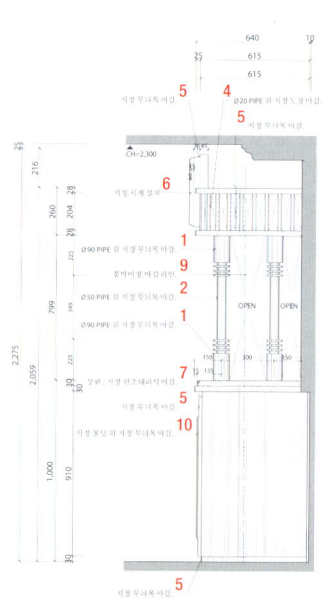

바 카운터 측면 G / bar counter side view G

1 Ø90 파이프 위 지정 무늬목　2 Ø50 파이프 위 지정 무늬목　3 상판 : 지정 인조대리석 / 하부 가구장 마감 라인　4 Ø20 파이프 위 지정 도장　5 지정 무늬목　6 시계 설치　7 상판 : 지정 인조대리석　8 지정 패브릭 / 지정 몰딩 위 지정 무늬목　9 붙박이장 마감 라인　10 지정 몰딩 위 지정 무늬목　11 지정 도장　12 지정 패브릭　13 3mm 줄눈

1 App. wood veneer on Ø90 pipe　2 App. wood veneer on Ø50 pipe　3 Top : App. composite marble / Lower cabinet finish line　4 App. wood veneer on Ø20 pipe　5 App. wood veneer　6 App. clock　7 Top : App. composite marble　8 App. fabric / App. wood veneer on moulding　9 Built-in cabinet finish line　10 App. wood veneer on moulding　11 App. painting　12 App. fabric　13 3mm reveal

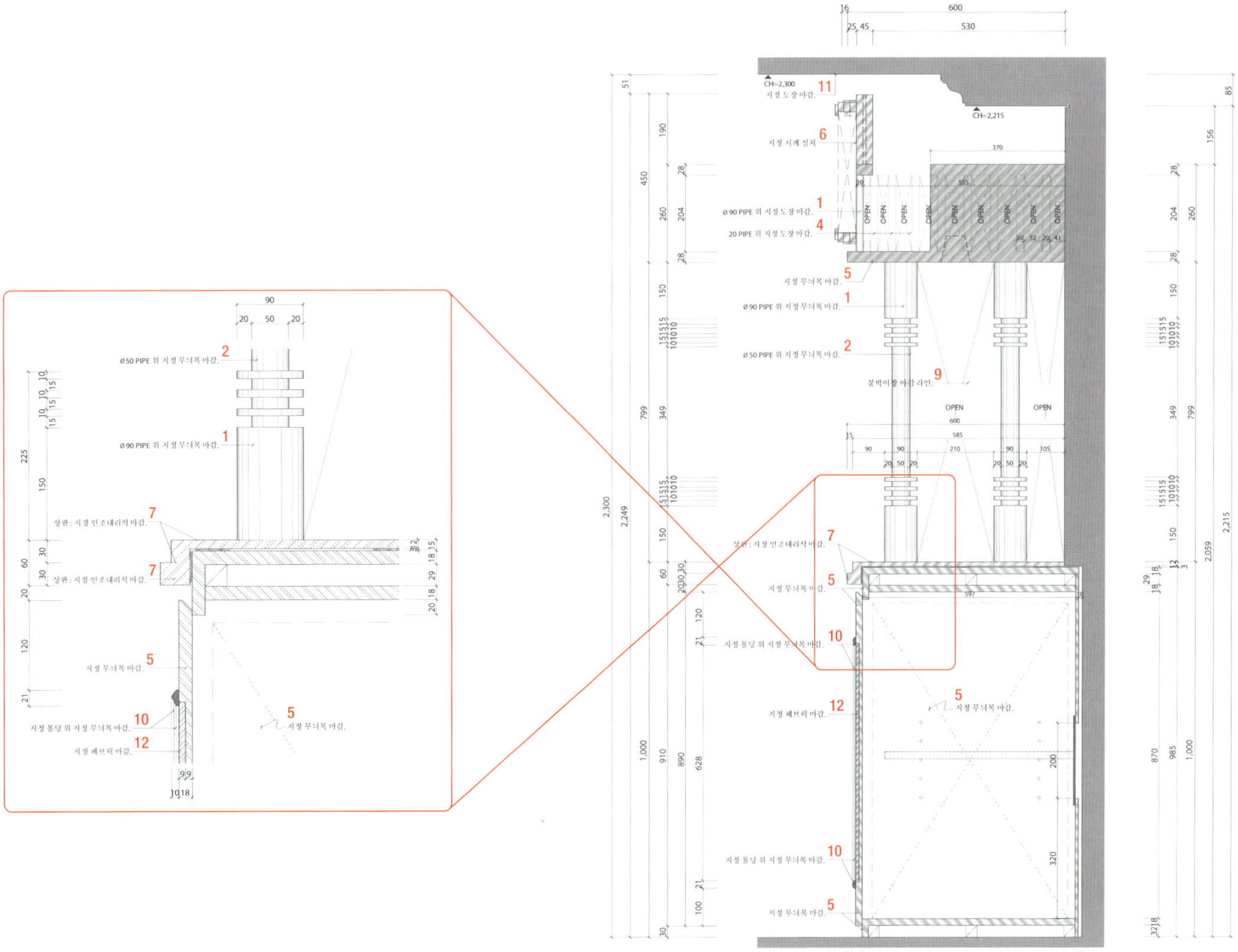

바 카운터 단면 상세 H / bar counter section detail H

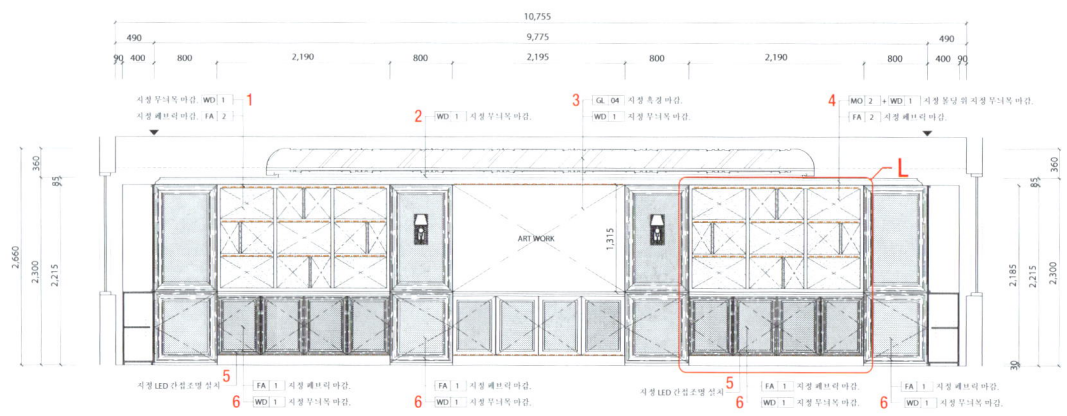

홀 입면 I / hall elevation I

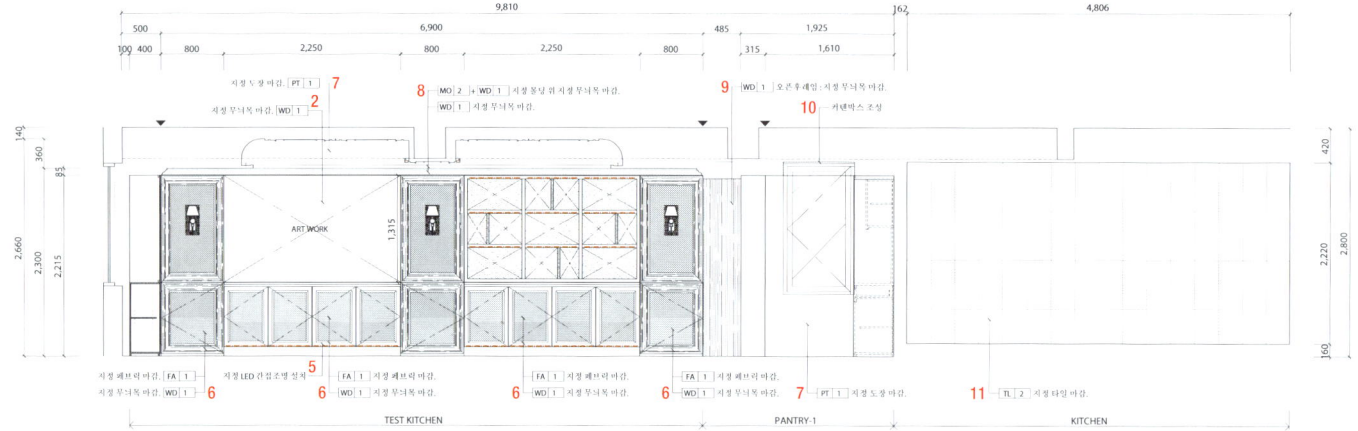

홀 입면 J / hall elevation J

1 지정 무늬목 / 지정 패브릭　2 지정 무늬목　3 지정 흑경 / 지정 무늬목　4 지정 몰딩 위 지정 무늬목 / 지정 패브릭　5 지정 LED 간접조명　6 지정 패브릭 / 지정 무늬목　7 지정 도장　8 지정 몰딩 위 지정 무늬목 / 지정 무늬목　9 오픈프레임 : 지정 무늬목　10 커튼박스　11 지정 타일　12 지정 패브릭　13 지정 무늬목 / 지정 LED 간접조명　14 지정 패브릭 / 지정 몰딩 위 지정 무늬목　15 지정 몰딩 위 지정 무늬목

1 App. wood veneer / App. fabric　2 App. wood veneer　3 App. black mirror / App. wood veneer　4 App. wood veneer on moulding / App. fabric　5 App. LED indirect lighting　6 App. fabric / App. wood veneer　7 App. painting　8 App. wood veneer on moulding / App. wood veneer　9 Opened frame : App. wood veneer　10 Curtain box　11 App. tile　12 App. fabric　13 App. wood veneer / App. LED indirect lighting　14 App. fabric / App. wood veneer on moulding　15 App. wood veneer on moulding

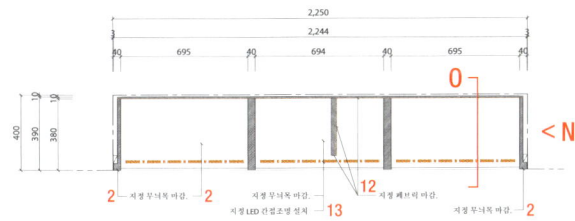

선반 평단면 K / shelves top section K

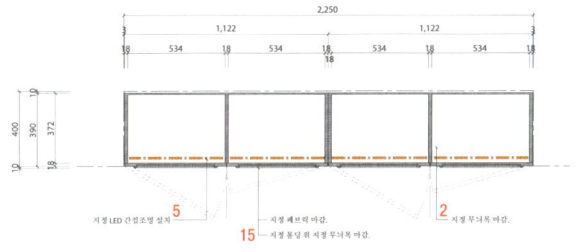

선반 평단면 M / shelves top section M

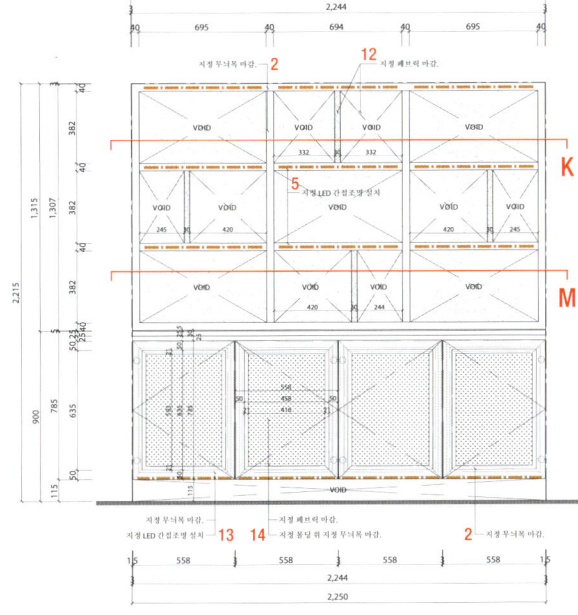

선반 정면 L / shelves front view L

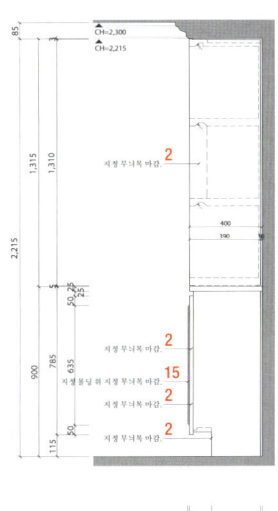

선반 측면 N / shelves side view N

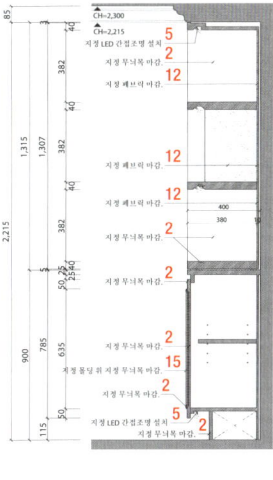

선반 단면 O / shelves section O

17

1 지정 몰딩 위 지정 무늬목 / 지정 몰딩 위 지정 무늬목 2 지정 무늬목 / T8 지정 강화유리 3 지정 도장 / 지정 도장 4 지정 무늬목 / 지정 몰딩 위 지정 무늬목 5 지정 블라인드 6 기존 우드 패널 / 기존 걸레받이 7 주출입구 : 지정 무늬목 8 지정 패브릭 9 오픈프레임 : 지정 무늬목 10 지정 패브릭 / 지정 무늬목 11 지정 무늬목 12 지정 LED 간접조명 13 지정 도장 14 소방발신기 : 지정 필름 15 지정 무늬목 / 지정 금속 16 붙박이장 월 패널 17 붙박이장 마감 라인 18 지정 몰딩 위 지정 무늬목 19 지정 패널 위 지정 도장 20 붙박이장 21 T1.6 갈바륨 위 지정 도장 22 ㅁ30X30 파이프 구조틀 @300 / T9.5 석고보드 / T9 합판 / 지정 패널 위 지정 도장 23 T1.6 갈바륨 위 지정 도장 / 지정 몰딩 위 지정 무늬목 24 T5 지정 흑경 25 지정 LED 조명 / T1.6 갈바륨 위 지정 도장 / 지정 패널 위 지정 도장 26 ㅁ30X30 파이프 구조틀 @300 / T9 합판 2겹 / 지정 패널 위 지정 도장

1 App. wood veneer on moulding / App. wood veneer on moulding 2 App. wood veneer / T8 tempered glass 3 App. painting / App. painting 4 App. wood veneer / App. wood veneer on moulding 5 App. blinds 6 Existing wooden panel / Existing baseboard finishing 7 Main entrance : App. wood veneer 8 App. fabric 9 Opened frame : App. wood veneer 10 App. fabric / App. wood veneer 11 App. wood veneer 12 App. LED indirect lighting 13 App. painting 14 App. film 15 App. wood veneer / App. metal 16 Built-in cabinet wall panel 17 Built-in cabinet finish line 18 App. wood veneer on moulding 19 App. painting on panel 20 Built-in cabinet 21 App. painting on T1.6 galvalume 22 ㅁ30X30 pipe structure @300 / T9.5 gypsum board / T9 plywood / App. painting on panel 23 App. painting on T1.6 galvalume / App. wood veneer on moulding 24 T5 black mirror 25 App. LED lighting / App. painting on T1.6 galvalume / App. painting on panel 26 ㅁ30X30 pipe structure @300 / T9 plywood 2ply / App. painting on panel

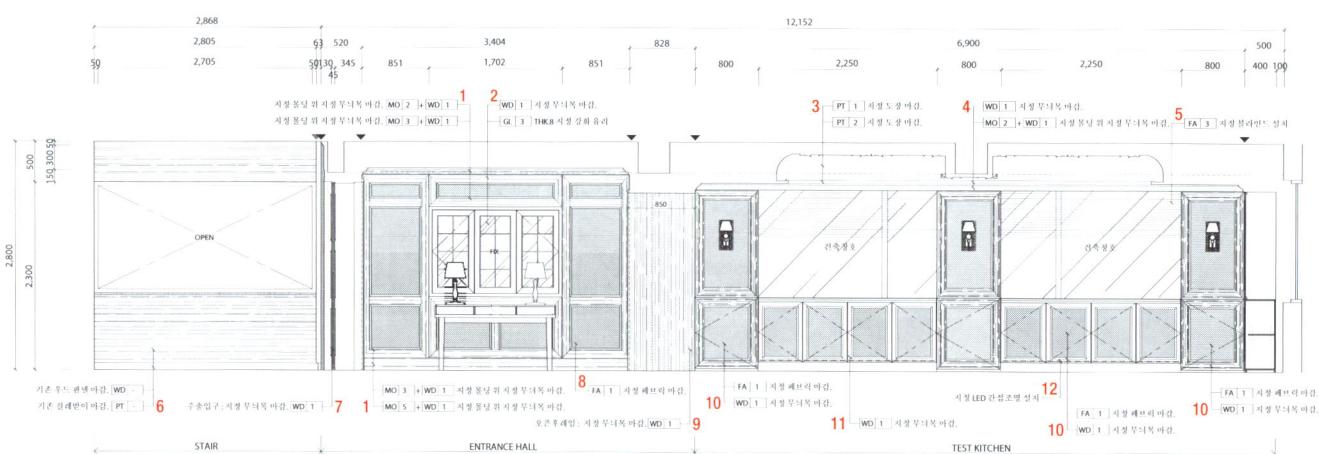

홀 입면 P / hall elevation P

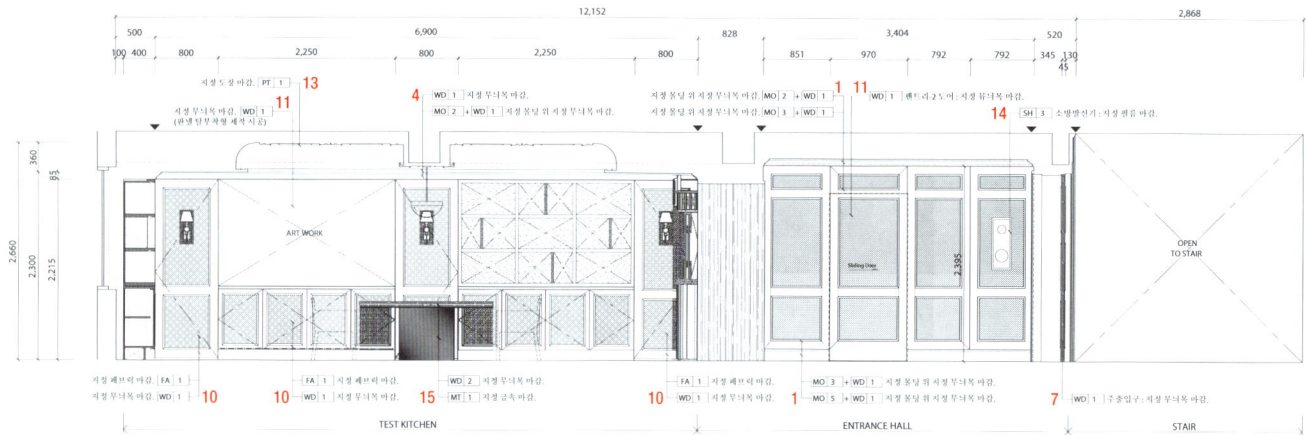

홀 입면 Q / hall elevation Q

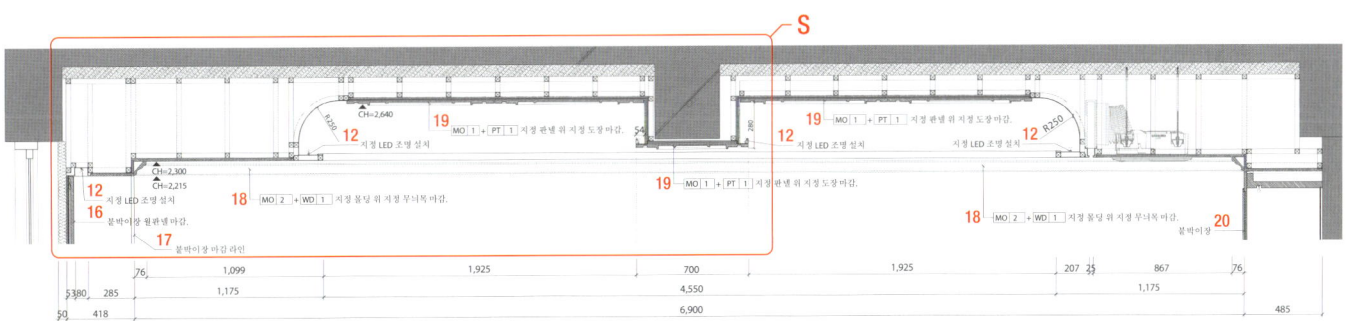

천장 단면 R / ceiling section R

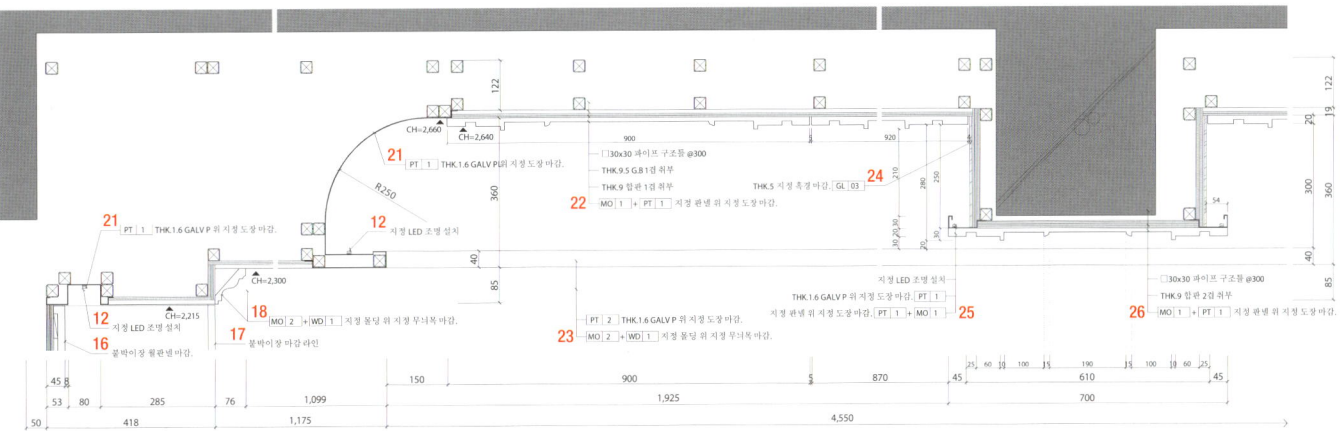

천장 단면 상세 S / ceiling detail S

IMOK SMOKE DINING

PROJECT MARK | Jaehong Son, Jeeseung Yang, Jihyung Song

이목의 첫 불씨는 청주시 낭성면 이목리에서 붙여졌다. 클라이언트가 지인들과 함께 붙인 불씨가 남영동 유용욱 바베큐 연구소를 거쳐 지금의 이목 스모크 다이닝에 이르렀다. 36년 된 목욕탕을 철거하고 나서 철거 현장 중간에 서서 공간을 바라보니 이형의 평면과 기준 없이 종횡을 지르는 기둥과 보들이 눈에 띄었다. 오랜 시간 습기와 열기가 만든 콘크리트의 분위기가 마치 비밀시설의 벙커 같다는 생각이 들었다. 태어날 때부터 물과 함께했던 공간에서 앞으로는 불을 쓰는 공간이 된다는 게 재밌게 느껴졌고, 우리의 첫 불씨가 눅눅한 이 공간을 밝혀주길 바라는 마음에 콘셉트는 Burn-ker로 지었다. 건물의 주출입구를 지나면 나타나는 이목의 첫 모습은 지하벙커를 연상케하는 녹슨 철문이다. 무거운 철문을 열어 불빛을 따라 한 걸음씩 내딛다 보면 지상으로부터 내려온 빛과 함께 로비를 마주하게 된다. 리셉션 데스크 뒤로 보이는 선큰가든은 이 공간이 지하에 있고 저 너머에 지상공간이 있다는 암시를 줘서 어두운 공간을 조금 더 비밀스럽고 안락하게 만들어준다. 동선을 따라 레스토랑의 홀로 들어가면 벽난로를 먼저 마주한다. 레스토랑 전체에 불에 관한 모티브를 각각 다르게 표현했는데 홀 복도에서는 붉은색 조명과 가습효과를 이용하여 불꽃을 표현했다. 전반적인 마감과 색감은 장작이 불에 타서 숯이 되고 재가 되는 과정에서 볼 수 있는 브라운, 차콜 블랙, 레드 브라운, 그레이 계열 마감재를 사용해서 벙커의 고독하고 웅장한 이미지와 뜨겁고 야성적인 바베큐의 이미지를 혼합하려 했다.

The first spark of Imok was lit in Imok-ri, Nangseong-myeon, Cheongju. From a small spark ignited by the client and his partners, it evolved to Yoo Yong-wook BBQ Laboratory in Nam-yeong-dong and has eventually become today's Imok Smoke Dining. When the designer visited the project site and stood in the middle of it after demolishing a 36-year-old bathhouse that used to be there, irregular floor plans and randomly positioned columns and beams caught the eye. The concrete that had endured years of moisture and heat reminded one of a secret bunker. The designer thought it was intriguing that a space originally built for water would become one using fire. This inspired the concept "Burn-ker," hoping that Imok's first spark would illuminate this once-damp space. Past the main entrance, a rusty steel door reminiscent of an underground bunker defines Imok's first impression.

As visitors open the heavy steel door and follow the light step by step, they reach the lobby illuminated by natural light from above. The sunken garden behind the reception desk gives a hint that this place is underground while the ground level is on the other side. This makes the dark space feel both mysterious and comfortable. Upon entering the restaurant hall, visitors are first greeted by a fireplace. Various fire-related elements are incorporated throughout the space. For example, red lighting and humidification effects are used together to represent flames in the hall corridor. As for finishes and colors, materials in brown, charcoal black, red-brown, and gray tones, which can be observed in the process of wood burning into charcoal and ash, are used to blend the solemn yet overwhelming image of a bunker with the hot, wild character of barbecue.

디자인 손재홍, 양지승, 송지형 / 프로젝트 마크
위치 서울특별시 강남구 압구정로2길 6, 지하 1층
용도 레스토랑
면적 468.7m²
마감 타일, 열연강판, 종석 미장, 페인트, 원목, 합판, 벽돌
디자인팀 김어진, 팽종인
사진 조동현

Location B1, Apgujeong-ro 2-gil, Gangnam-gu, Seoul
Use Restaurant
Area 468.7m²
Finishing Tile, Hot rolled steel plate, Plaster, Wood, Plywood, Brick
Photographer Donghyun Cho

평면도 / floor plan

1 입구 2 로비 3 직원실 4 사무실 5 선큰 가든 6 와이셀러 7 홀 8 프라이빗 룸 9 오픈 키친 10 주방 11 창고 12 화장실

1 Entrance 2 Lobby 3 Staff room 4 Office 5 Sunken garden 6 Wine cellar 7 Hall 8 Private room 9 Open kitchen 10 Back kitchen 11 Storage 12 Restroom

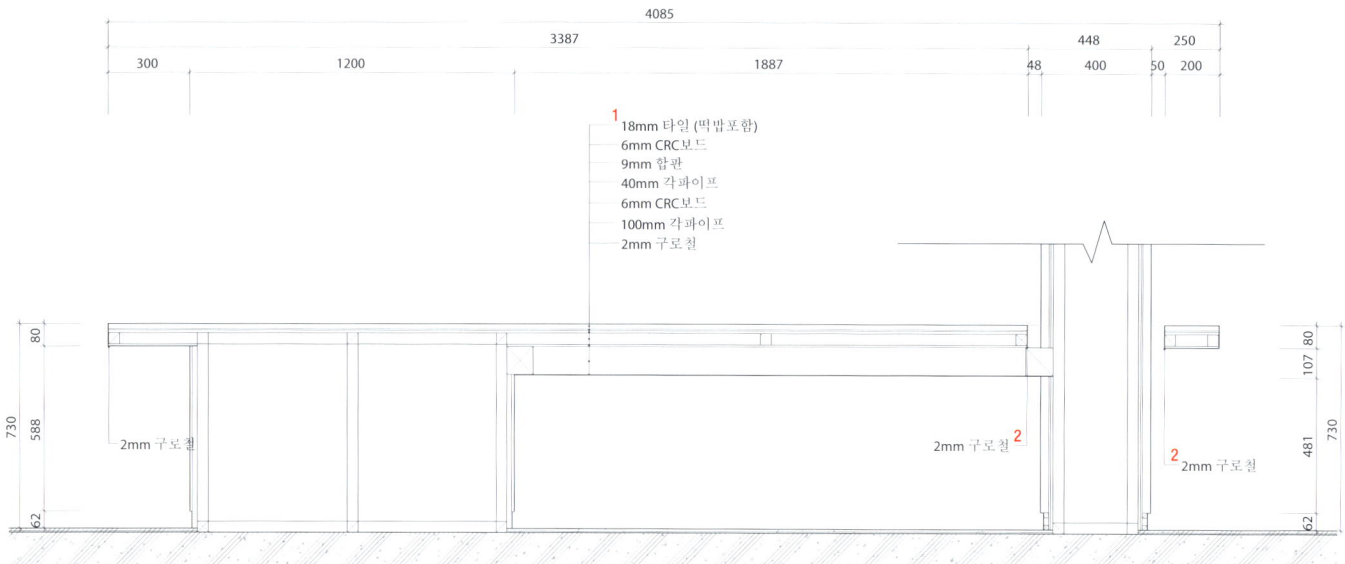

1. 18mm 타일 (떡밥포함)
   6mm CRC보드
   9mm 합판
   40mm 각파이프
   6mm CRC보드
   100mm 각파이프
   2mm 구로철
2. 2mm 구로철

테이블 단면 A / table section A

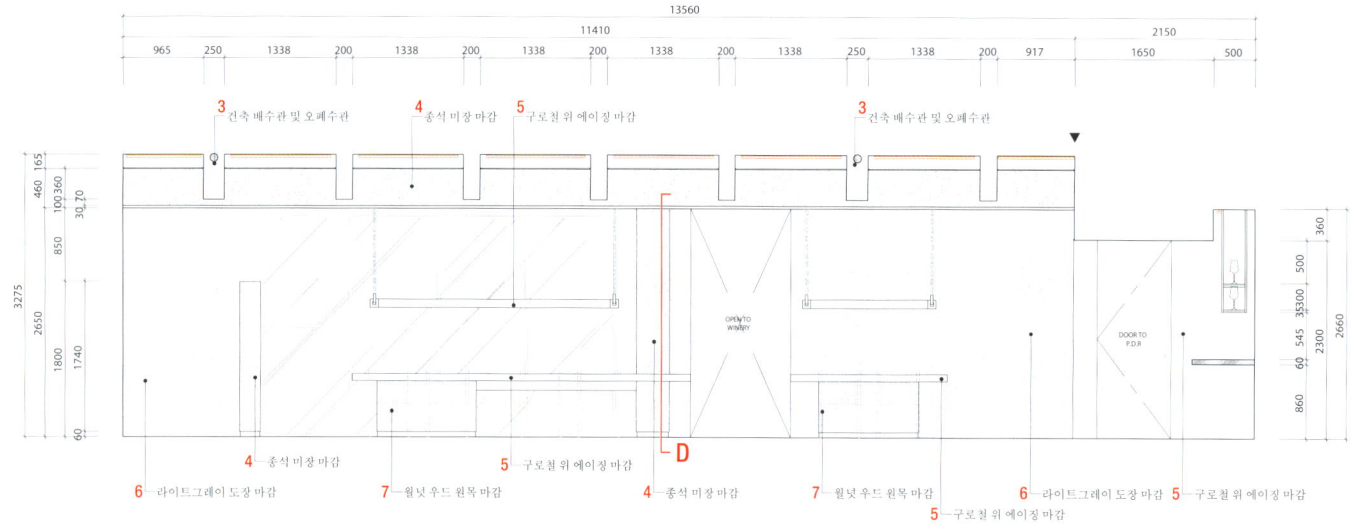

3. 건축 배수관 및 오폐수관
4. 종석 미장 마감
5. 구로철 위 에이징 마감
6. 라이트그레이 도장 마감
7. 월넛 우드 원목 마감

홀 입면 B / hall elevation B

1 T12 박판 타일 / T9 석고보드 / T9 합판 / □40X40 각파이프 / T9 석고보드 2 T2 열연강판 3 건축 배수관 및 오폐수관 4 종석 미장 5 열연강 위 에이징 마감 6 연회색 도장 7 월넛 원목 8 배기공조 그릴 9 후드 위 열연강 10 타일 11 Ø50 열연강파이프 12 벽난로 연출 13 바리솔 14 시멘트 보드 위 메쉬망 삽입 후 미장 15 기존 건축 배수관

1 T12 thin tile / T9 gypsum board / T9 plywood / □40X40 square pipe / T9 gypsum board 2 T2 hot rolled steel sheet 3 Drain pipe and sewer pipe 4 Stone plaster 5 Hot rolled steel rust effect 6 Light gray painting 7 Walnut wood 8 Exhaust air grille 9 Hot rolled steel on hood 10 Tile 11 Ø50 hot rolled steel pipe 12 Fireplace 13 Barrisol 14 Insert wired mesh on cement board, Plastering 15 Existing drain pipe

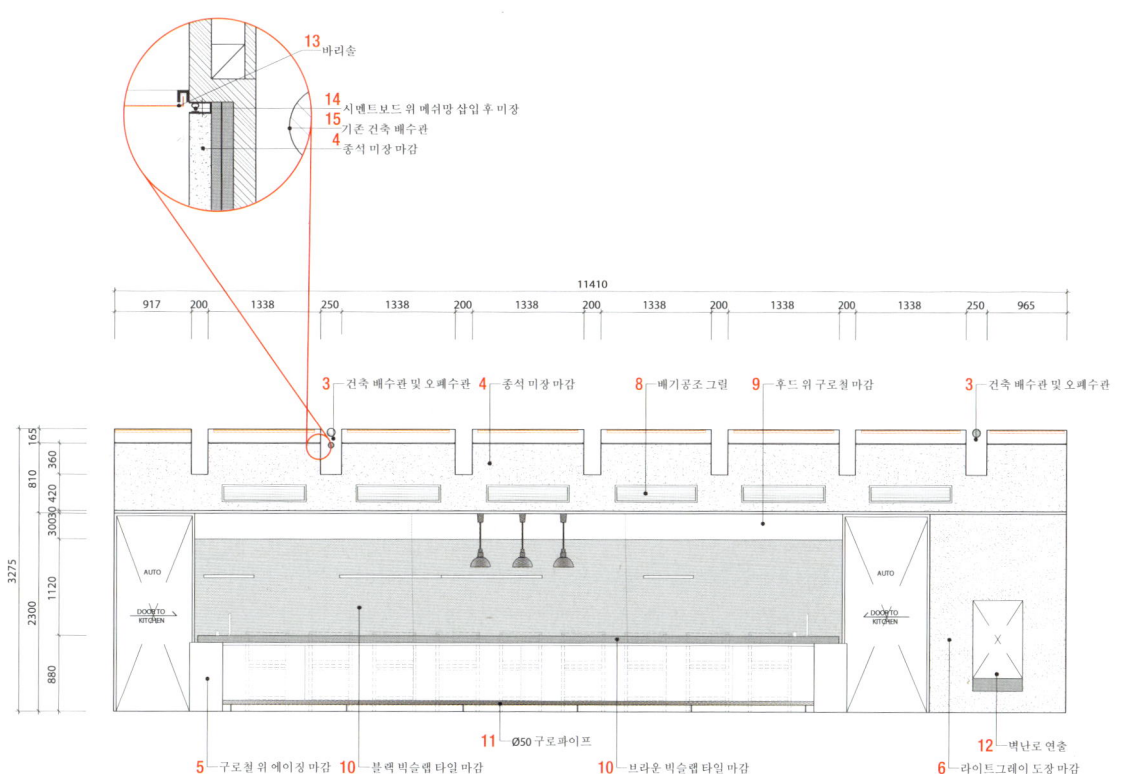

홀 입면 C / hall elevation C

1 LED 2 T1.2 갈바륨 3 □40X40 각파이프 4 T5 아크릴 5 열연강 절곡 6 □20X20 각파이프 7 종석 마감 8 에어컨

1 LED 2 T1.2 galvalume 3 □40X40 square pipe 4 T5 acrylic 5 Hot rolled steel bending 6 □20X20 square pipe 7 Stone finishing 8 Air conditioner

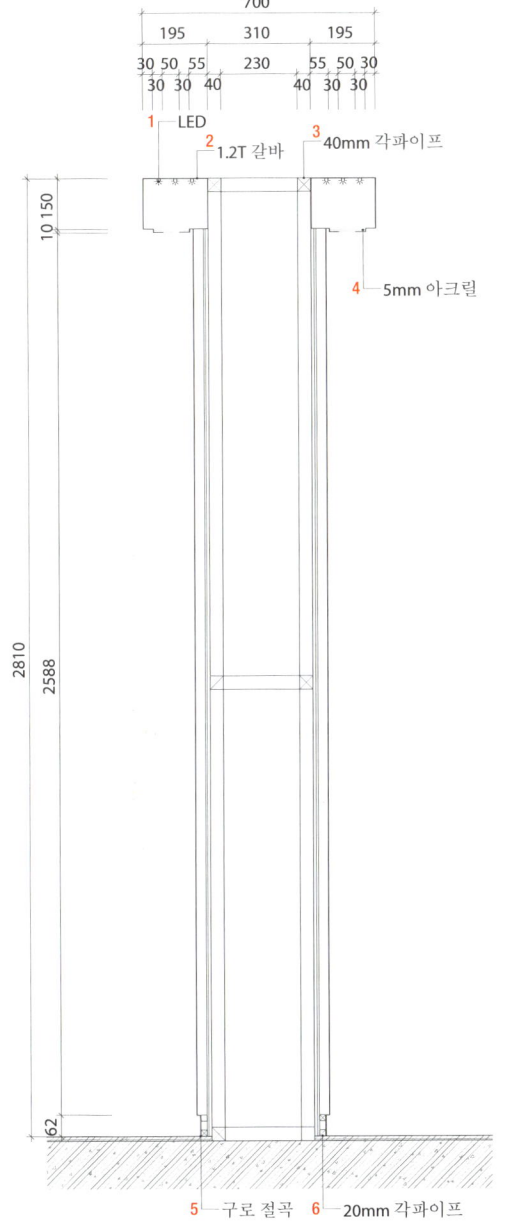

기둥 단면 D / pillar section D

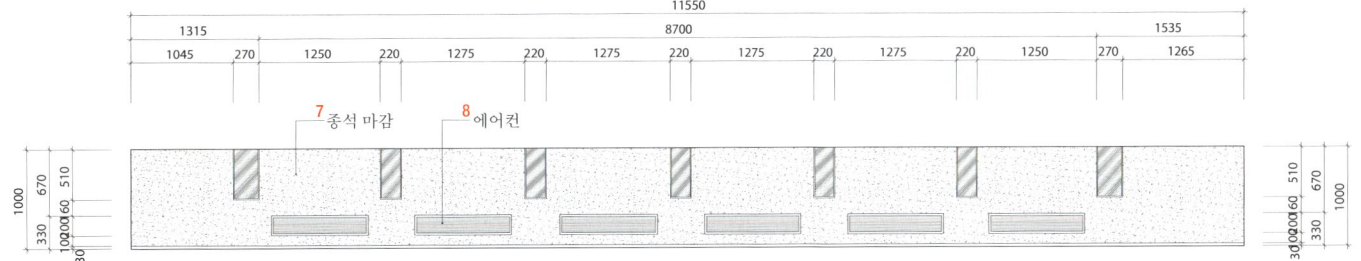

천장 단면 상세 / ceiling section detail

1 후드 2 급기 & 냉난방 라인디퓨저 3 라인조명 매립 4 바리솔 5 종석 미장 6 냉난방 라인디퓨저 7 커튼박스 8 열연강 위 에이징 마감 9 연회색 도장 10 LED 바 11 월넛 원목 12 시멘트 벽돌 13 갈바륨 절곡 위 회색 분체도장 14 제작 손잡이 15 T15 투명 강화유리 16 SUS 바이브레이션 17 열연강 18 T12 박판 타일 / T9 석고보드 / T9 합판 / □40X40 각파이프 / T9 석고보드 19 T3 스테인리스 스틸 헤어라인 / T9 합판 / □40X40 각파이프 20 T18 박판 타일 / T6 CRC 보드 / T9 합판 / □40X40 각파이프 / T5 합판 21 T3 스테인리스 스틸 헤어라인

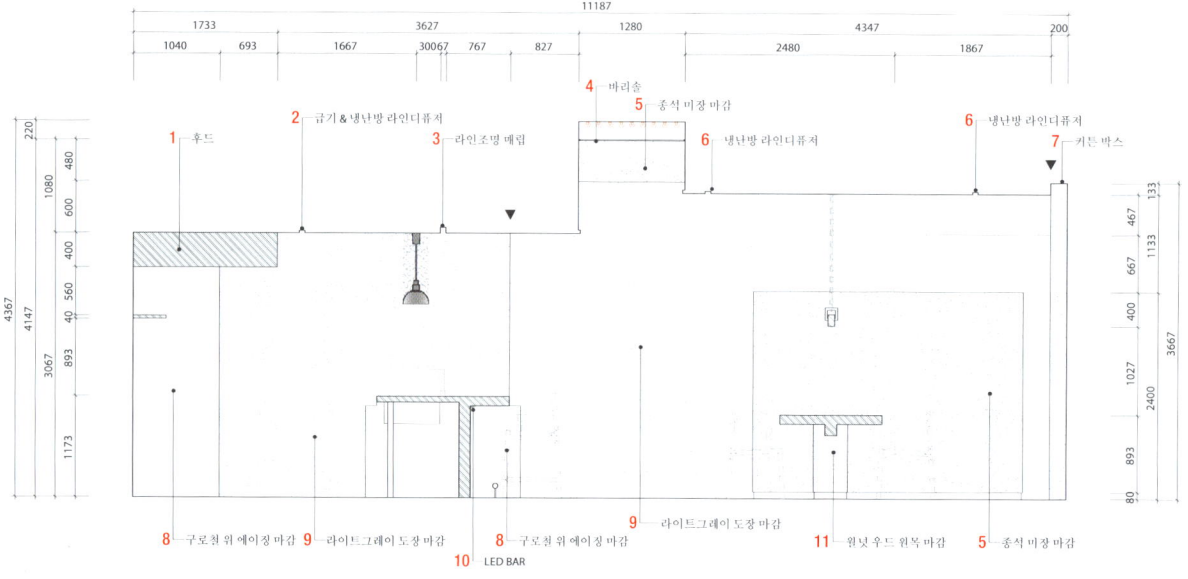

홀 입면 E / hall elevation E

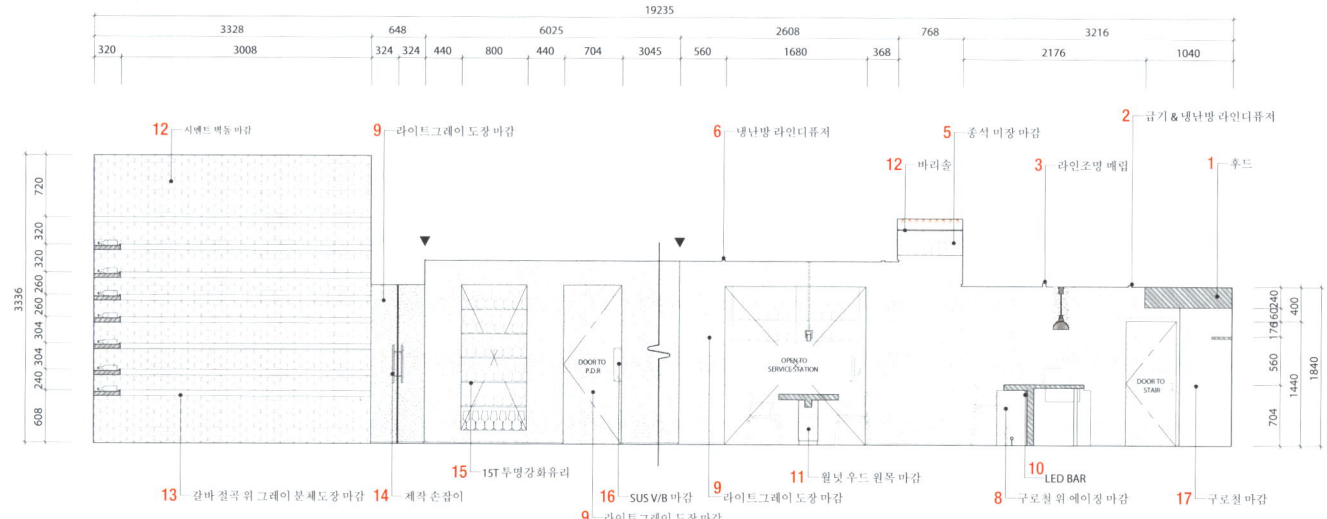

홀 입면 F / hall elevation F

1 Hood **2** Ventilation & HVAC diffuser **3** Embedding linear lighting **4** Barrisol **5** Stone plastering **6** HVAC diffuser **7** Curtain box **8** Hot rolled steel rust effect **9** Light gray painting **10** LED bar **11** Walnut wood **12** Cement brick **13** Gray powder coating on galvalume bending **14** Custom-made handle **15** T15 clear tempered glass **16** SUS vibration **17** Hot rolled steel **18** T12 thin tile / T9 gypsum board / T9 plywood / □40X40 square pipe / T9 gypsum board **19** T3 stainless steel hairline / T9 plywood / □40X40 square pipe **20** T18 thin tile / T6 CRC board / T9 plywood / □40X40 square pipe / T5 plywood **21** T3 stainless steel hairline

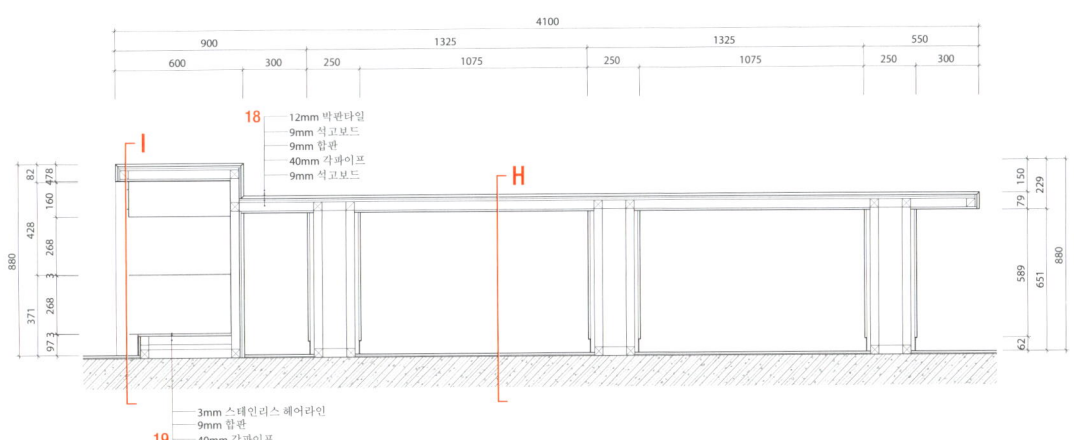

테이블 단면 G / table section G

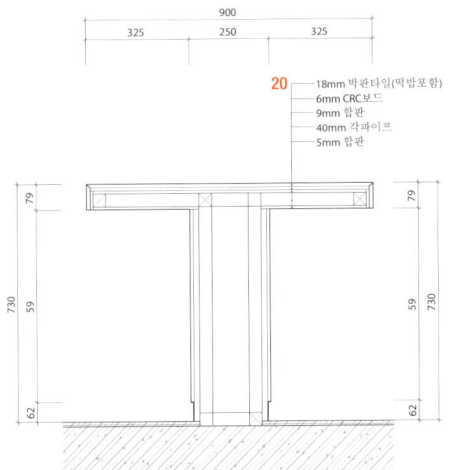

테이블 단면 H / table section H

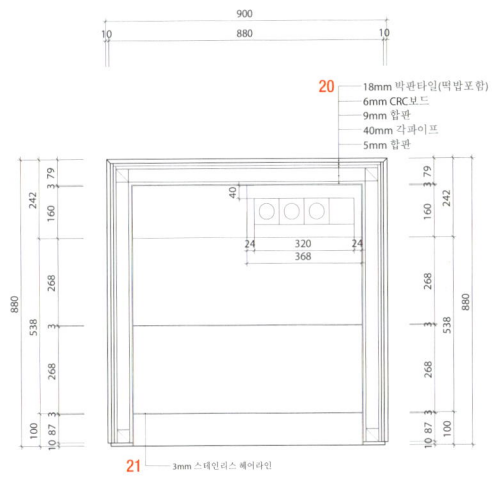

테이블 단면 I / table section I

OBALTAN SAMSUNG BRANCH

INTOEX | Yunjun Yang

강남구 삼성로에 위치한 오발탄은 2개 층으로 구성되어 있으며 지난 2018년도, 인투익스에서 2층 인테리어 리뉴얼을 진행 후 이번에 1층과 파사드를 새롭게 인테리어 리뉴얼을 하게 되었다. 이번에는 기존의 절제된 공간에 삼성동이라는 우리나라 최고 도심의 풍경을 담아내려 했다. 한여름의 활기차고 뜨거운 하늘 아래 도심의 마천루 같은 풍경, 밤이 되어도 꺼지지 않는 도심의 빛. 그런 모습을 공간에 넣어 삼성동의 이야기, 그리고 그와 어우러진 도심의 삶, 우리의 모습을 담아내고자 했다. 메인 입구를 들어서면 어두운 철재 면 위 레이저 커팅을 반복적으로 적용해 연출한 격자무늬가 돋보인다. 은은한 조도를 제공하기 위하여 이 천장 구조물에 직선의 빛 설치하였으며 데스크와 홀의 기둥, 천장에도 직선 빛을 이용해 다양한 선이 겹쳐지게 보이는 효과를 만들어냈다. 전체적으로 어두운 면과 여러 직선들, 블랙서스 마감, 레드 노출 천장 등으로 인해 차갑고 날카롭게 느껴질 수 있는 공간에 매스감 있는 아치 게이트를 이용해 메인 입구 공간을 감싸 안는 조화를 이루도록 하였다. 평면 조닝은 단순한 구성이지만 오발탄만의 분위기에 맞는 무드를 조성하였으며 과도한 장식적 요소는 없지만 분명히 대비되는 컬러와 본질적인 구조감을 강조하며 차분하고 우아한 공간을 연출해 고객에게 특별한 디자인과 특별한 시간을 선사한다.

Obaltan, located on Samsung-ro in Gangnam-gu, consists of two floors. In 2018, INTOEX renovated the interior design of the second floor. This time, we were responsible for revamping the interior design of the first floor and façade. In this scenario, designer's goal was to capture the essence of Samsung-dong, one of the representative urban areas in our country, within our understated established narrative. They aimed to accentuate the space with the lively and dynamic city skyline against the backdrop of a hot summer sky, as well as the incessant glow of city lights at night. By incorporating these visual elements into the space, they wished to portray the unique narrative of Samsung-dong, showcasing its harmonious urban lifestyle and our multifaceted identity within that context. Upon entering the main entrance, a striking lattice pattern, formed by repeatedly laser-cutting a dark iron surface, immediately catches the eye.

To achieve subdued lighting, straight lights were installed within the ceiling structure. This incorporation of straight lines on the desk, pillars, and ceiling throughout the hall creates a visually dynamic effect with overlapping lines. To enhance the harmonious effect and soften the potentially cold and sharp atmosphere created by the overall dark surfaces, numerous straight lines, black SUS finishes, and exposed LED ceilings, we installed a massive arched gate enveloping the main entrance. The flat zoning has a simple composition, yet it creates a mood that aligns with Obaltan's distinctive atmosphere. Despite the presence of intricate decorative elements, the design highlights contrasting colors and essential structural features, imparting a sense of calm and elegance to the space. This approach provides customers with a unique design and an unforgettable experience.

디자인 양윤준 / 인투익스
위치 서울특별시 강남구 삼성로 606
용도 레스토랑
면적 355.07㎡
마감 바닥 – 타일 / 벽 – 스타코, 벽돌, 인테리어 필름, 유리 블록, SUS, 타일, 시멘트 보드 / 천장 – 페인트, SMC 패널
완공 2023. 10
디자인팀 강신영, 이다솜
시공팀 김태하
사진 강명국

Location 606, Samsung-ro, Gangnam-gu, Seoul
Use Restaurant
Area 355.07m²
Finishing Floor - Tile / Wall - Stucco, Brick, Interior film, Glass block, SUS, Tile, Cement board / Ceiling - Paint, SMC panel
Completion 2023. 10
Design team Shinyoung Kang, Dasom Lee
Construction team Taeha Kim
Photographer Myeongguk Gang

평면도 / floor plan

1 로비 2 입구 3 홀 4 단체룸 5 준비실 6 주방 7 화장실 8 창고

1 Lobby 2 Entrance 3 Hall 4 Group room 5 Preparation room 6 Kitchen 7 Washroom 8 Storage

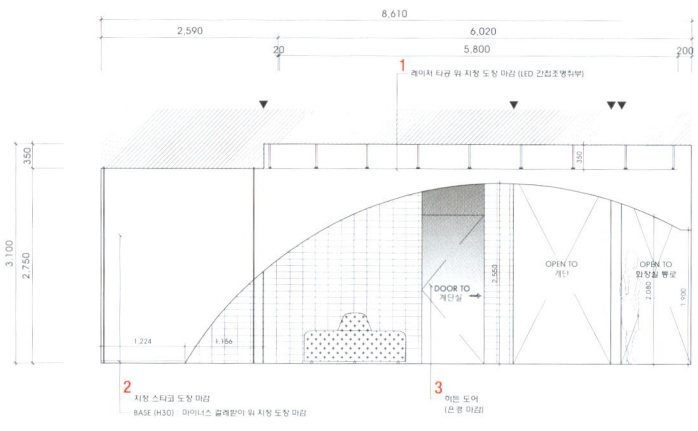

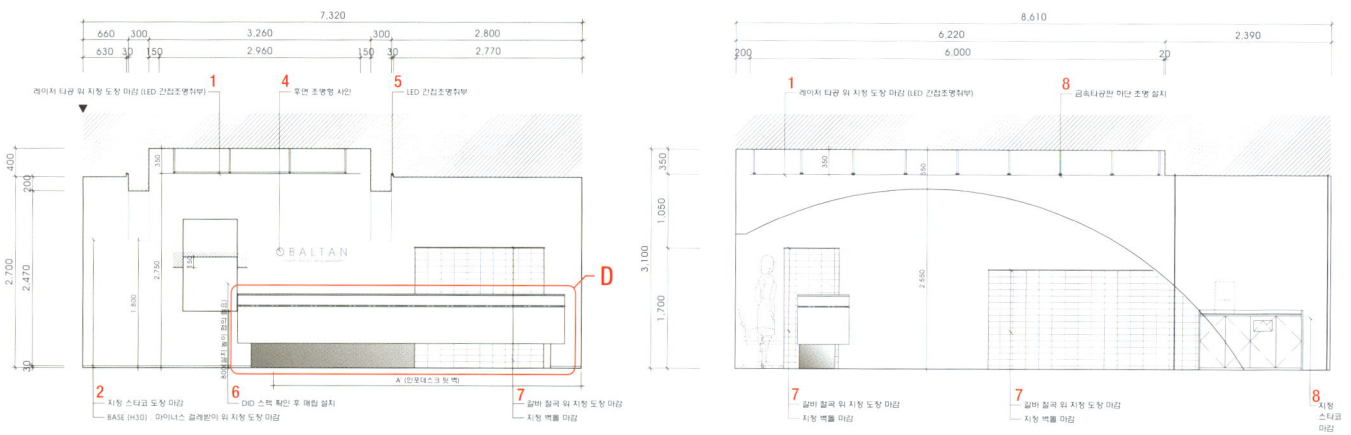

로비 입면 A / lobby elevation A

로비 입면 B / lobby elevation B

로비 입면 C / lobby elevation C

1 레이저 타공 위 지정 도장 2 지정 스투코 도장 / 걸레받이 : 마이너스 걸레받이 위 지정 도장 3 히든도어, 은경 마감 4 조명형 사인 5 LED 간접조명 6 DID 매입 7 갈바륨 절곡 위 지정 도장 / 지정 벽돌 8 지정 스투코 도장 9 데스크 상판 : 지정 시트, T5 강화유리 / 상판 : 지정 인조대리석 10 벽돌 벽체 + 상하부장 11 상판 : 지정 인조대리석 12 지정 검은색 SUS / 은경 13 LED 바 14 지정 시트 / 데스크 상판 : 지정 시트, T5 강화유리 15 서랍 : 지정 PB / 선반 : 지정 PB

1 App. painting on laser perforation 2 App. stucco painting / Base : App. painting on minus baseboard 3 Hidden door, Silver mirror 4 Illuminated sign 5 LED indirect lighting 6 Embedding DID 7 App. painting on galvalume bending / App. brick 8 App. stucco painting 9 Desk top : App. sheet, T5 tempered glass / Top : App. engineered marble 10 Brick wall + Upper/lower cabinet 11 Top : App. engineered marble 12 App. black SUS / Mirror 13 LED bar 14 App. sheet / Desk top : App. sheet, T5 tempered glass 15 Drawer : App. PB board / Shelf : App. PB board

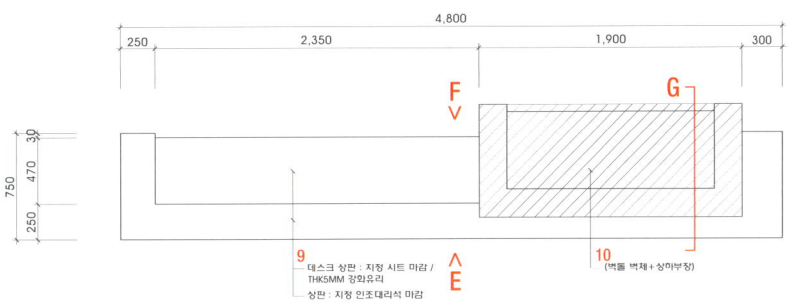

리셉션 데스크 평면 D / reception desk top view D

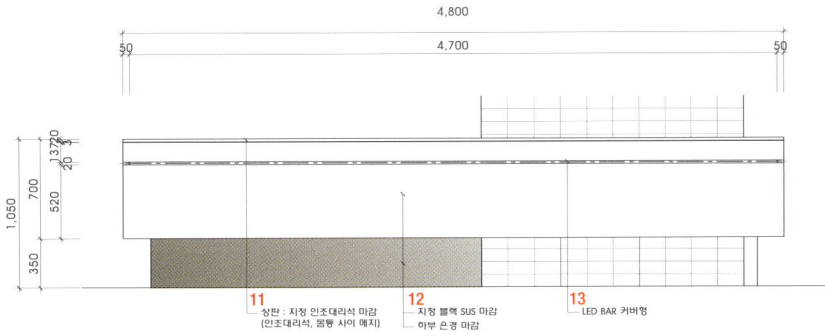

리셉션 데스크 정면 E / reception desk front view E

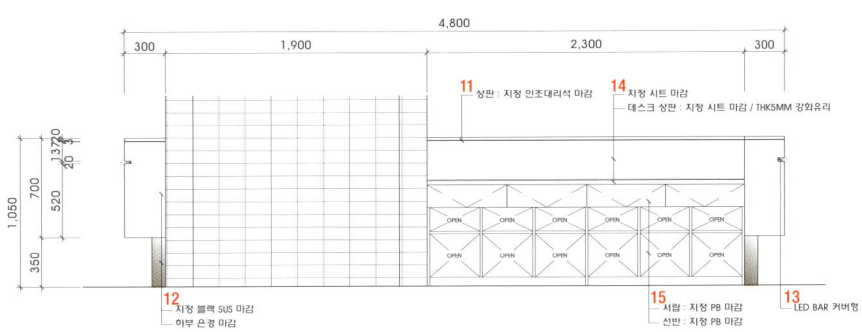

리셉션 데스크 후면 F / reception desk rear view F

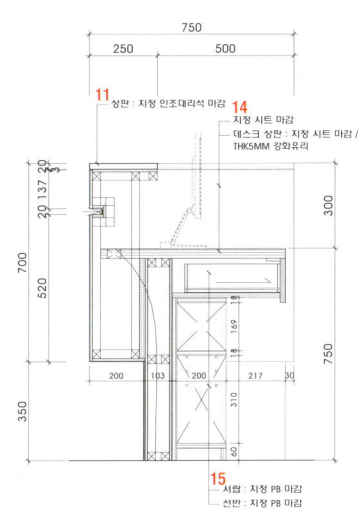

리셉션 데스크 단면 G / reception desk section G

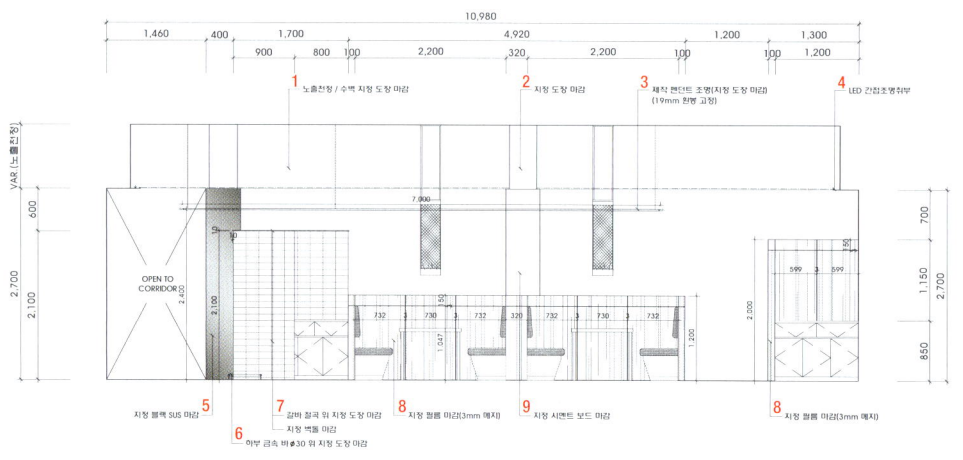

홀 입면 H / hall elevation H

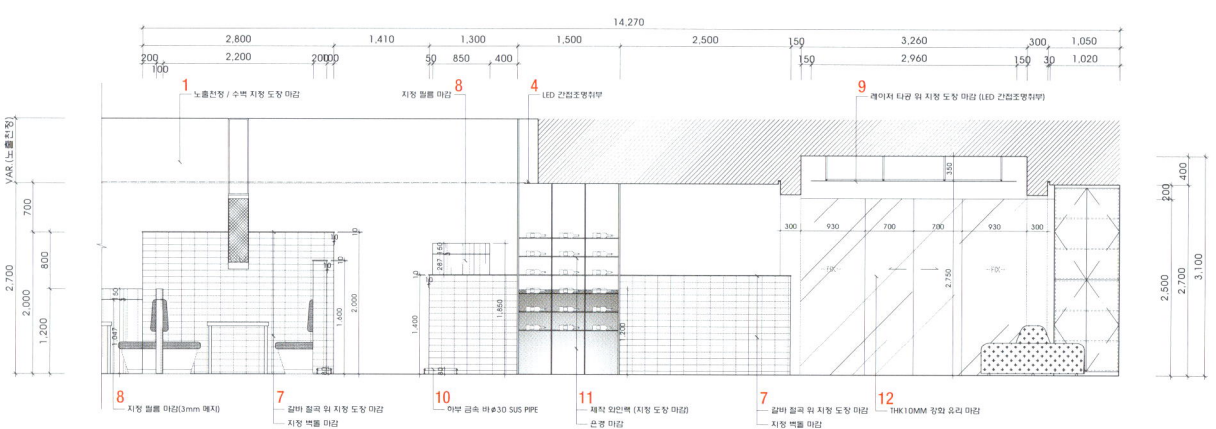

홀 입면 I / hall elevation I

1 노출천장, 수벽 지정 도장 2 지정 도장 3 제작 펜던트 조명 4 LED 간접조명 5 지정 검은색 SUS 6 하부 Ø30 금속 바 위 지정 도장 7 갈바륨 절곡 위 지정 도장 / 지정 벽돌 8 지정 필름 9 지정 시멘트 보드 10 하부 금속 바 Ø30 SUS 파이프 11 제작 와인랙 : 지정 도장 / 은경 12 지정 스투코 도장 13 몰딩 : 갈바륨 절곡 위 지정 도장 14 게이트 : 갈바륨 절곡 위 지정 도장 15 은경 16 와이어 고정 17 T1.2 갈바륨 절곡 위 지정 도장

1 Exposed ceiling, Reveal : App. painting 2 App. painting 3 Custom-made pendant light 4 LED indirect lighting 5 App. black SUS 6 App. painting on Ø30 metal bar 7 App. painting on galvalume bending / App. brick 8 App. film 9 App. cement board 10 Lower metal bar : Ø30 SUS pipe 11 Custom-made wine rack : App. painting / Mirror 12 App. stucco painting 13 Moulding : App. painting on galvalume bending 14 Gate : App. painting on galvalume bending 15 Mirror 16 Join with steel wire rope 17 App. painting on T1.2 galvalume bending

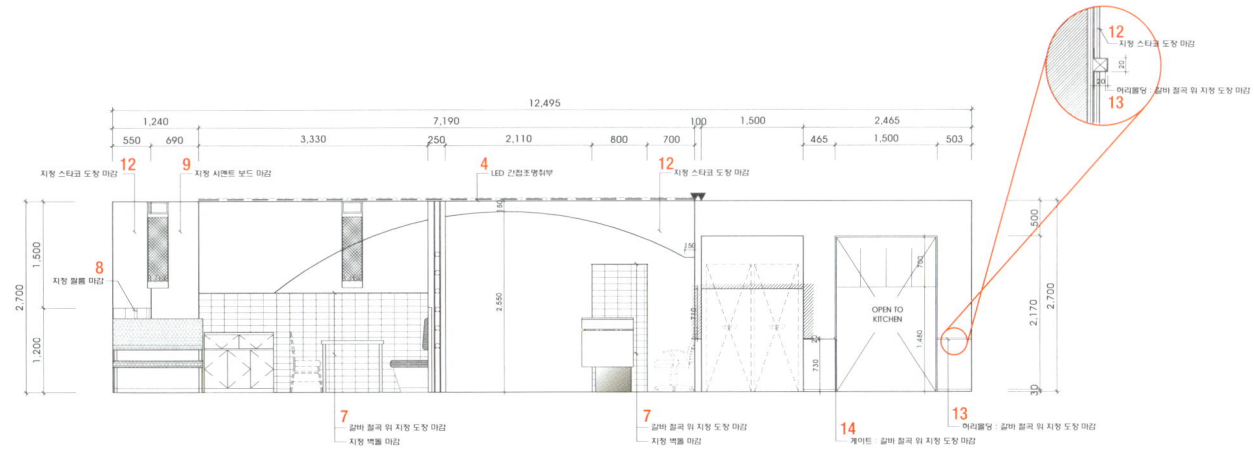

홀 입면 J / hall elevation J

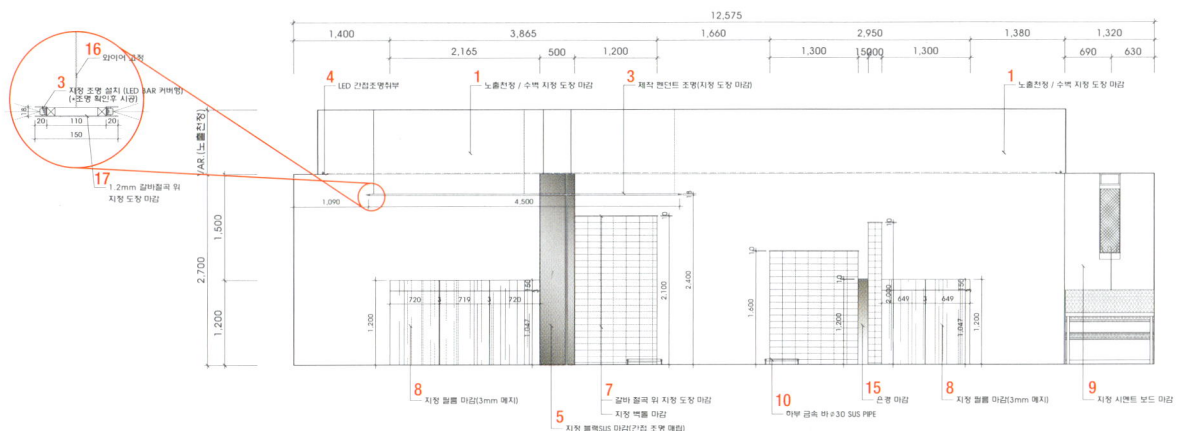

홀 입면 K / hall elevation K

33

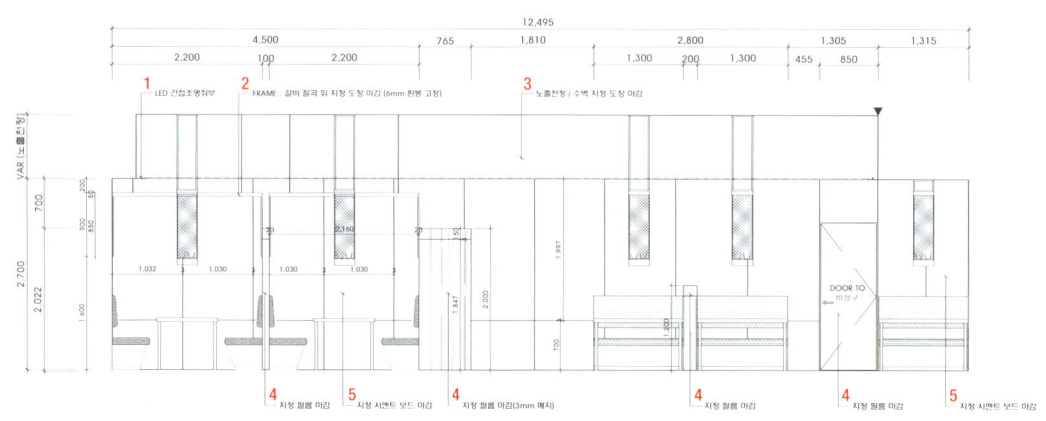

홀 입면 L / hall elevation L

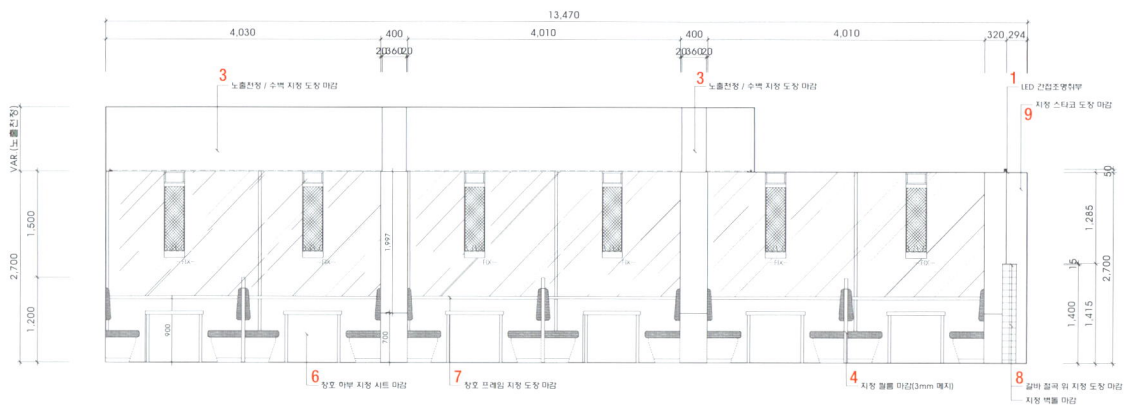

홀 입면 M / hall elevation M

1 LED 간접조명 2 프레임 : 갈바륨 절곡 위 지정 도장 3 노출천장, 수벽 지정 도장 4 지정 필름 5 지정 시멘트 보드 6 창호 하부 : 지정 시트 7 창호 프레임 : 지정 도장 8 갈바륨 절곡 위 지정 도장 / 지정 벽돌 9 지정 스투코 도장 10 간접조명 매입 11 프레임 : 지정 필름 12 문, 프레임 : 지정 필름 13 지정 유리 블록 14 T10 강화유리 15 몰딩 : 갈바륨 절곡 위 지정 도장 16 POS 설치 / 지정 스투코 도장 17 망입유리 도어 18 갈바륨 절곡 위 지정 도장

1 LED indirect lighting 2 Frame : App. painting on galvalume bending 3 Exposed ceiling, Reveal : App. painting 4 App. film 5 App. cement board 6 Below the window : App. sheet 7 Window frame : App. painting 8 App. painting on galvalume bending / App. brick 9 App. stucco painting 10 Embedding indirect lighting 11 Frame : App. film 12 Door, frame : App. film 13 App. glass block 14 T10 tempered glass 15 Moulding : App. painting on galvalume bending 16 POS installation / App. stucco painting 17 Wired sheet glass door 18 App. painting on galvalume bending

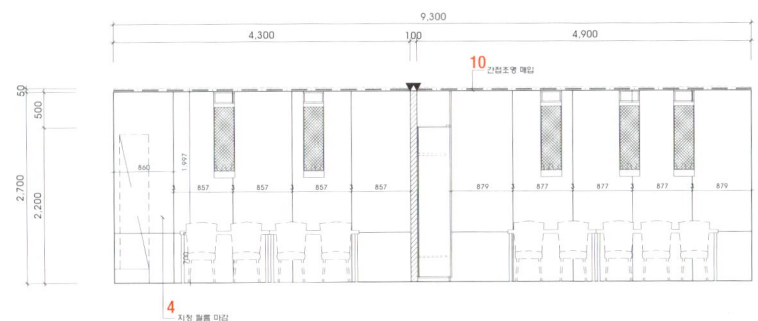

단체 룸 입면 N / group room elevation N

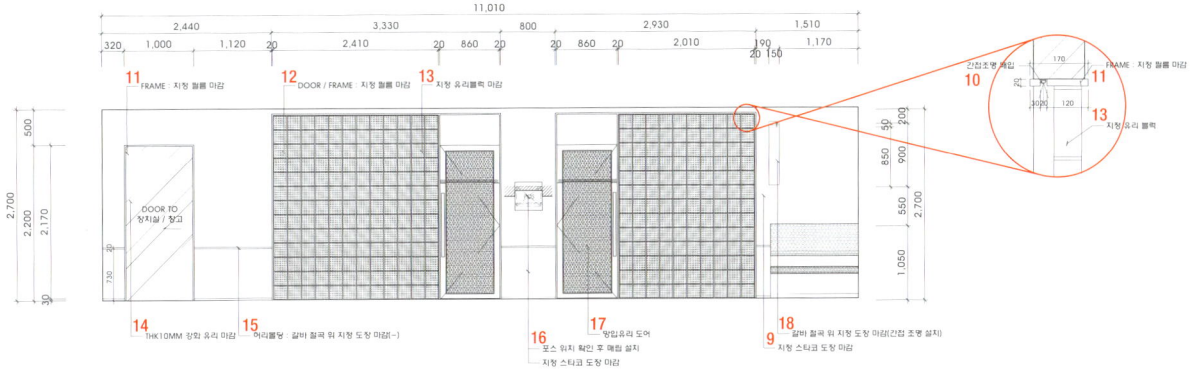

홀 입면 O / hall elevation O

CHEESEWAVE

BY SEOG BE SEOG | Junwoong Seog

사람들이 인생에서 자신이 누구인지 알아가는 데 많은 시간을 보내듯 모든 브랜드가 그 시작부터 정체성을 확립하기는 쉽지 않은 일이다. '치즈웨이브'라는 브랜드 또한 정체성을 발견해 나아가는 과정에 있었기 때문에 실체화되지 않은 감성적 이미지를 시각적으로 표현하는 것이 이번 프로젝트의 주안점이었다. 하지만 그 분기점이 될 송도 현장은 높은 천장고를 가지고 있어 브랜드의 이미지를 표현하기에 적절한 상황은 아니었다. 긴 고민과 논의 끝에 디자이너는 큰 공간을 배경삼아 작고 소박한 집을 한 채 짓기로 했다. 경제적이고 따뜻함이 베어 있는 소재들을 중심으로 벽을 세우고 지붕을 올려 집의 형태를 구축했고, 브랜드 컬러를 곳곳에 무심히 두어 고객들이 브랜드 이미지를 어렵지 않게 파악할 수 있도록 했다.

감성 중심의 경험을 할 수 있는 내부와는 다르게 외부에서는 성장과 확장이 필요한 브랜드의 상황을 고려해 보다 직관적으로 인지될 수 있는 포인트 요소가 필요하다고 생각했다. 파사드가 그러한 역할을 부여 받았고 'wave'라는 단어의 의미를 직관적으로 연출했다. 이로써 외부의 직관적인 표현과 내부의 감성적 경험의 조화를 통해 고객들에게 '치즈웨이브'의 아이덴티티가 명확하게 전달될 수 있도록 했다.

People spend much of their lives discovering who they are. Likewise, it is not easy for a brand to establish its identity from the beginning. 'Cheesewave' is still in the process of discovering its identity. Therefore, the biggest challenge of this project was to express an intangible emotional image through a visual language. However, the project site in Songdo, which marks an important milestone in this journey, had unusually high ceilings that made it difficult to express the brand's image. After extensive contemplation and discussion, the designer decided to build a small, modest house that would stand against the backdrop of the large space. For the house, economical and warm materials are used to form the basic structure of a house with walls and a roof. Brand colors are randomly placed throughout, allowing customers to easily recognize the brand's image. While the interior was designed to offer an emotional experience, the exterior required more intuitive design elements to match the brand's need for growth and expansion. This role was assigned to the facade, which was crafted to embody the concept of 'wave' in an intuitive way. As a result, the combination of intuitive language on the exterior and emotional experience in the interior helps effectively convey the brand's identity to customers.

디자인 석준웅 / 바이석비석
위치 인천광역시 연수구 송도과학로16번길 33-2, B동 126호
용도 식음공간
면적 87.25m²
마감 바닥 – 셀프레벨링 위 무광 코팅, 테라코타 타일 / 벽 – 페인트, 라왕 합판, 아연도금골판, 테라코타 타일 / 천장 – 노출천장 위 페인트
디자인팀 김미영
사진 김동규

Location B-126, 33-2, Songdogwahak-ro 16beon-gil, Yeonsu-gu, Incheon
Use F&B space
Area 87.25m²
Finishing Floor - Matt coat on self-leveling, Terracotta tile / Wall - Paint, Lauan plywood, Galvanized corrugated metal sheet, Terracotta tile / Ceiling - Paint on exposed ceiling
Photographer Donggyu Kim

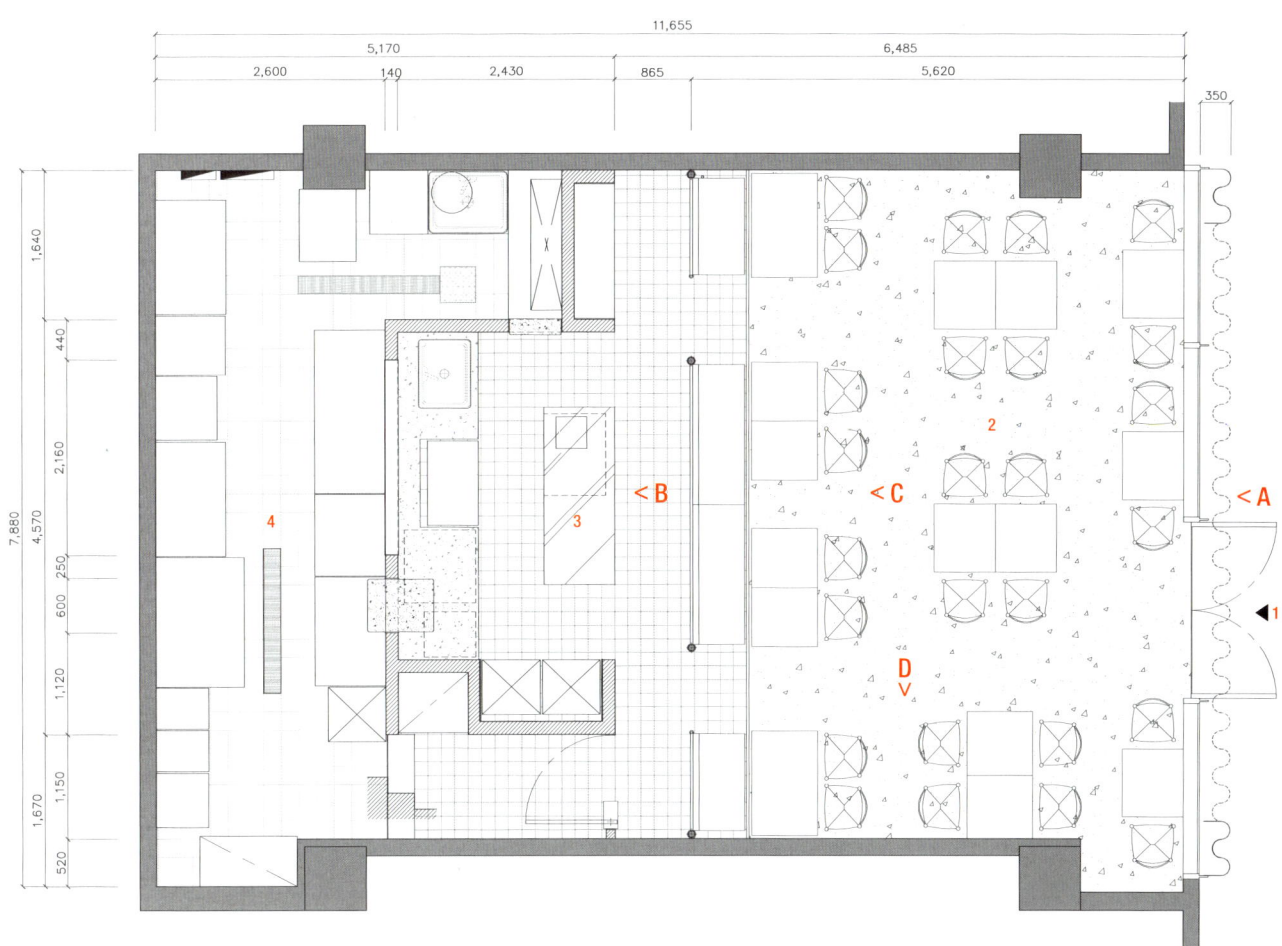

평면도 / floor plan

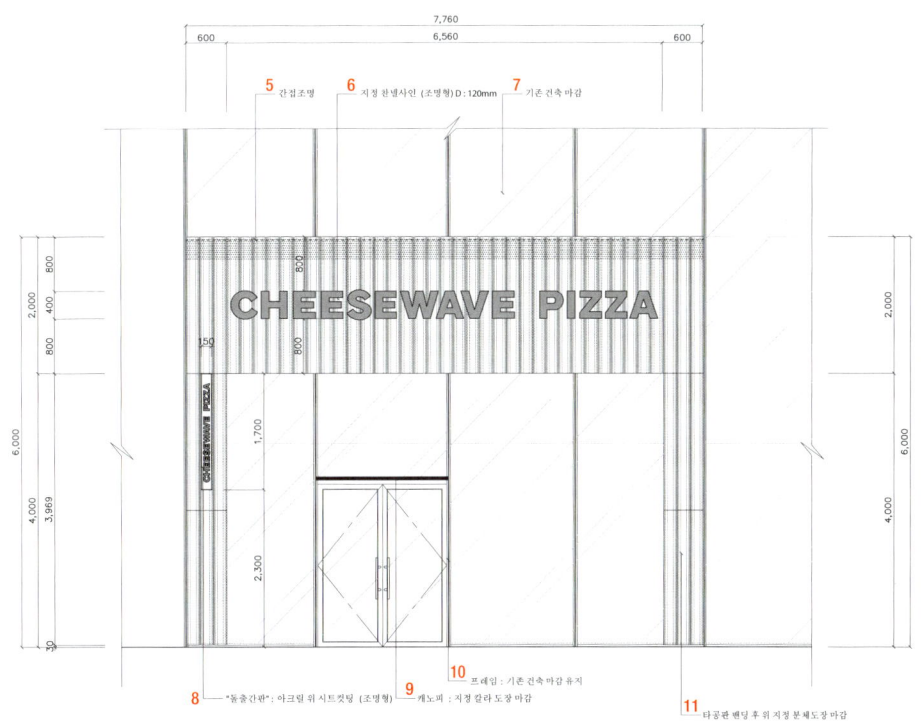

파사드 A / facade A

1 입구 2 홀 3 카운터 4 주방 5 간접조명 6 지정 조명형 채널사인 7 기존 건축 마감 8 간판 : 아크릴 위 시트 커팅(조명형) 9 캐노피 : 지정색 도장 10 프레임 : 기존 건축마감 11 타공판 절곡 후 지정 분체 도장

1 Entrance 2 Hall 3 Counter 4 Kitchen 5 Indirect lighting 6 App. illumated channel sign 7 Existing architecture finishing 8 Sign : Sheet cutting on acrylic (illuminated sign) 9 Canopy : App. color painting 10 Frame : Existing architecture finishing 11 App. powder coating after bending perforated sheet

1 지붕 : 금속 골판 위 지정색 도장   2 벽 : 지정색 스타코 도장 / 기존 레이스웨이 위 지정색 도장   3 간접조명   4 프레임 : 지정 분체도장   5 큐링 합판 위 지정 오일 스테인, 투명 도장   6 스테인리스 스틸 헤어라인   7 측면 : 지정 테라코타 타일   8 간접조명 / 상판 : T5 스테인리스 스틸 바이브레이션 / 지정 테라코타 타일   9 벽 : 큐링 합판 위 지정 오일 스테인, 투명 도장 / 걸레받이 : 큐링 합판 위 지정 오일 스테인, 투명 도장   10 선반 : 큐링 합판 위 지정 오일 스테인, 투명 도장   11 타공판 위 지정 분체도장   12 기존 레이스웨이 위 지정색 도장   13 지정 사인 : 지정 분체도장   14 지붕 : 금속 골판 위 지정색 도장 / 간접조명   15 캡 : 지정색 도장   16 Ø30 원형 파이프 위 지정색 도장 / 쿠션 : 지정 패브릭 / 좌판 : 지정 인조가죽   17 큐링 합판 위 지정 오일 스테인, 투명 도장 / Ø30 원형 파이프 위 지정색 도장   18 Ø80 원형 파이프 위 지정색 도장

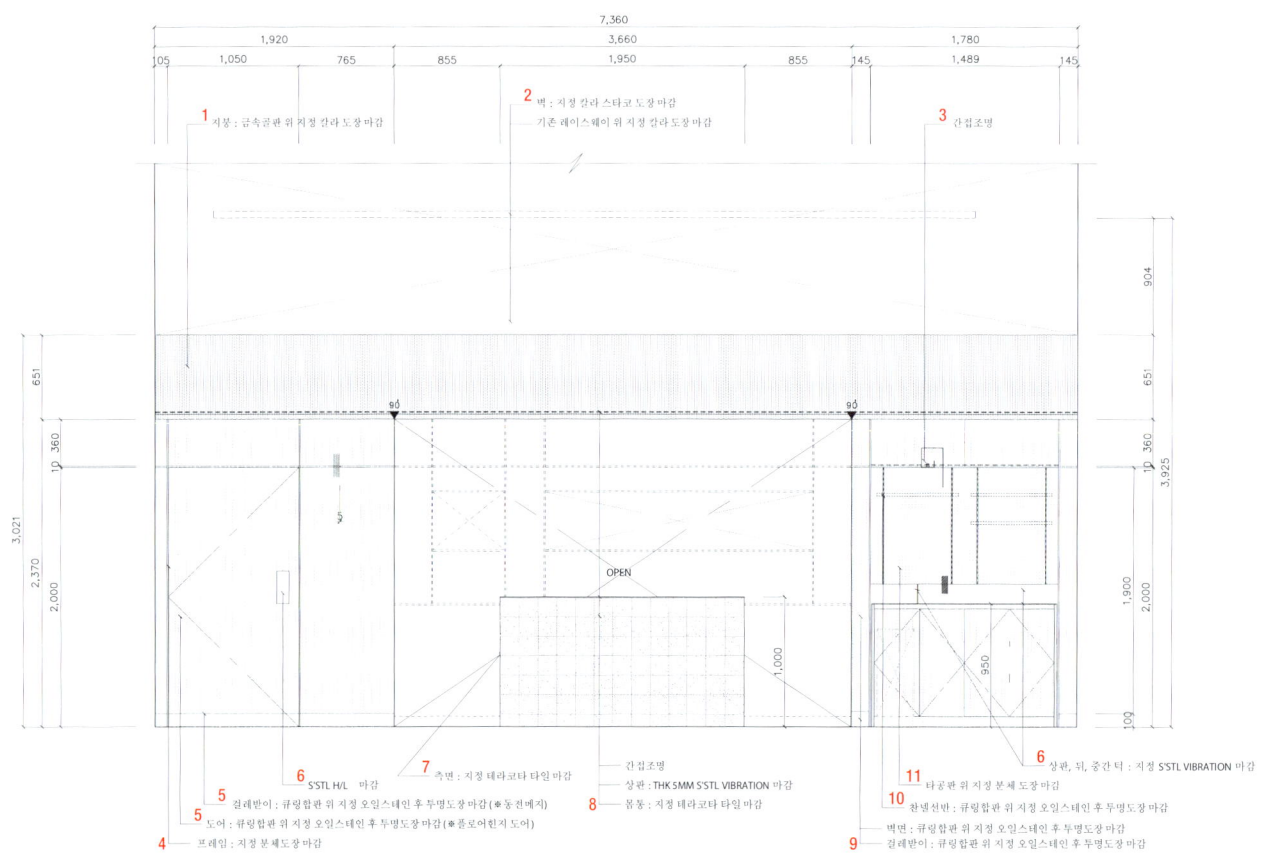

홀 입면 B / hall elevation B

1 Roof : App. color painting on corrugated sheet 2 Wall : App. color stucco painting / App. color painting on existing wire duct 3 Indirect lighting 4 Frame : App. powder coating 5 App. oil stain on keruing plywood / App. color painting on existing wire duct 6 Stainless steel hairline 7 Side : App. teracotta tile 8 Indirect lighting / Top : T5 stainless steel vibration / App. teracotta tile 9 Shelf : App. oil stain on keruing plywood / Baseboard : App. oil stain on keruing plywood, Clear coat 10 Shelf : App. oil stain on keruing plywood, Clear coat 11 App. powder coating on perforated sheet 12 App. color painting on existing wire duct 13 App. sign : App. powder coating 14 Roof : App. color painting on corrugated sheet / Indirect lighting 15 Cap : App. color painting 16 App. color painting on Ø30 round pipe / Cushion : App. fabric / Seat : App. artificial leather 17 App. oil stain on keruing plywood, Clear coat / App. color painting on Ø30 round pipe 18 App. color painting on Ø80 round pipe

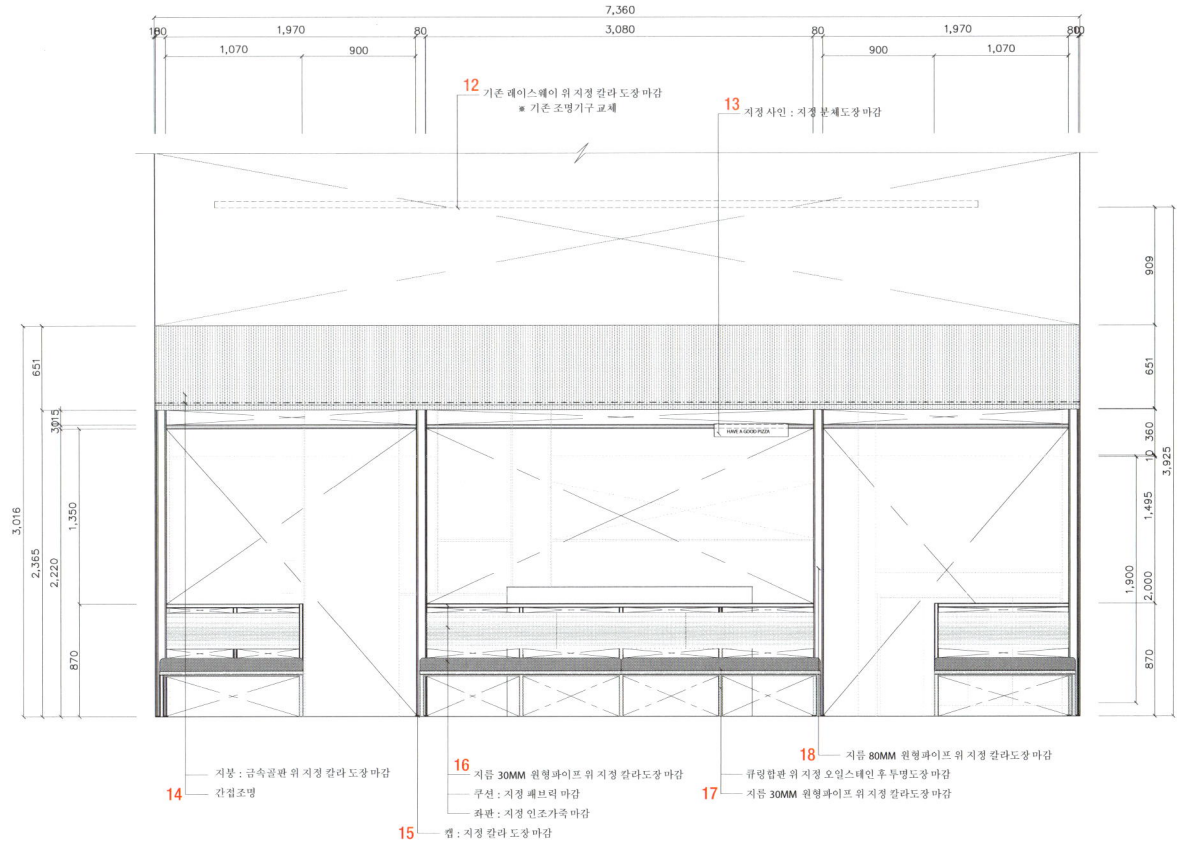

홀 입면 C / hall elevation C

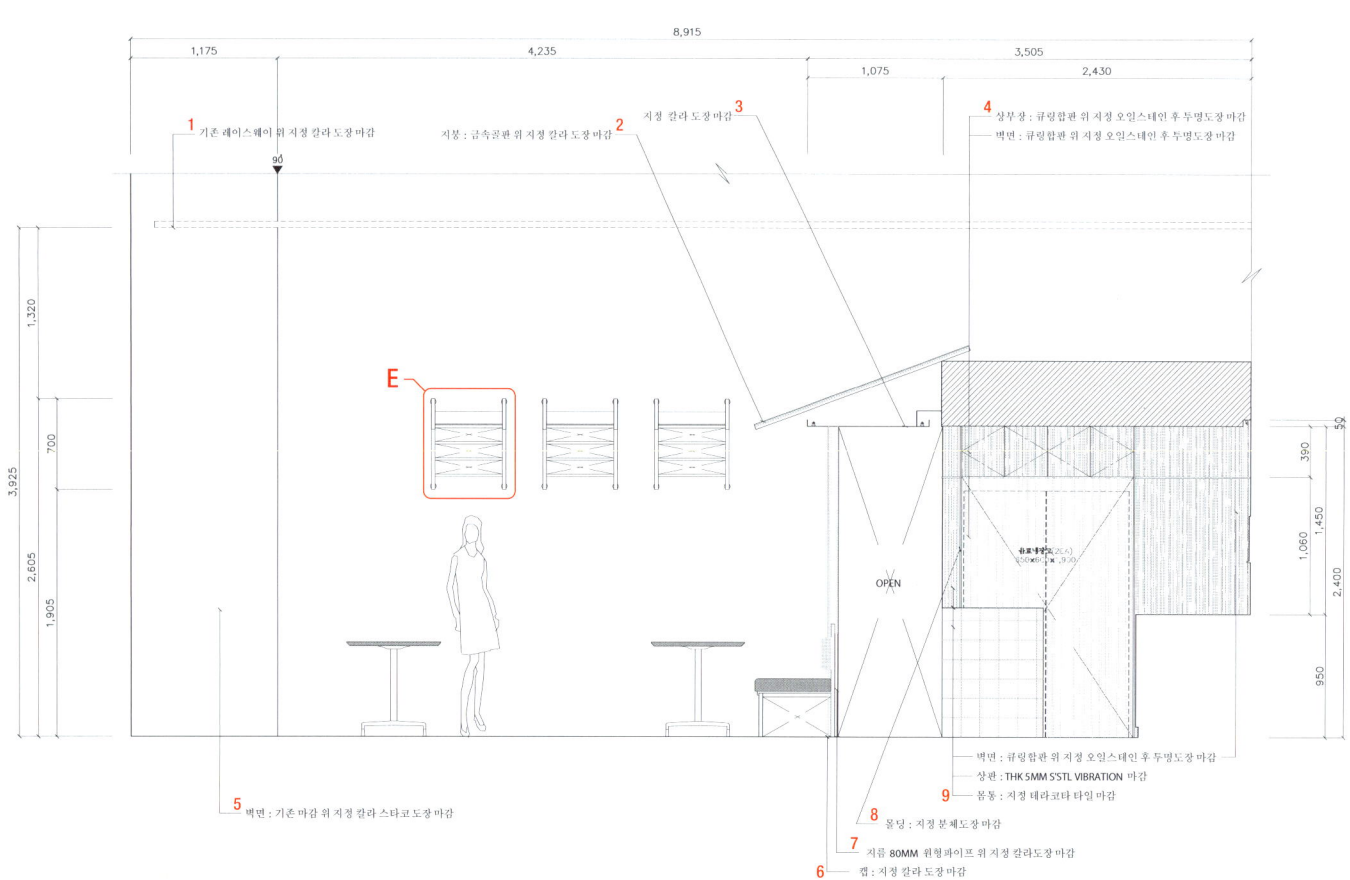

1 기존 레이스웨이 위 지정색 도장 2 지붕 : 금속 골판 위 지정색 도장 3 지정색 도장 4 상부장 : 큐링 합판 위 지정 오일 스테인, 투명 도장 / 벽 : 큐링 합판 위 지정 오일 스테인, 투명 도장 5 벽 : 기존 마감 위 지정색 스타코 도장 6 캡 : 지정색 도장 7 Ø80 원형파이프 위 지정색 도장 8 몰딩 : 지정 분체도장 9 벽 : 큐링 합판 위 지정 오일 스테인, 투명 도장 / 상판 : T5 스테인리스 스틸 바이브레이션 / 몸통 : 지정 테라코타 타일 10 Ø30 원형파이프 위 지정 분체 도장 11 Ø20 원형파이프 위 지정 분체 도장 12 T9 큐링 합판 위 지정 오일 스테인, 투명 도장 13 T9 큐링 합판 위 지정 오일 스테인, 투명 도장 / Ø20 원형파이프 위 지정 분체 도장

홀 입면 D / hall elevation D

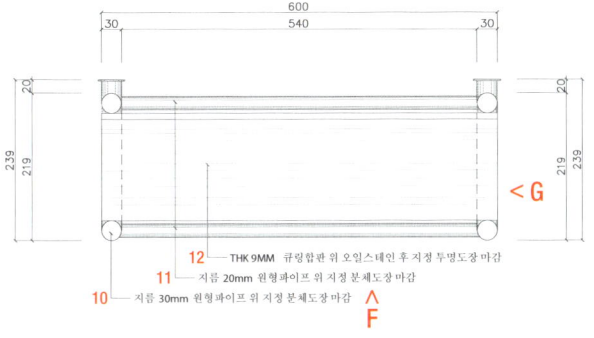

선반 평면 E / shelf top view E

1 App. color painting on existing wire duct   2 Roof : App. color painting on corrugated sheet   3 App. color painting   4 Upper cabinet : App. oil stain on keruing plywood, Clear coat / Wall : App. oil stain on keruing plywood, Clear coat   5 Wall : App. stucco painting on existing finishing   6 Cap : App. color painting   7 App. color painting on Ø80 round pipe   8 Moulding : App. powder coating   9 Wall : App. oil stain on keruing plywood, Clear coat / Top : T5 stainless steel vibration / Base : App. teracotta tile   10 App. powder coating on Ø30 round pipe   11 App. powder coating on Ø20 round pipe   12 App. oil stain on T9 keruing plywood, Clear coat   13 App. oil stain on T9 keruing plywood, Clear coat / App. powder coating on Ø20 round pipe

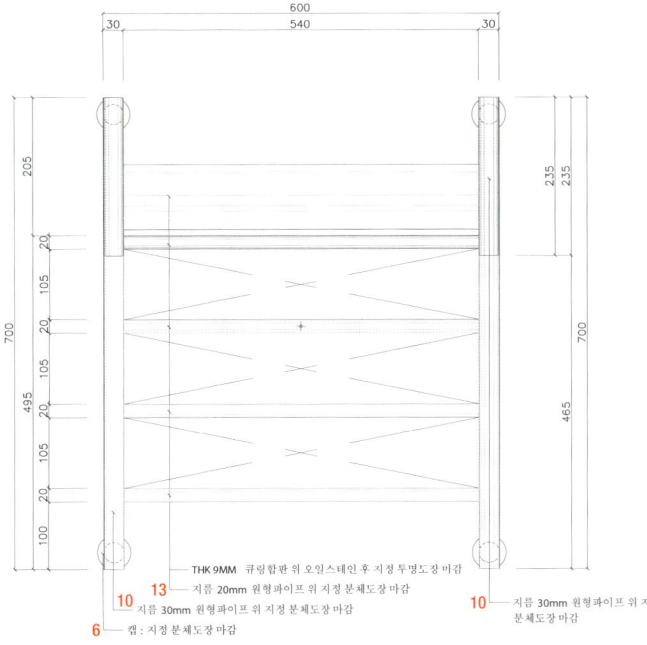

선반 정면 F / shelf front view F

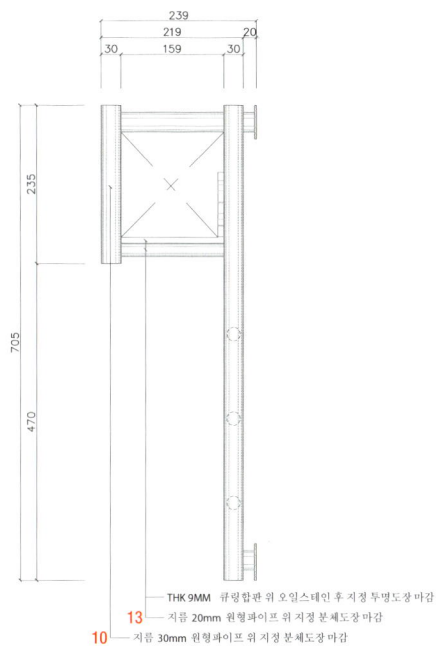

선반 측면 G / shelf side view G

STINKY BACON TRUCK

PROJECT MARK | Jaehong Son, Jeeseung Yang, Jihyung Song

스팅키 베이컨 트럭은 아메리칸 델리 문화의 느낌으로 남녀노소 모두가 편하게 이용할 수 있는 캐주얼 다이닝으로 기획되었다. "Garage Party"에서 영감을 받아 "Delicious Garage"로 콘셉트를 설정하고, 모던한 형태와 레트로한 마감, 그리고 펑키한 그래픽을 결합하여 스팅키 베이컨 트럭만의 독특한 분위기를 자아낸다. 전반적인 공간 분위기는 모던한 형태 속 레트로 포인트들을 활용하여 재해석된 빈티지를 연출하려 했다. 불규칙하게 조합된 펑키한 그래픽과 자유로운 디스플레이의 조합을 통해 델리의 분위기를 연출하였고, 바의 표면은 복고풍의 타일로 마감하여 한층 더 따뜻한 느낌을 가미해주었다. 펜던트 조명은 옛날 차고에서 흔히 볼 수 있는 산업용 형광등을 모티브로 현대적으로 재해석하여 디자인했다. 공간을 둘러싸고 있는 부스 소파는 안락함을 위해 쿠션을 두텁게 얹어주고, 이동식 의자는 미니멀한 형태와 소지품을 수납할 수 있는 가죽 해먹을 추가하여 빈티지한 포인트로 활용했다. 마지막으로 이러한 브랜딩의 바이브 속 스팅키 베이컨 트럭만의 아메리칸 메뉴들은 이 공간을 완성시키며 활력을 불어넣는다.

Stinky Bacon Truck is a casual dining space with an American delicatessen concept, welcoming guests of all ages. Inspired by "Garage Party," the design concept "Delicious Garage" is developed to combine modern forms with retro-style finishes and funky graphics and to create Stinky Bacon Truck's unique atmosphere. The space is characterized by a vintage sensibility, reinterpreted with retro-style elements within the modern framework. The delicatessen-like atmosphere is enhanced by irregularly combined funky graphics and casually arranged displays, while the bar surface is finished with retro tiles to add warmth. The pendant lights are a modern reinterpretation of industrial fluorescent lights commonly found in old garages. The booth sofas surrounding the space are topped with thick cushions for comfort, while the movable chairs are designed to have minimalist forms and leather hammocks for personal items, adding to the vintage aesthetic. Finally, Stinky Bacon Truck's American menu items perfectly complement the space's narrative, adding vibrancy to the overall atmosphere.

디자인 손재홍, 양지승, 송지형 / 프로젝트 마크
위치 경기도 수원시 장안구 수성로 75
용도 레스토랑
면적 178.8㎡
마감 타일, 페인트, 무늬목, 스테인리스 스틸
완공 2024. 1
디자인팀 김어진, 팽종인
사진 조동현

Location 75, Suseong-ro, Jangan-gu, Suwon-si, Gyeonggi-do
Use Restaurant
Area 178.8㎡
Finishing Tile, Paint, Wood veneer, Stainless steel
Completion 2024. 1
Photographer Donghyun Cho

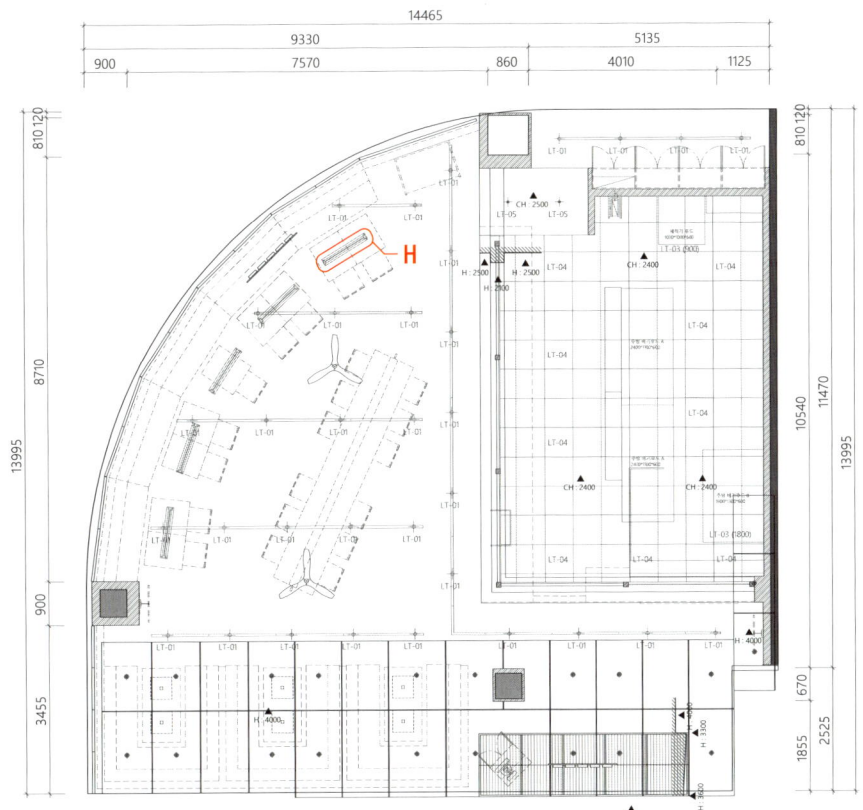

천장도 / ceiling plan

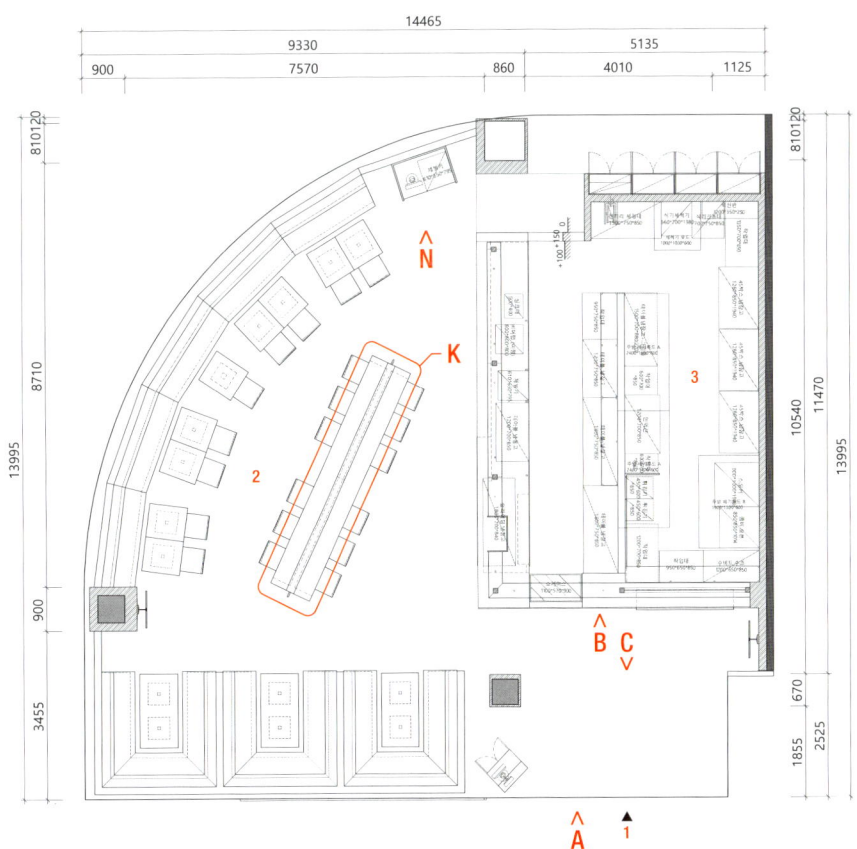

평면도 / floor plan

1 입구 2 홀 3 주방

1 Entrance 2 Hall 3 Kitchen

1 사인 2 컬러 강판 3 지정 조명 4 지정 가죽 / 지정 도장 / 걸레받이 : 지정 도장 위 코팅 5 난간 6 지정 도장 7 지정 타일 8 유리 블록 9 지정 금속 10 지정 도장 / 걸레받이 : 지정 도장 위 코팅 11 지정 금속 / 지정 금속

1 Sign   2 Color steel plate   3 App. lighting   4 App. leather / App. painting / Baseboard : App. painting, Coating   5 Balustrade   6 App. painting   7 App. tile   8 Glass block   9 App. metal   10 App. painting / Baseboard : App. painting, Coating   11 App. metal / App. metal

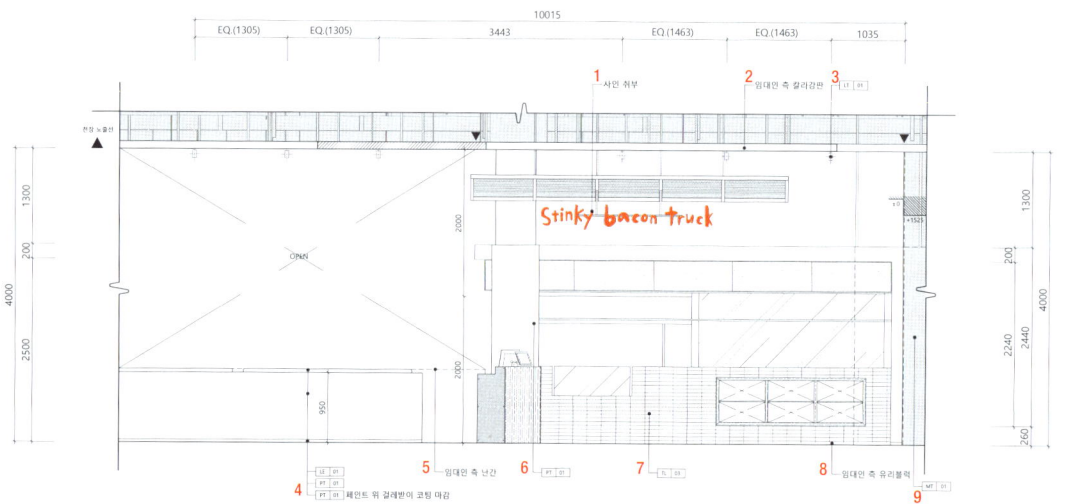

홀 입면 A / hall elevation A

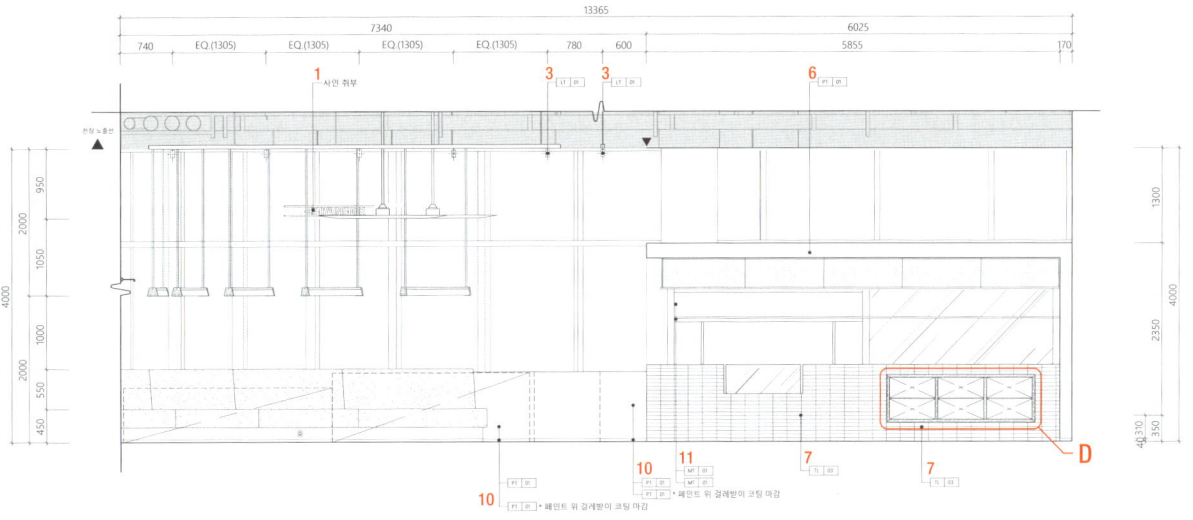

홀 입면 B / hall elevation B

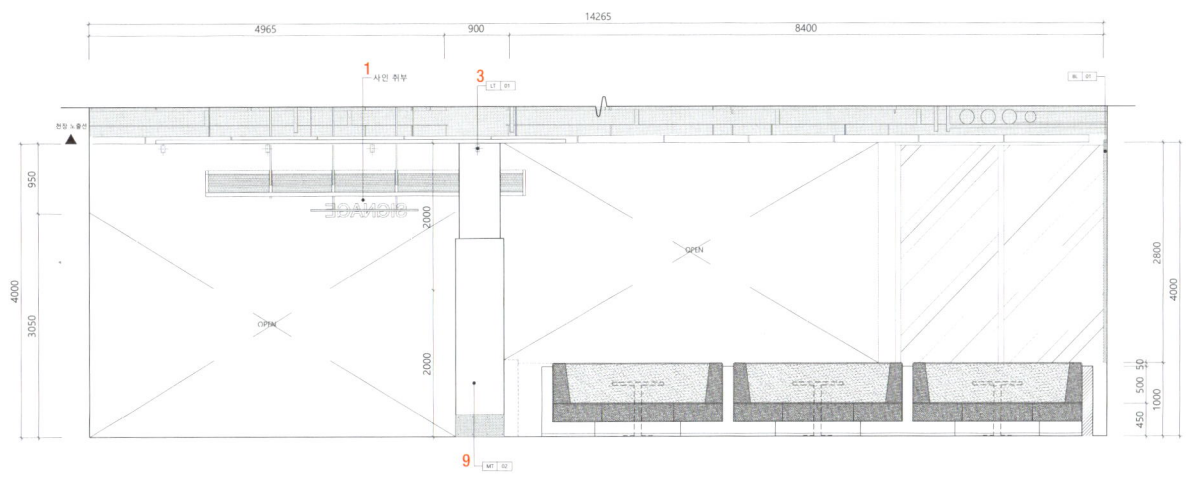

홀 입면 C / hall elevation C

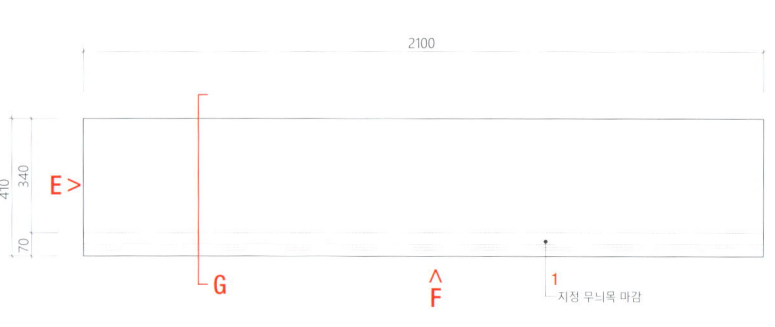

선반장 평면 D / shelf cabinet top view D

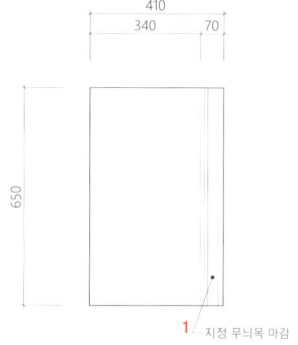

선반장 측면 E / shelf cabinet side view E

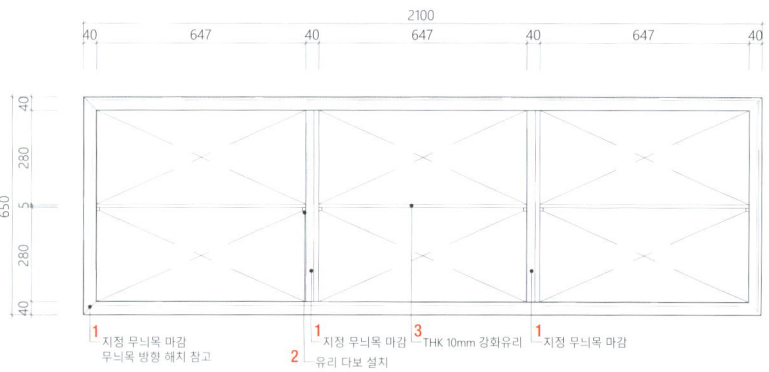

선반장 정면 F / shelf cabinet front view F

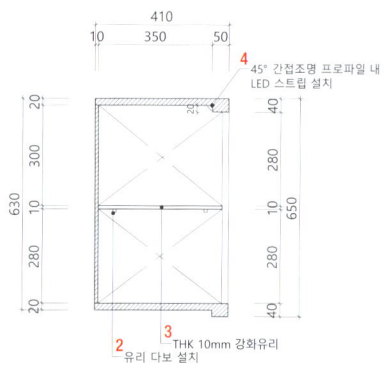

선반장 단면 G / shelf cabinet section G

1 지정 무늬목 2 유리 다보 설치 3 T10 강화유리 4 LED 스트립 설치 5 Ø10 접시머리 렌치 볼트 6 T2 SUS 실버 바이브레이션 / T10 새틴 유리 7 지정 와이어 8 T2 SUS 실버 바이브레이션 9 전선 구멍 타공 10 T2 SUS 실버 바이브레이션 / 3,000K 조명 / T10 새틴 유리 11 T10 새틴 유리 12 지정 조명

1 App. wood veneer 2 Glass standoff installation 3 T10 tempered glass 4 LED strip installation 5 Ø10 flat head cap screw 6 T2 SUS silver vibration / T10 satin glass 7 App. wire 8 T2 SUS silver vibration 9 Hole for cable 10 T2 SUS silver vibration / 3,000K lighting / T10 satin glass 11 T10 satin glass 12 App. lighting

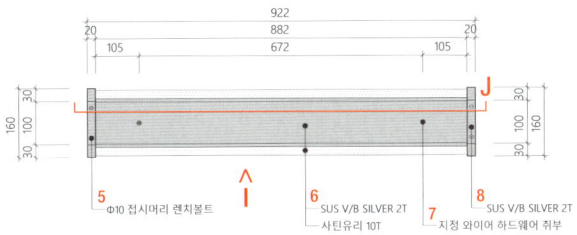

조명 평면 H / pendant light top view H

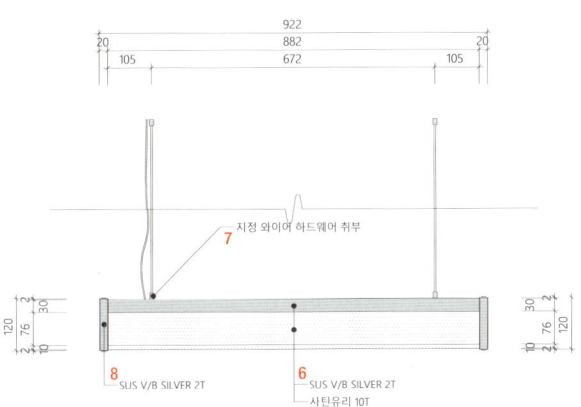

조명 정면 I / pendant light front view I

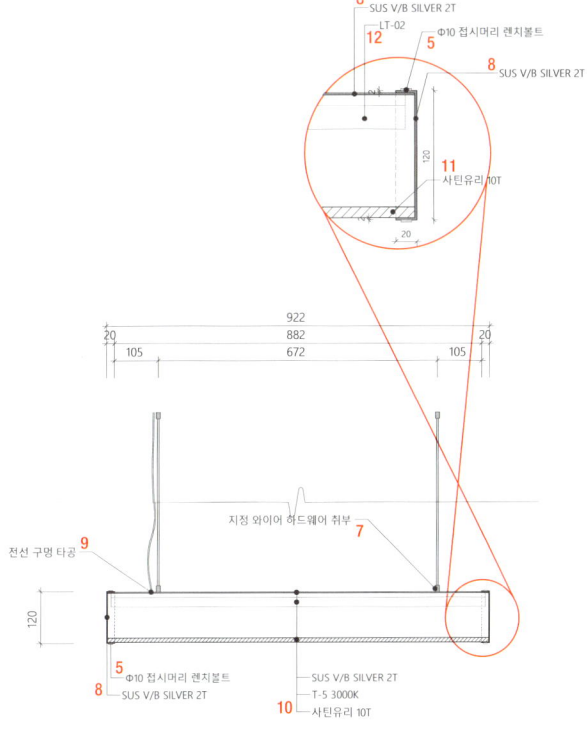

조명 단면 J / pendant light section J

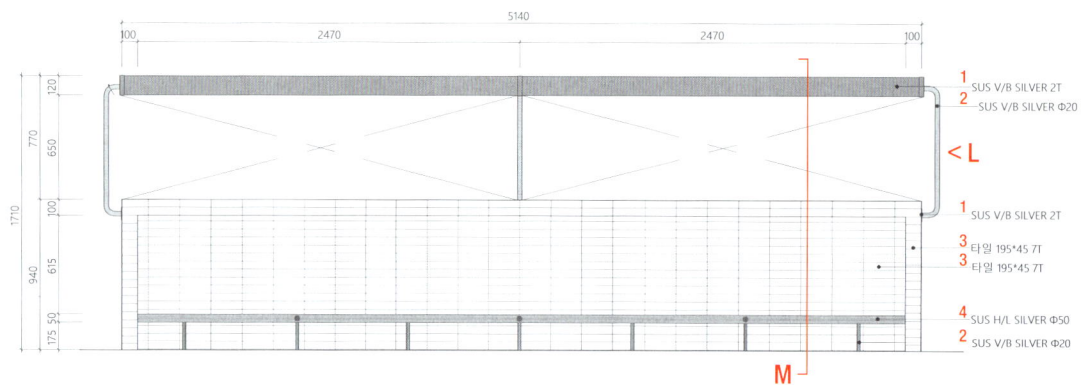

테이블 정면 K / table front view K

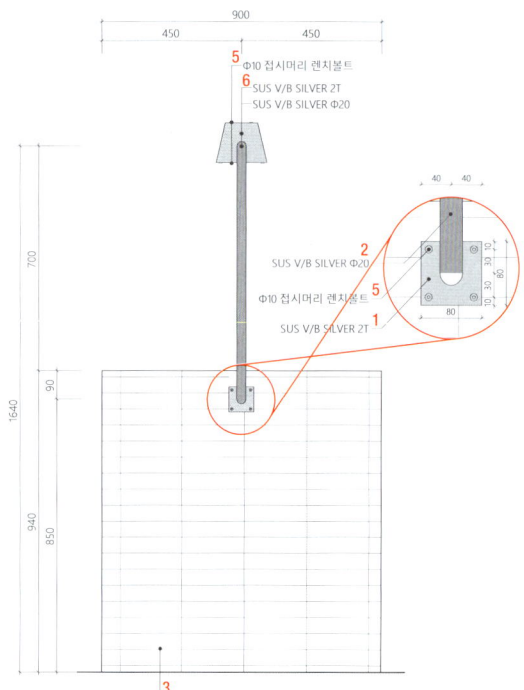

테이블 측면 L / table side view L

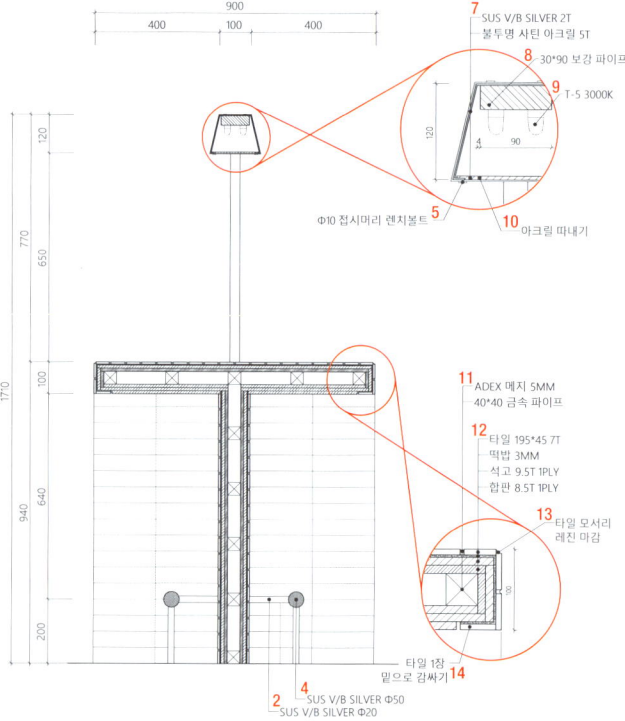

테이블 단면 M / table section M

1 T2 SUS 실버 바이브레이션 2 Ø20 SUS 실버 바이브레이션 3 T7 지정 타일 195X45 4 Ø50 SUS 실버 바이브레이션 5 Ø10 접시머리 렌치 볼트 6 T2 SUS 실버 바이브레이션 / Ø20 SUS 실버 바이브레이션 7 T2 SUS 실버 바이브레이션 / T5 새틴 아크릴 8 30X90 보강 파이프 9 3,000K 조명 10 아크릴 따내기 11 줄눈 5mm / 40X40 금속 파이프 12 T7 지정 타일 195X45 / 3mm 시멘트 / T9.5 석고보드 / T8.5 합판 13 타일 모서리 레진 마감 14 타일 15 사인 16 지정 조명 17 난간 18 지정 도장 / 걸레받이 : 지정 도장 위 코팅 19 지정 금속 20 지정 콘센트

1 T2 SUS silver vibration 2 Ø20 SUS silver vibration 3 T7 app. tile 195X45 4 Ø50 SUS silver vibration 5 Ø10 flat head cap screw 6 T2 SUS silver vibration / Ø20 SUS silver vibration 7 T2 SUS silver vibration / T5 satin acrylic 8 30X90 reinforced pipe 9 3,000K lighting 10 Cutting acrylic 11 Joint reveal 5mm / 40X40 square pipe 12 T7 app. tile 195X45 / 3mm cement / T9.5 gypsum board / T8.5 plywood 13 Resin finishing to tile edge 14 Tile 15 Sign 16 App. lighting 17 Balustrade 18 App. painting / Baseboard : App. painting, Coating 19 App. metal 20 App. outlet

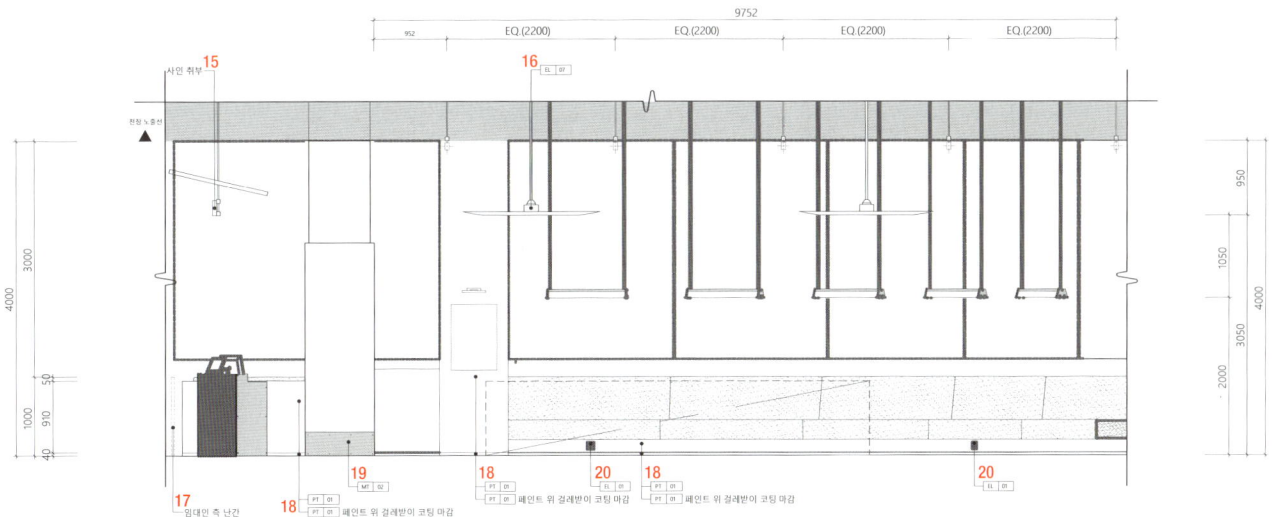

홀 입면 N / hall elevation N

DEOKBOON KOREAN DINING & GRILL

FLYMINGO | Junho Kim, Jiyeon Hwang

'덕분'은 베풀어 준 은혜나 도움이라는 뜻으로 긍정적 맥락에서 사용해온 우리네 정서가 담겨있는 단어다. 따로 '감사하다'라는 말을 하지 않아도 '덕분'이라는 단어에는 온기가 담겨있다. '덕분'은 그런 온기가 가득 담겨있는 공간이다. 디자이너가 생각하는 공간은 결과가 아닌 경험의 과정이다. 나열된 경험들은 하나의 이야기가 된다. 공간은 발단, 전개, 절정, 결말, 여운의 이야기로 구성되었다. 공간에 들어가 되돌아 나오는 과정 동안 전체적인 밝기와 톤앤무드가 그에 맞추어 변화하는 시퀀스를 통해 이야기가 진행되고 있음을 경험할 수 있다. 이야기는 마을 어귀에서 시작된다. 공간의 입구는 담으로 구성한 파사드가 마을 어귀를 연상케 하고, 높고 낮은 담장 너머로 쏟아지는 천창의 빛은 따스한 밥 짓는 연기를 형상화한 천장 오브제에 겹겹이 맺혀 환대의 감정을 전달한다. 담장을 지나 입구로 들어오면 마을의 고목 아래 바람에 흔들리는 잎 사이로 떨어져 산란하는 햇빛같이 살아 숨쉬는 온기가 담긴 안뜰을 마주하게 된다. 마당을 표현한 메인 홀에는 계절마다 바뀌는 제철음식의 다양한 식재료들이 레진 속에서 벽면을 가득 채우고 있다. 라이브그릴에서 피어오르는 불꽃과 마주앉아 식사하는 사람들의 온기, 레진 오브제 속에 담아낸 계절 등 저마다의 다양한 온기가 서로를 빛내는 긍정적 에너지로 아우러져 공간을 가득 채운다. 가장 깊은 공간인 룸은 온기를 가득 담은 숯의 물성을 담아내고자 했다. 숯은 제조과정에 따라 탄화된 나무의 표면에 하얀 재가 붙은 백탄과 새까만 검탄으로 나뉜다. 룸 역시 백탄룸과 검탄룸으로 나누어 공간을 다채롭게 구성했다. 룸으로 가는 복도의 끝엔 벼루로 벽을 마감하고 온기를 형상화한 오브제를 매달아 아늑한 공간 안에서 더욱 따뜻하게 짙어지는 온기를 표현했다.

'Deokboon' is a word that carries the meaning of grace or help bestowed upon someone, and it is used in a positive context. This word embodies the unique emotional sentiments of Korean people. Even without explicitly saying "thank you," the word 'deokboon' can convey a sense of warmth and gratitude. 'Deokboon' is a space filled with this kind of warmth. For the designer, space is not about the end result, but about a journey of experience. The accumulated experiences should become a story. The entire space is structured in a sequence of exposition, development, climax, resolution, and epilogue. As visitors enter and exit the space, the overall brightness and atmosphere change according to the stage of their journey, making them aware that a story is unfolding. The story begins at the entrance of a village. The façade, which is made of walls, evokes the entrance to a traditional village. Light streaming through the skylight over the high and low walls embraces a ceiling object that symbolizes the warm smoke from cooking rice, conveying a message of welcome. If visitors walk past the walls and through the entrance, they can find a courtyard where sunlight scatters through leaves swaying in the wind under an old village tree, filling the space with vivid warmth. The main hall, representing a traditional courtyard, features walls filled with seasonal ingredients preserved in resin, showcasing the variety of foods that change with each season. The warmth of people dining in front of the flickering flames of the live grill, and the seasons captured within the resin objects - these various forms of warmth complement each other as positive energy and enrich the narrative of space. In the deepest part of the space, private rooms are designed to capture the material properties of charcoal and its inherent warmth. Charcoal is classified into two types based on its manufacturing process: white charcoal, which has white ash on its carbonized wooden surface, and black charcoal. Likewise, the rooms are divided into white and black charcoal rooms, adding variety to the space. At the end of the corridor leading to these rooms, the walls are finished with inkstones, and objects symbolizing warmth are hung from above to express how warmth intensifies within this cozy space.

디자인 김준호, 황지연 / 플라이밍고
위치 서울특별시 용산구 장문로 23, 1층
용도 레스토랑
면적 674㎡
마감 바닥 - 포세린 타일 / 벽 - 디자인 패널, 페인트 / 천장 - 스페셜 페인트
완공 2024. 7
디자인팀 전민선, 김태훈, 류우진
사진 최용준

Location 1F, 23, Jangmun-ro, Yongsan-gu, Seoul
Use Restaurant
Area 674㎡
Finishing Floor - Porcelain tile / Wall - Design panel, Paint / Ceiling - Special paint
Completion 2024. 7
Photographer Yongjoon Choi

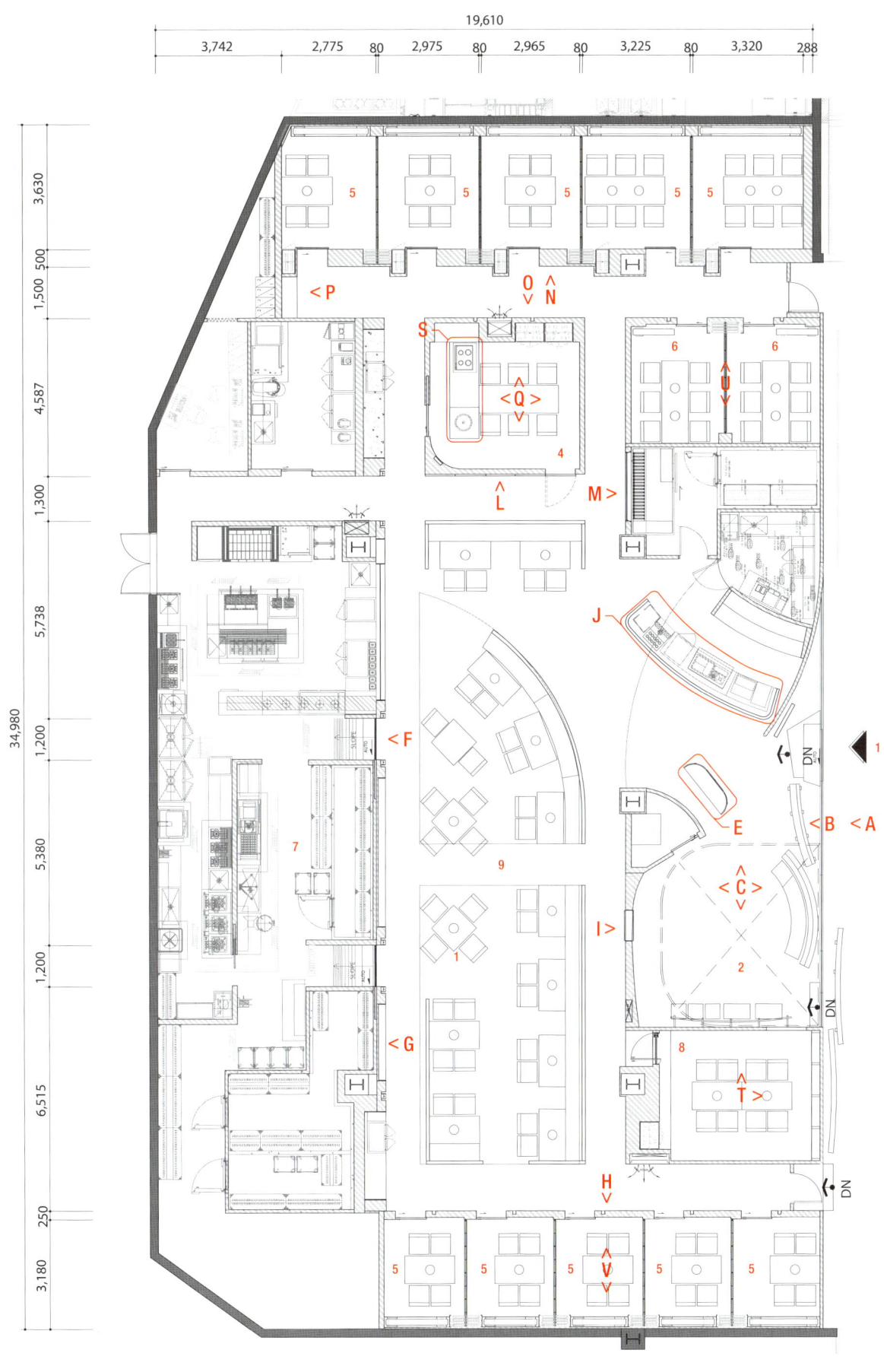

평면도 / floor plan

1 입구 2 안뜰 3 바 4 오마카세 룸 5 백탄룸 6 검탄룸 7 주방 8 VIP 룸 9 홀

1 Entrance 2 Courtyard 3 Bar 4 Omakase room 5 White room 6 Black room 7 Kitchen 8 VIP room 9 Hall

1 지정 횐색 스페셜 페인트   2 지정 디자인 패널 / 문 : 지정 디자인 패널   3 지정 투명 FRP   4 지정 투명 FRP / 지정 갈색 SUS 바이브레이션   5 자동문 : 투명 유리   6 클레이 로고 디자인   7 지정 콘크리트 보드 / 걸레받이 : 지정 스톤 타일   8 지정 콘크리트 보드 / 문 : 지정 콘크리트 보드   9 스폿라이트 : 지정 갈색 SUS 바이브레이션, Ø28 스폿라이트   10 문 : 지정 오크 고재   11 문 : 지정 콘크리트 보드   12 지정 디자인 패널   13 지정 오크 고재 / 지정 디자인 패널   14 카운터 : 지정 갈색 SUS 바이브레이션   15 지정 횐색 크랙 페인트   16 문 : 지정 디자인 패널   17 자동문 : 지정 투명 유리 / 계단 : 오크 고재 / 계단 : 지정 연회색 화강석   18 계단 : 지정 연회색 화강석   19 상부 간접조명   20 프레임 : 지정 갈색 SUS 바이브레이션   21 옷걸이 : 지정 갈색 SUS 바이브레이션 / 지정 디자인 패널

1 App. white special paint   2 App. design panel / Door : App. design panel   3 App. clear FRP   4 App. clear FRP / App. brown SUs vibration   5 Automatic door : Clear glass   6 Clay logo design   7 App. concrete board / Baseboard : App. stone tile   8 App. concrete board / Door : App. concrete board   9 Spot light : App. brown SUS vibration, Ø28 spot light   10 Door : App. old oak wood   11 Door : App. concrete board   12 App. design panel   13 App. old oak wood / App. design panel   14 Counter : App. brown SUS vibration   15 App. white crack paint   16 Door : App. design panel   17 Automatic door : App. clear glass / Stairs : Old oak wood / Stairs : App. light gray granite stone   18 Stairs : App. light gray granite stone   19 Upper indirect lighting   20 Frame : App. brown SUS vibration   21 Hanger : App. brown SUS vibration / App. design panel

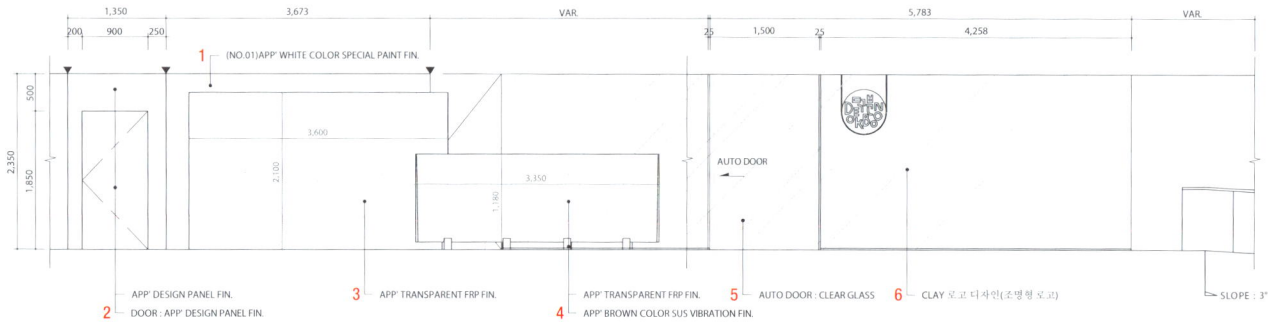

파사드 A / facade A

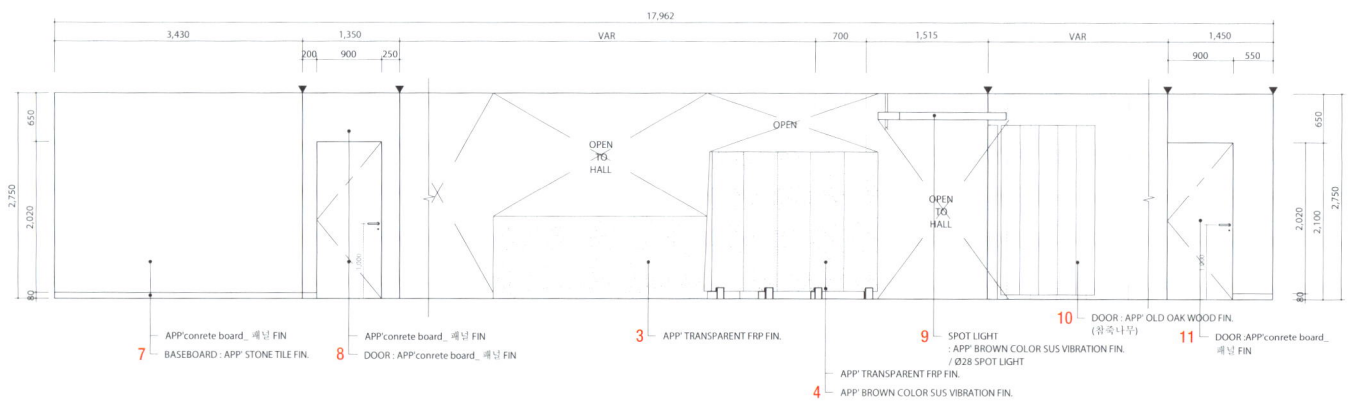

파사드 B / facade B

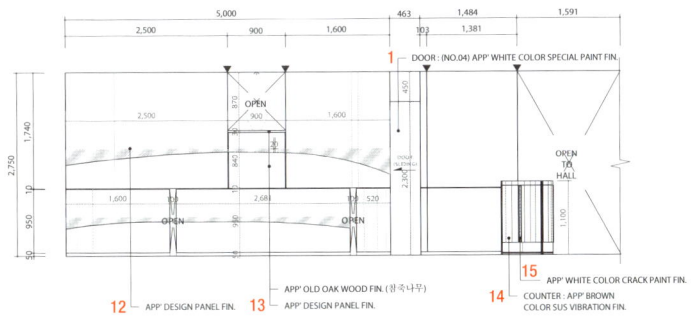

안뜰 입면 C1 / courtyard elevation C1

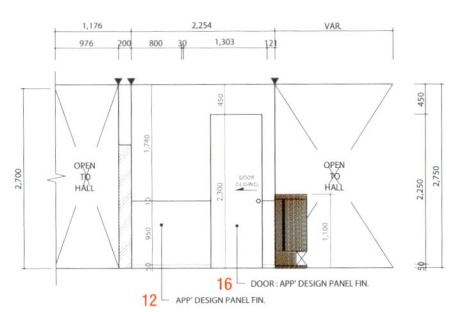

안뜰 입면 C2 / courtyard elevation C2

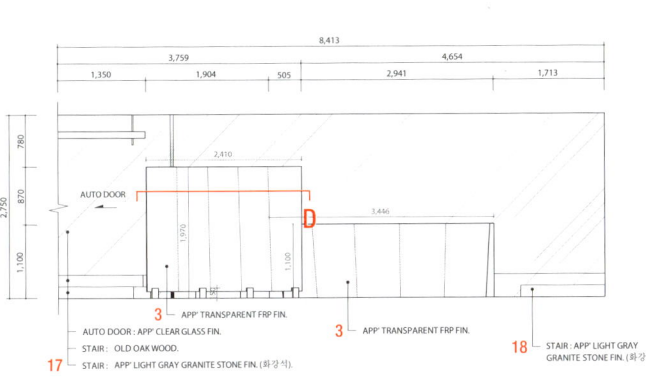

안뜰 입면 C3 / courtyard elevation C3

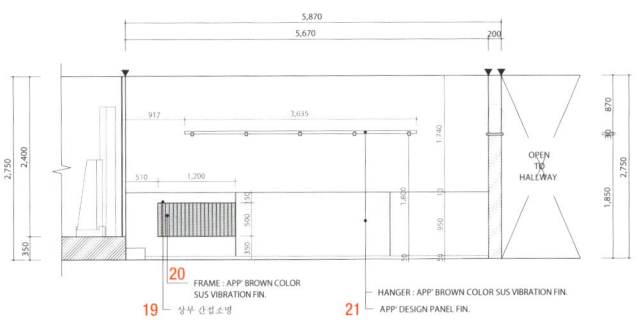

안뜰 입면 C4 / courtyard elevation C4

1 지정 투명 FRP   2 지지대 : 지정 갈색 SUS 바이브레이션   3 지정 투명 FRP / 지지대 : 지정 갈색 SUS 바이브레이션   4 지정 갈색 우레탄 페인트 / T1.2 지정 갈색 SUS 바이브레이션   5 T1.2 지정 갈색 SUS 바이브레이션   6 지정 검은색 우레탄 페인트 / T1.2 지정 갈색 SUS 바이브레이션   7 지정 흰색 크랙 페인트 / 지정 검은색 우레탄 페인트 / T1.2 지정 갈색 SUS 바이브레이션

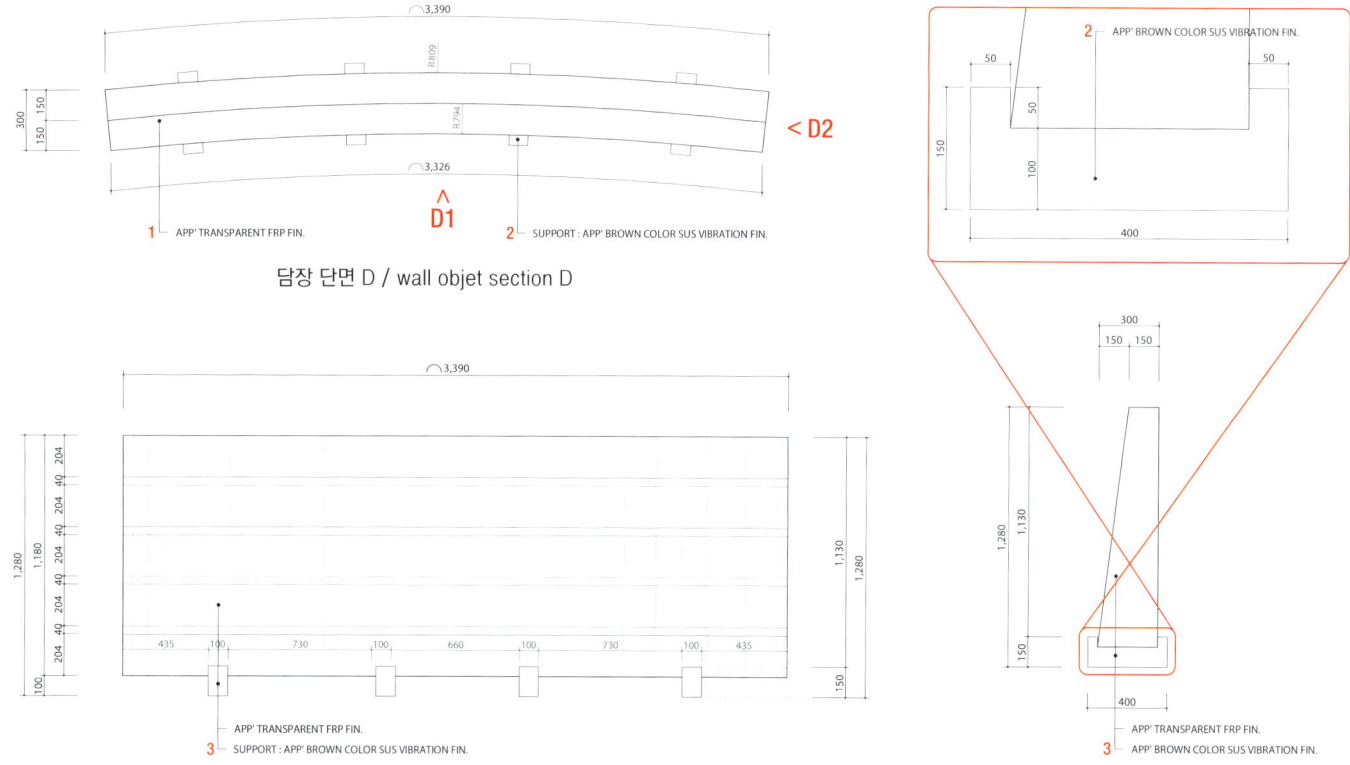

담장 단면 D / wall objet section D

담장 정면 D1 / wall objet front view D1

담장 측면 D2 / wall objet side view D2

1 App. clear FRP 2 Support : App. brown SUS vibration 3 App. clear FRP / Support : App. brown SUS vibration 4 App. brown urethane paint / T1.2 app. brown SUS vibration 5 T1.2 app. brown SUS vibration 6 App. black urethane paint / T1.2 app. brown SUS vibration 7 App. white crack paint / App. black urethane paint / T1.2 app. brown SUS vibration

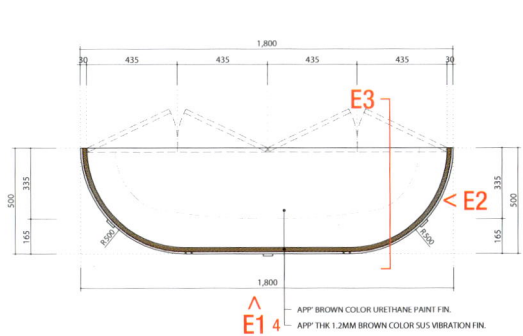

카운터 평면 E / counter top view E

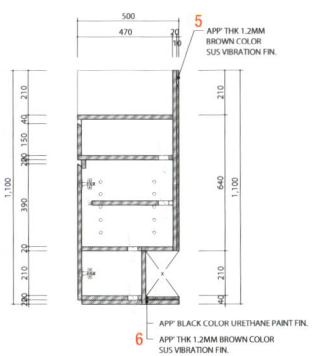

카운터 단면 E3 / counter section E3

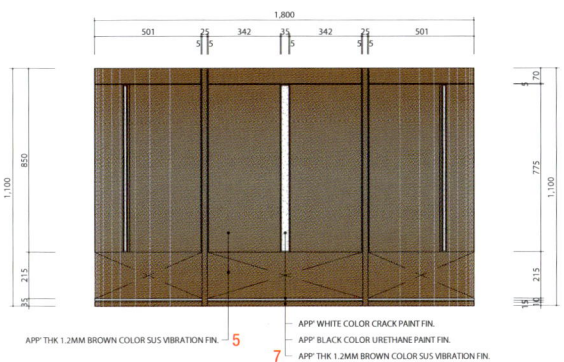

카운터 정면 E1 / counter front view E1

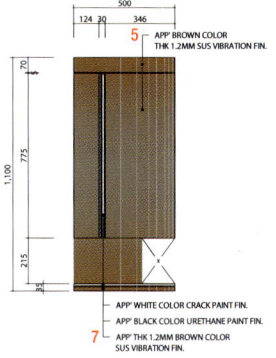

카운터 측면 E2 / counter side view E2

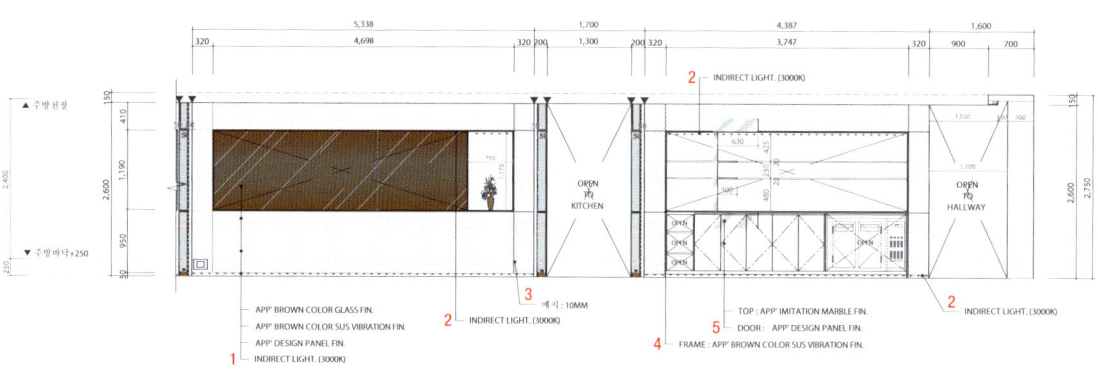

홀 입면 F / hall elevation F

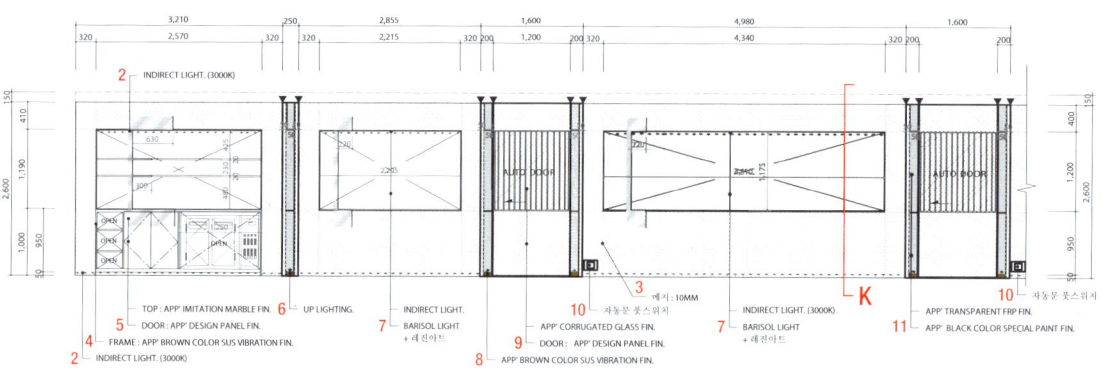

홀 입면 G / hall elevation G

1 지정 갈색 유리 / 지정 갈색 SUS 바이브레이션 / 지정 디자인 패널 / 간접조명 2 간접조명 3 10mm 줄눈 4 프레임 : 지정 갈색 SUS 바이브레이션 5 상판 : 지정 인조대리석 / 문 : 지정 디자인 패널 6 업라이트 7 간접조명 / 바리솔 조명 + 레진 아트 8 지정 갈색 SUS 바이브레이션 9 지정 골판유리 / 문 : 지정 디자인 패널 10 자동문 풋 스위치 11 지정 투명 FRP / 지정 검은색 스페셜 페인트 12 지정 디자인 패널 / 걸레받이 : 지정 회색 스페셜 페인트 13 지정 검은색 스페셜 페인트 / 오브제 조명 14 문 : 지정 디자인 패널 15 제작 손잡이 16 지정 디자인 패널 / 간접조명 17 지정 오크 고재 / 지정 디자인 패널 18 주물 제작 손잡이

1 App. brown glass / App. brown SUS vibration / App. design panel / Indirect lighting 2 Indirect lighting 3 10mm joint reveal 4 Frame : App. brown SUS vibration 5 Top : App. imitation marble / Door : App. design panel 6 Uplight 7 Indirect lighting / Barrisol + Resin art 8 App. brown SUS vibration 9 App. corrugated glass / Door : App. design panel 10 Automatic foot switch 11 App. clear FRP / App. black special paint 12 App. design panel / Baseboard : App. gray special paint 13 App. black special paint / Objet light 14 Door : App. design panel 15 Custom-made handle 16 App. design panel / Indirect lighting 17 App. old oak wood / App. design panel 18 Custom-made casting handle

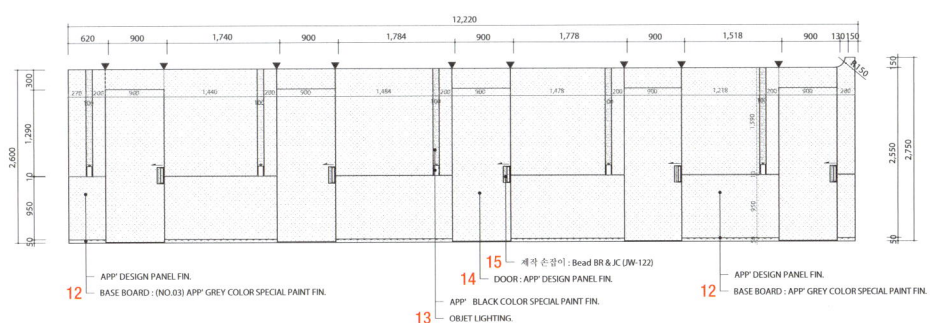

복도 입면 H / hallway elevation H

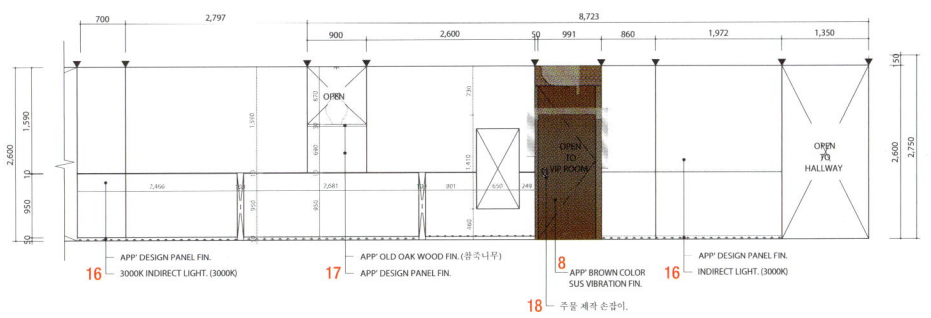

복도 입면 I / hallway elevation I

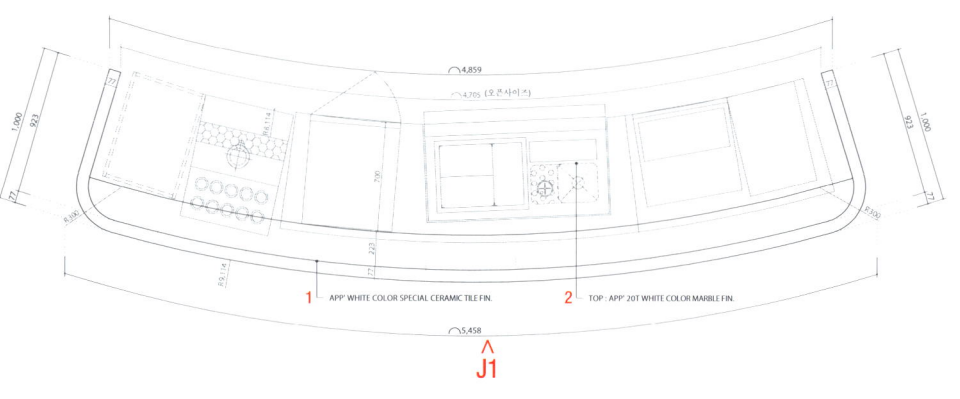

바 평면 J / bar top view J

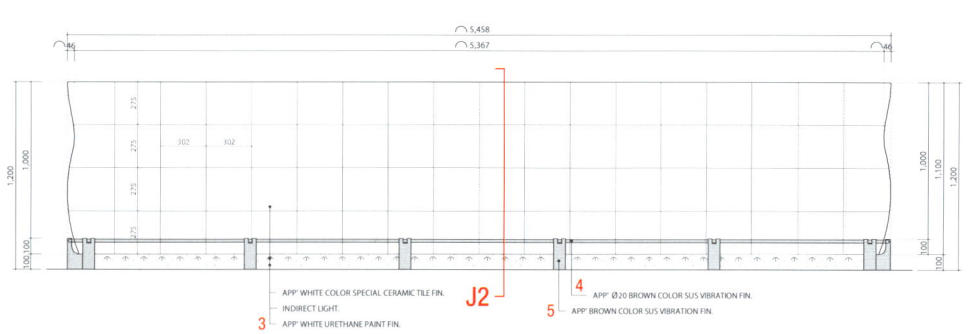

바 정면 J1 / bar front view J1

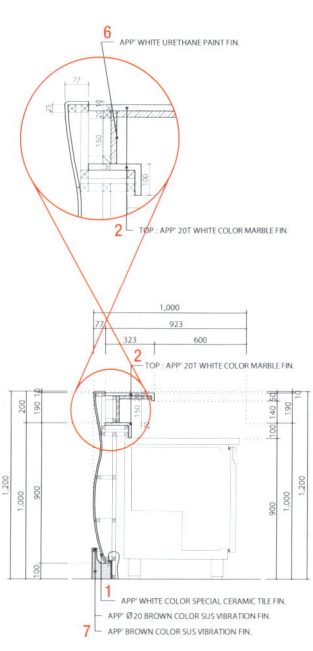

바 단면 J2 / bar section J2

1 지정 흰색 세라믹 타일 2 상판 : T20 지정색 대리석 3 지정 흰색 세라믹 타일 / 간접조명 / 지정 흰색 우레탄 페인트 4 Ø20 지정 갈색 SUS 바이브레이션 5 지정 갈색 SUS 바이브레이션 6 지정 흰색 우레탄 페인트 7 Ø20 지정 갈색 SUS 바이브레이션 / 지정 갈색 SUS 바이브레이션 8 T1.6 아연도금강판 위 지정 스페셜 페인트 9 다운라이트 10 LED 바 11 T9.5 석고보드 2겹 위 지정 스페셜 페인트 12 T1.2 갈색 SUS 바이브레이션 13 레진 오브제 14 패브릭 조명 15 RGB 조명 16 T9 합판 위 지정 흰색 도장 17 □30X30 철제 파이프 구조 18 T9 연회색 타일

1 App. white ceramic tile 2 Top : T20 app. color marble 3 App. white ceramic tile / Indirect lighting / App. white urethane paint 4 Ø20 app. brown SUS vibration 5 App. brown SUS vibration 6 App. white urethane paint 7 Ø20 app. brown SUS vibration / App. brown SUS vibration 8 App. special paint on T1.6 galvanized steel 9 Downlight 10 LED bar 11 App. special paint on T9.5 gypsum board 2ply 12 T1.2 brown SUS vibration 13 Resin objet 14 Fabric light 15 RGB light 16 App. white paint on T9 plywood 17 □30X30 steel pipe structure 18 T9 light gray tile

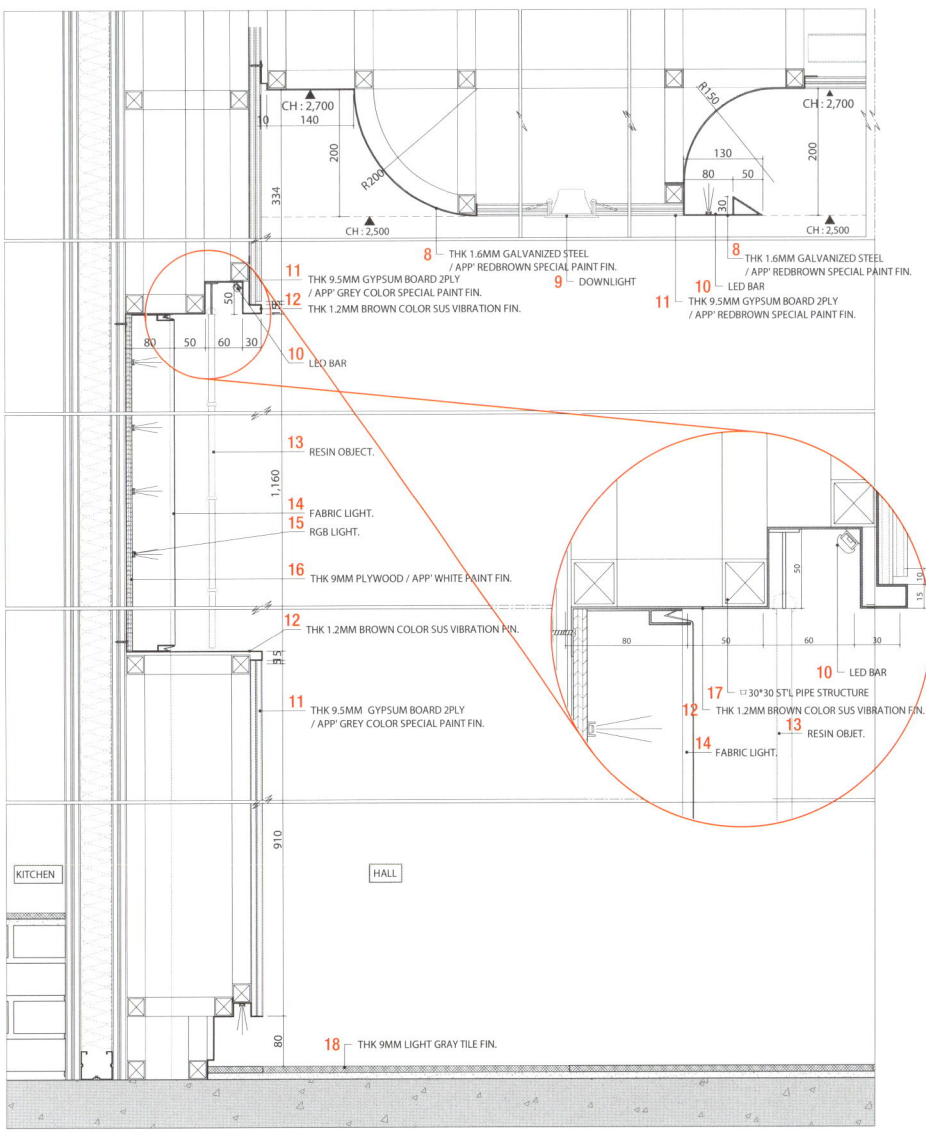

홀 벽 단면 상세 K / hall wall section detail K

1 그래픽 : 지정 금속 사인 2 지정 디자인 패널 3 프레임 : 지정 갈색 SUS 바이브레이션 4 문 : 지정 디자인 패널 5 지정 디자인 패널 / 간접조명 / 걸레받이 : 지정 회색 스페셜 페인트 6 와인셀러 시창 : 지정 투명 유리 / 프레임 : 지정 갈색 SUS 바이브레이션 7 지정 디자인 패널 / 걸레받이 : 지정 회색 타일 8 식물 오브제 9 간접조명 10 지정 디자인 패널 / 문 : 지정 디자인 패널 11 줄눈 10mm 12 지정 흰색 크랙 페인트 / 포인트 조명 / 지정 회색 스페셜 페인트 / 걸레받이 : 지정 회색 타일 13 지정 디자인 패널 / 간접조명 14 벼루 부착

1 Graphic : App. metal sign 2 App. design panel 3 Frame : App. brown SUS vibration 4 Door : App. design panel 5 App. design panel / Indirect lighting / Baseboard : App. gray special paint 6 Wine cellar window : App. clear glass / Frame : App. brown SUS vibration 7 App. design panel / Baseboard : App. gray tile 8 Plant objet 9 Indirect lighting 10 App. design panel / Door : App. design panel 11 Joint reveal 10mm 12 App. white crack paint / Point light / App. gray special paint / Baseboard : App. gray tile 13 App. design panel / Indirect lighting 14 Ink stone

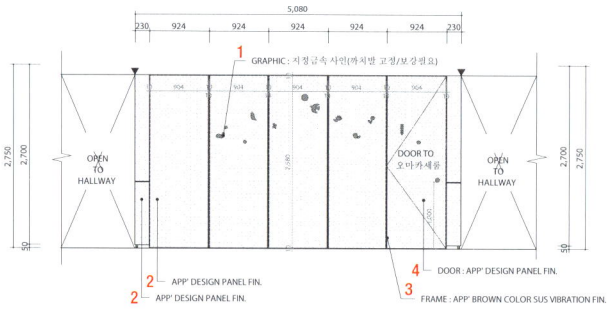

복도 입면 L / hallway elevation L

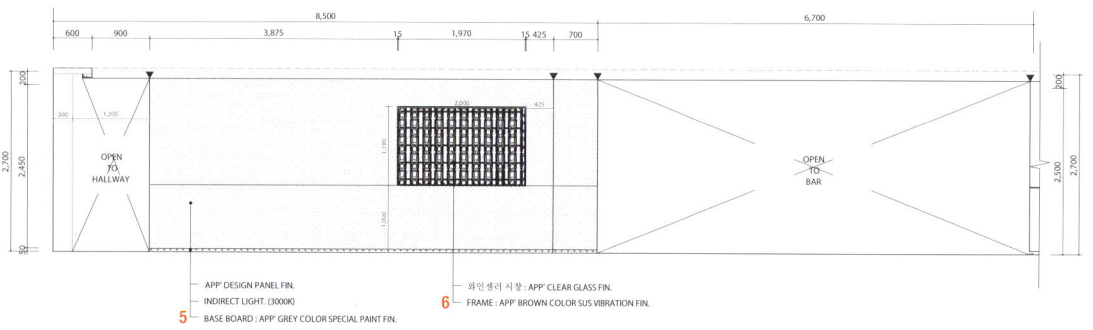

복도 입면 M / hallway elevation M

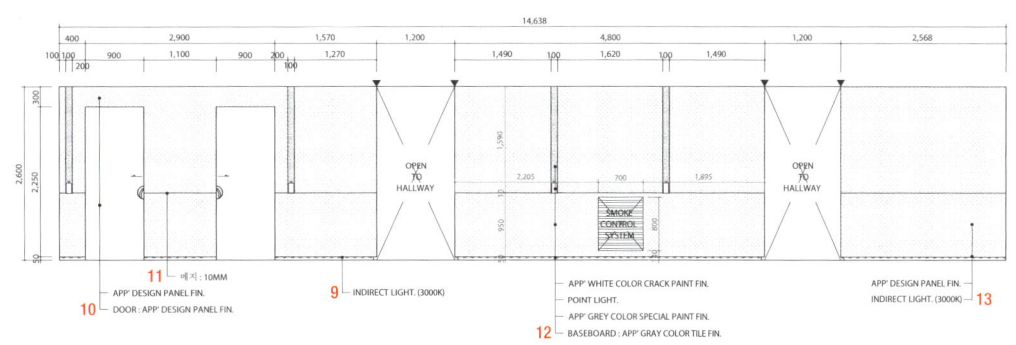

복도 입면 N / hallway elevation N

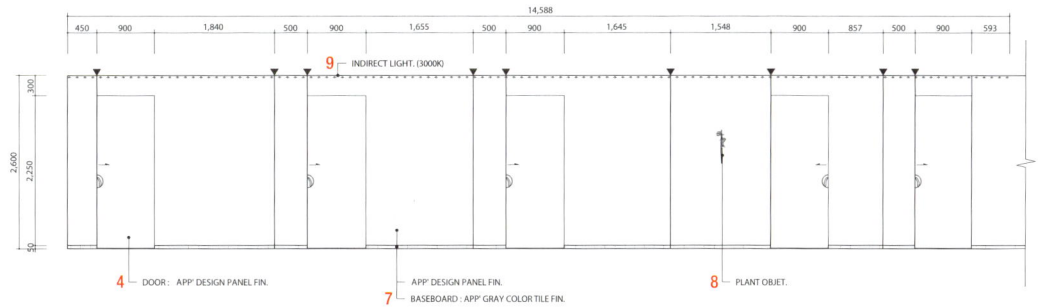

복도 입면 O / hallway elevation O

복도 입면 P /
hallway elevation P

1 간접조명 2 지정 검은색 스페셜 페인트 / 지정 검은색 탄화목 3 지정 검은색 탄화목 / 손잡이 : 지정 검은색 우레탄 4 선반 : 검은색 무늬목 5 선반 : 검은색 무늬목 / 문 : 검은색 무늬목 6 후드 : 지정 검은색 SUS 헤어라인 / 카운터 : 지정 화강석 7 문 : 검은색 무늬목 / 프레임 : T10 지정 검은색 SUS 헤어라인 8 옷장 : 검은색 무늬목 9 지정 검은색 흡음뿜칠 / 지정 포인트 대리석 10 문 : 지정 검은색 탄화목 / 프레임 : 지정 갈색 SUS 바이브레이션 11 지정 검은색 탄화목 / 걸레받이 : 지정 검은색 스페셜 페인트 12 T9.5 석고보드 2겹 위 검은색 흡음뿜칠 13 T1.6 아연도금강판 위 지정 검은색 흡음뿜칠 14 트랙조명 15 T20 포인트 대리석 16 □30X30 철제 파이프 구조 17 T9 진회색 타일 18 T12.5 석고보드 2겹, T9 MDF, 지정 회색 스페셜 페인트 19 Ø500 타공 / 지정 화강석 20 지정 화강석 21 4구 인덕션 타공 22 인덕션 23 지정 연회색 우레탄 페인트 / 지정 연회색 우레탄 페인트 24 지정 화강석 / 지정 연회색 우레탄 페인트 25 □30X30 각파이프 위 20T 합판 보강 26 지정 화강석, □30X30 각파이프 위 20T 합판 보강

1 Indirect lighting 2 App. black special paint / App. black color burned wood panel 3 App. black color burned wood panel / Handle : App. black urethane 4 Shelf : Black wood veneer 5 Shelf : Black wood veneer / Door : Black wood veneer 6 Hood : App. black SUS hairline / Counter : App. granite 7 Door : App. wood veneer / Frame : T10 app. black SUS hairline 8 Closet : Black wood veneer 9 App. black cotton spray paint / App. point marble 10 Door : App. black color burned wood panel / Frame : App. brown SUS vibration 11 App. black color burned wood panel / Baseboard : App. black special paint 12 Black cotton spray paint on T1.6 galvanized steel 2ply 13 Black cotton spray paint on T1.6 galvanized steel 14 Track light 15 T20 point marble 16 □30X30 steel pipe structure 17 T9 dark gray tile 18 T12.5 gypsum board 2ply, T9 MDF, App. gray special paint 19 Ø500 hole / App. granite 20 App. granite 21 4-element induction cooktop hole 22 Induction cooktop 23 App. light gray urethane paint / App. light gray urethane paint 24 App. granite / App. light gray urethane paint 25 20T plywood reinforcement on □30X30 square pipe 26 App. granite, 20T plywood reinforcement on □30X30 square pipe

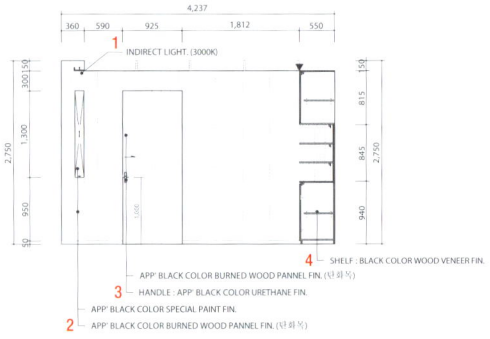

오마카세 룸 입면 Q1 / omakase room elevation Q1

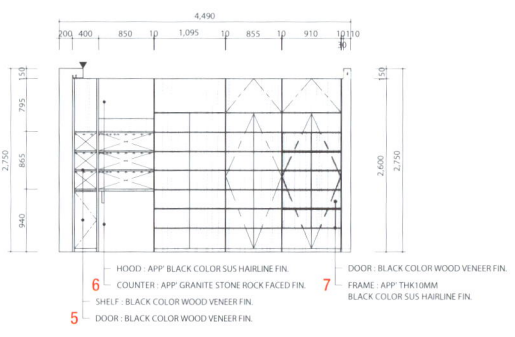

오마카세 룸 입면 Q2 / omakase room elevation Q2

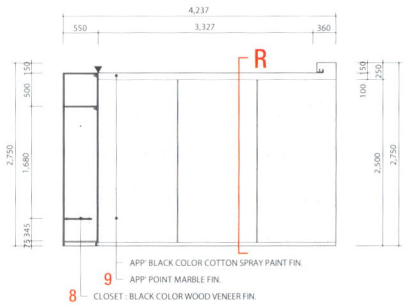

오마카세 룸 입면 Q3 / omakase room elevation Q3

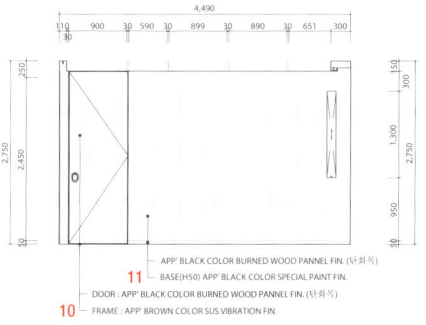

오마카세 룸 입면 Q4 / omakase room elevation Q4

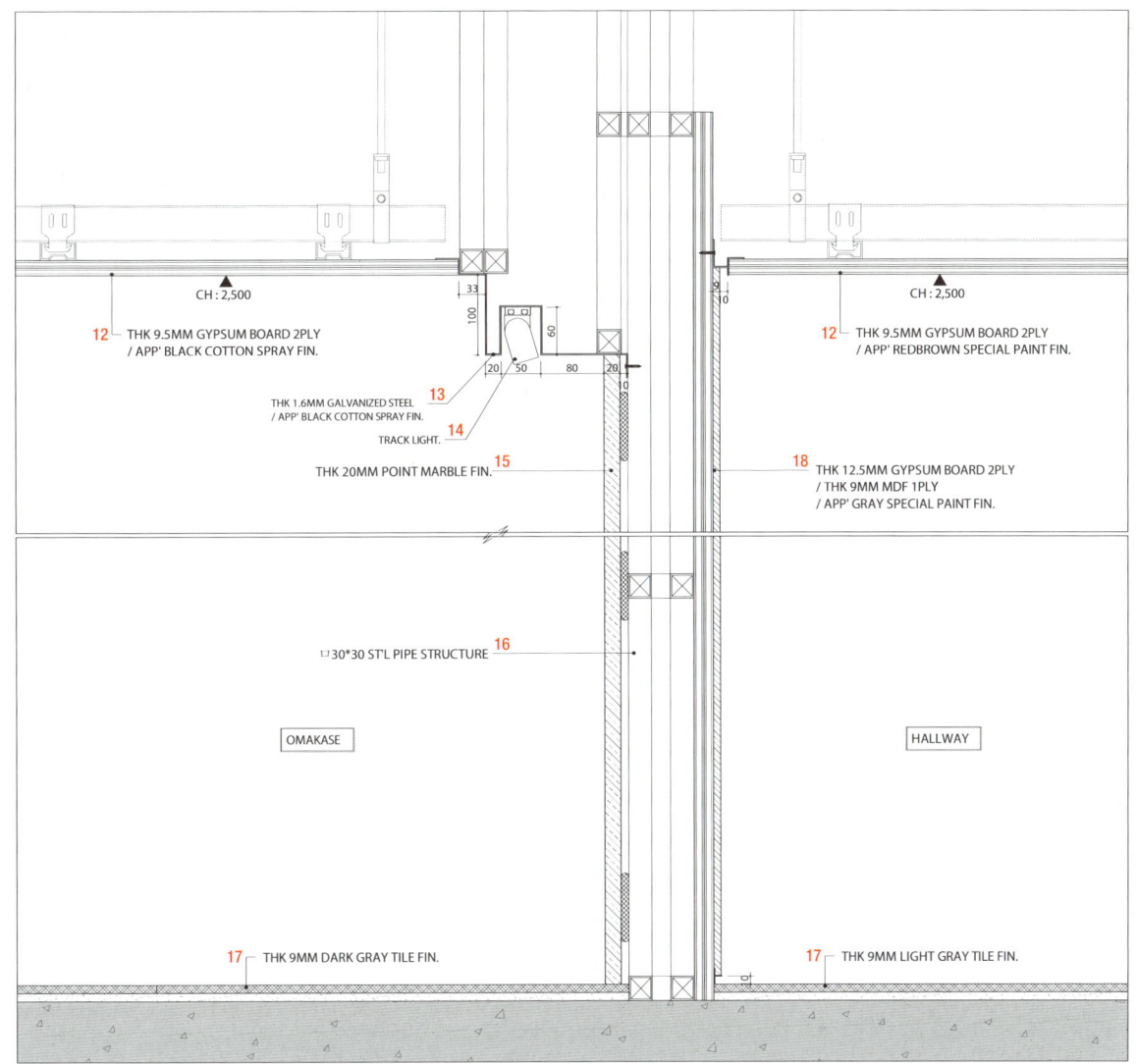

오마카세 룸 벽 단면 상세 R / omakase room wall section detail R

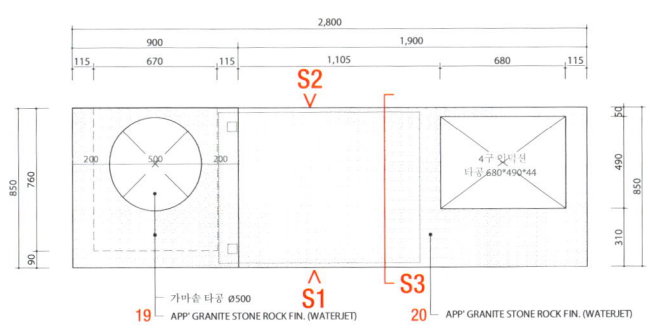

조리대 평면 S / cooking table top view S

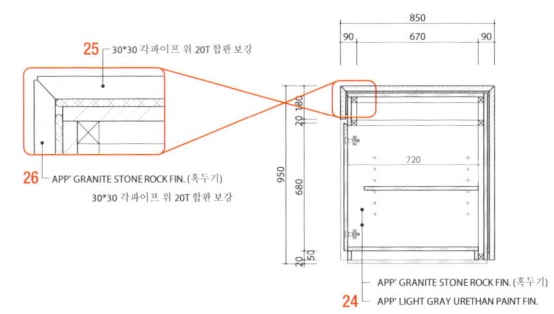

조리대 단면 S3 / cooking table section S3

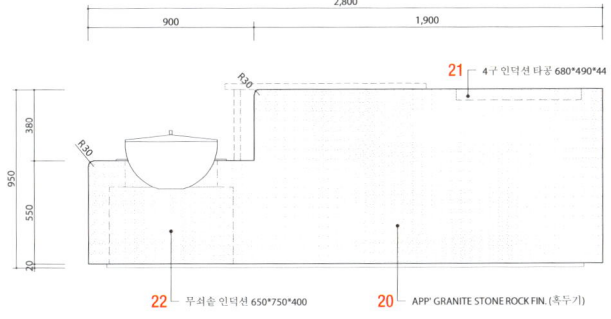

조리대 정면 S1 / cooking table front view S1

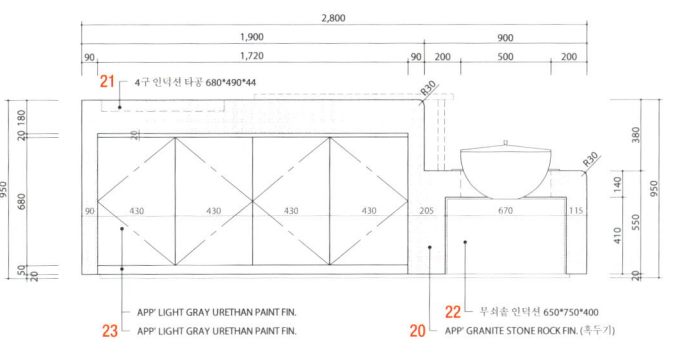

조리대 후면 S2 / cooking table rear view S2

1 간접조명 / 지정 검은색 탄화목 2 후면 벽 : 지정 포인트 대리석 3 지정 검은색 탄화목 4 지정 검은색 스페셜 페인트 / 걸레받이 : 지정 검은색 스페셜 페인트 5 프레임 : 부식 구리 6 지정 오리엔탈 종이 7 간접조명 8 지정 검은색 스페셜 페인트 9 지정 검은색 디자인 패널 10 문 : 지정 검은색 디자인 패널 11 지정 검은색 디자인 패널 / 걸레받이 : 지정 검은색 스페셜 페인트 12 지정 검은색 무니크 패널 13 파티션 : 지정 디자인 패널 14 레일 박스 15 지정 흰색 무니크 패널 16 지정 디자인 패널 17 아크릴 스툴 18 지정 흰색 무니크 패널 / 지정 디자인 패널

1 Indirect lighting / App. black color burned wood panel 2 Back wall : App. point marble 3 App. black color burned wood panel 4 App. black special paint / Baseboard : App. black special paint 5 Frame : Corroisn copper 6 App. oriental paper 7 Indirect lighting 8 App. black special paint 9 App. black design panel 10 Door : App. black design panel 11 App. black design panel / Baseboard : App. black special paint 12 App. black muniq panel 13 Partition : App. design panel 14 Rail box 15 App. white muniq panel 16 App. design panel 17 Acrylic stool 18 App. white muniq panel / App. design panel

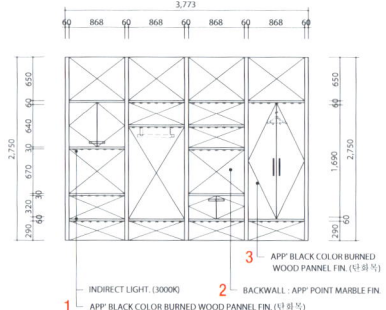

VIP 룸 입면 T1 / VIP room elevation T1

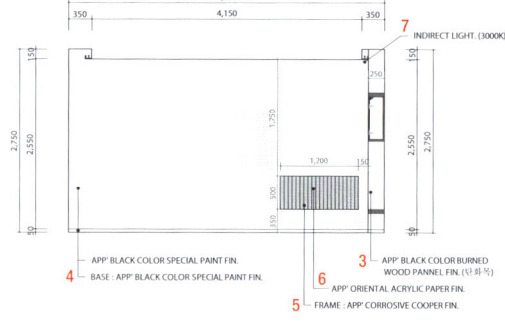

VIP 룸 입면 T2 / VIP room elevation T2

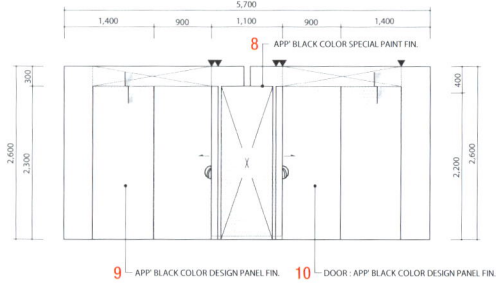

검탄룸 입면 U1 / black room elevation U1

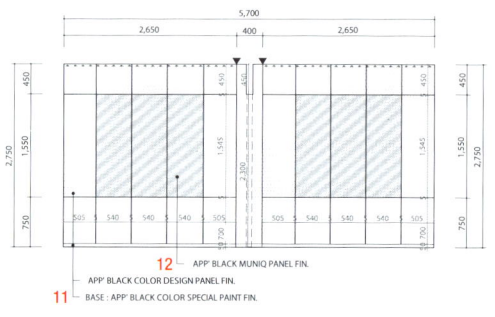

검탄룸 입면 U2 / black room elevation U2

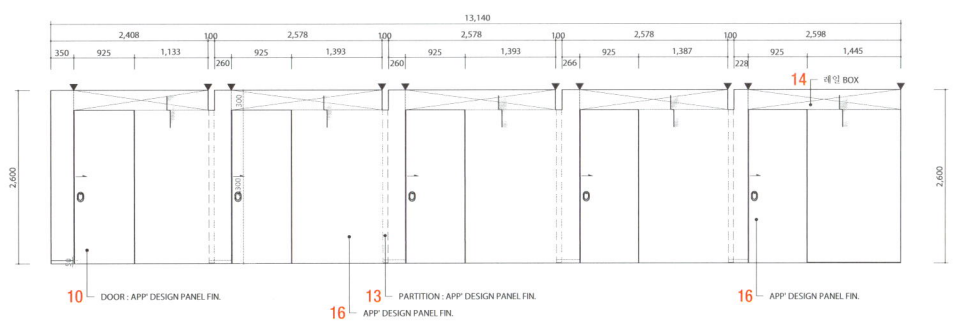

백탄룸 입면 V1 / white room elevation V1

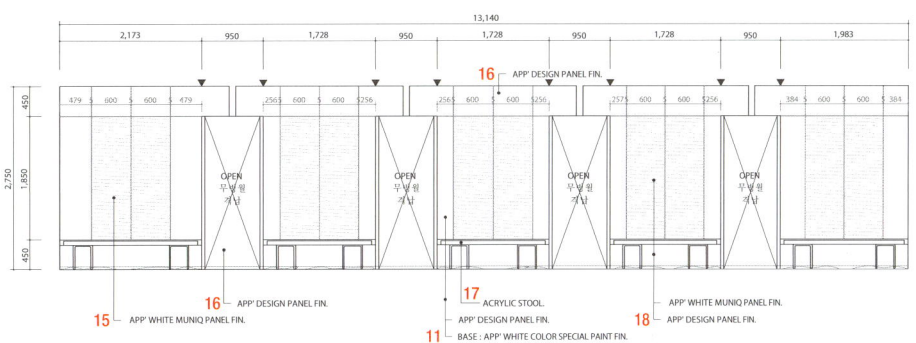

백탄룸 입면 V2 / white room elevation V2

RAINBOW BRIDGE GWANGGYO BRANCH

INTOEX | Yunjun Yang

레인보우 브릿지는 3개의 브랜드가 통합되어 새롭게 론칭한 하나의 브랜드이다. 아메리칸 정통 스테이크하우스, 한식, 디저트카페로 하나의 공간에 3개의 브랜드가 있는 통합된 공간을 디자인해야 한다는 것이 첫 번째 과제였다. 디자이너는 모두 본연의 개성이 도드라지는 각각의 브랜드를 하나로 아우르는 '레인보우 브릿지'의 이름과 같이 다채로운 색상의 하나의 공간을 만들었다.

입구에는 낮은 건물의 테이크아웃 카페 NODH가 위치하여 시작점을 알림과 동시에 내부 공간에 대한 호기심을 일으킨다. 오른쪽 방향의 입면에 순서대로 애견 동반 식사가 가능한 박공 형태의 유리 온실과 한식 레스토랑인 호감당, 그리고 제일 안쪽엔 놉스 스테이크하우스가 마치 골목과 같이 다른 높이와 다른 마감, 다른 형태로 줄지어 있으며, 길 건너로 테라스와 오솔길 그리고 야외공간으로 조성되었다. 호감당은 흙벽의 거친 질감이 느껴지는 스타코 소재와 우드 포인트, 문살의 형태를 차용한 디자인 블록으로 한국적인 이미지를 그렸으며, 맞은편 길 건너 마당과 같은 공간에서 식사가 가능하도록 하였다. 호감당 마당의 끝에 있는 붉은 벽돌의 놉스 건물은 시야를 차단함으로써 새로운 공간을 예고한다. 박공 형태의 벽의 입구를 지나서 들어가면 붉은 벽돌과 시그니처 컬러인 버건디로 구성된 놉스는 조금 더 프라이빗한 야외 테라스와 같은 식사공간이 펼쳐진다. 마지막 공간에는 고객을 생각하는 클라이언트의 마음을 담아 가족 단위 고객을 위한 어린이 놀이 공간을 마련해두었고, 시야의 끝이 머무는 곳엔 프로젝션 맵핑을 통해 변화 가능한 자연의 확장성을 보여주었다.

Rainbow Bridge is a fresh brand born from the fusion of three established brands. Designer's priority was crafting a unified environment that incorporates elements from each: an authentic American steakhouse, a Korean dining experience, and a dessert café, all within a single location. Each brand possessed a unique identity, each with its own strong personality. As a result, their goal was to forge a cohesive environment characterized by vibrant colors, aligning with the name "Rainbow Bridge," which embraces the collective essence of all these brands.

The low-rise building with the takeout café NODH at the entrance serves as a starting point, sparking curiosity about the interior space. To the right of the entrance, you'll find a gabled glass greenhouse designed for dog-friendly dining, followed by the Korean restaurant Hogamdang, and NOP'S Steakhouse at the far end. These establishments are arranged like an alley, featuring a variety of heights, finishes, and shapes. It consists of an outdoor dining area located across the street from the building, a trail-like ambiance, and an exterior dining setup enveloped by nature. Hogamdang carries a Korean aesthetic with its earth walls made of rough-textured stucco, wood accents, and design blocks inspired by traditional door lattice patterns. This allows diners to enjoy the ambiance of the courtyard located across the street. The red brick NOP's building at the end of the courtyard obstructs the view, enhancing the vibrancy of the façade and hinting at a new space beyond. Upon entering through the gabled wall entrance, NOP's reveals a more private dining area resembling an outdoor terrace, characterized by its red brick construction and signature burgundy color. The final space embodies the client's vision of a child-friendly playground. At the end of the sightline, they wished to bring a variety of natural expanses through projection mapping.

디자인 양윤준 / 인투익스
위치 경기도 수원시 영통구 센트럴타운로 85, 지하 1층
용도 레스토랑
면적 196㎡
마감 바닥 – LVT, 카펫 타일, 유크리트 / 벽 – 브릭 타일, 페인트, 스투코, 모노타일 / 천장 – 비닐페인트
완공 2024. 1
디자인팀 김유나, 강신영
시공팀 구광현
사진 강명국

Location B1, 85, Centraltown-ro, Yeongtong-gu, Suwon-si, Gyeonggi-do
Use Rastaurant
Area 196m²
Finishing Floor - LVT, Carpet tile, Ucrete / Wall - Brick tile, Paint, Stucco, Mono tile / Ceiling - Vinyl paint
Completion 2024. 1
Design team Yuna Kim, Syinyoung Kang
Construction team Gyanghyeon Ku
Photographer Myungguk Kang

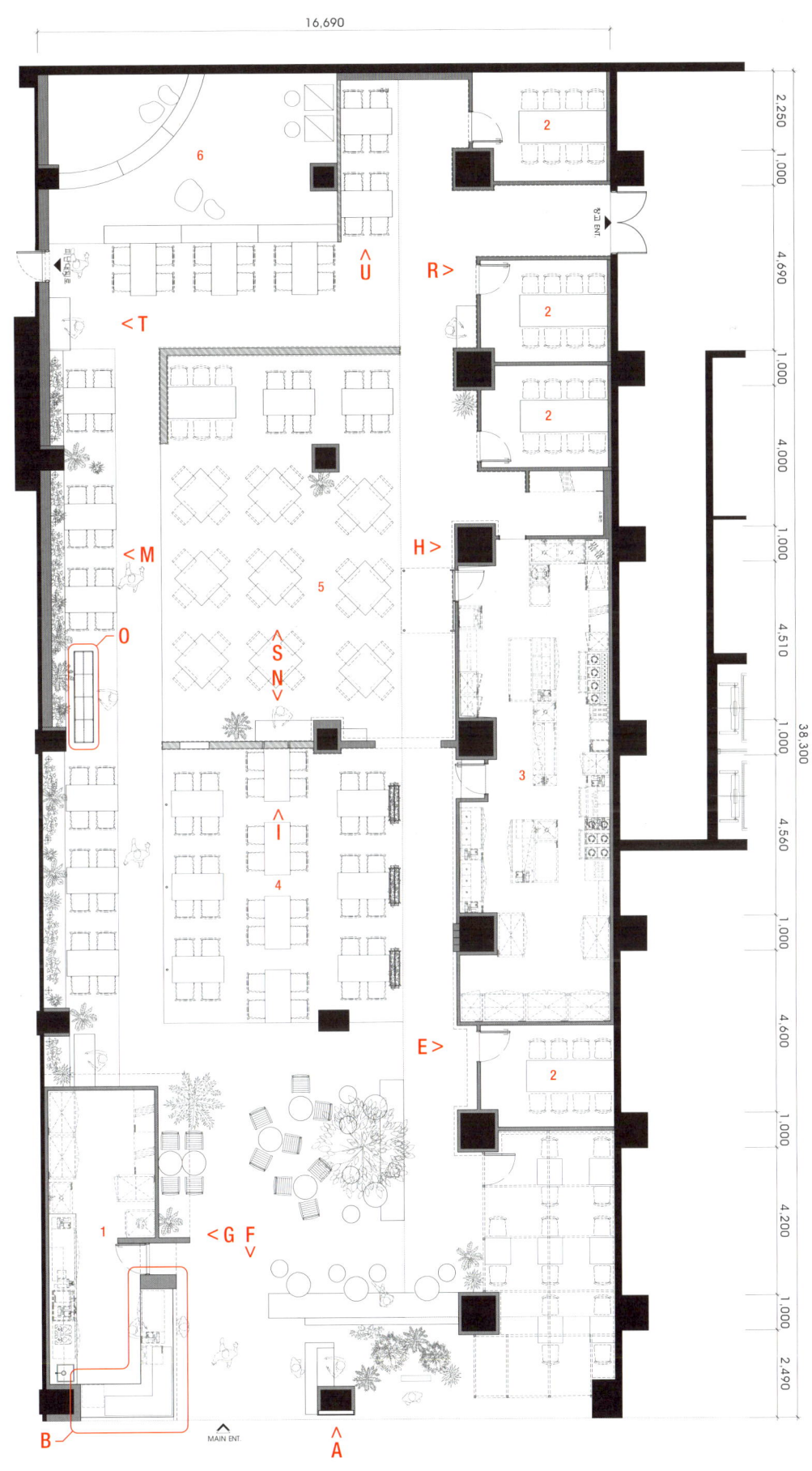

평면도 / floor plan

1 NODH 2 룸 3 주방 4 호감당 5 놉스 6 유아 놀이공간

1 NODH 2 Room 3 Kitchen 4 Hogamdang 5 NOP'S 6 Children's playground

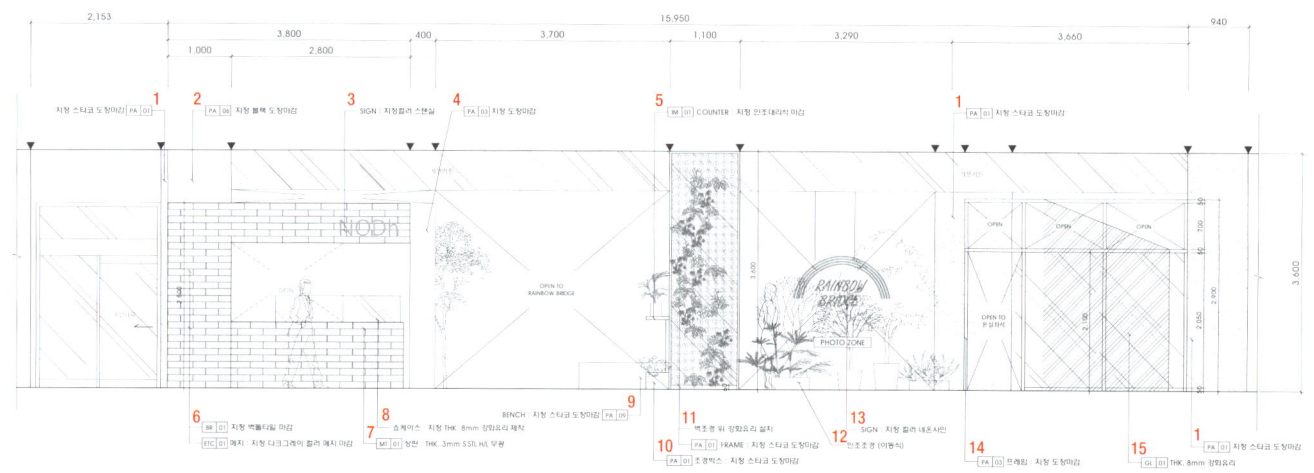

파사드 A / facade A

1 지정 스타코 도장　2 지정 검은색 도장　3 사인 : 지정색 스텐실　4 지정 도장　5 카운터 : 지정 인조대리석　6 지정 벽돌 / 줄눈 : 지정 진회색　7 상판 : T3 스테인리스 스틸 헤어라인　8 쇼케이스 : T8 강화유리　9 벤치 : 지정 스타코 도장　10 조경박스 : 지정 스타코 도장　11 벽 조경 위 강화유리 설치　12 인조 조경(이동식)　13 사인 : 지정색 네온사인　14 프레임 : 지정 도장　15 T8 강화유리　16 □390X90 지정 모노타일 위 지정 도장　17 □30X30 각파이프 보강　18 하부장 : 지정 LPM

1 App. stucco painting　2 App. black painting　3 Sign : App. color stencil　4 App. painting　5 Counter : App. engineered marble　6 App. brick / App. dark gray joint reveal　7 Top : T3 stainless steel hairline　8 Showcase : T8 tempered glass　9 Bench : App. stucco painting　10 Landscaping box : App. stucco painting　11 Tempered glass installation above wall landscaping　12 Artificial landscaping (movable)　13 Sign : App. color neon sign　14 Frame : App. painting　15 T8 tempered glass　16 App. painting on □390X90 app. mono tile　17 □30X30 square pipe reinforcement　18 Lower cabinet : App. LPM

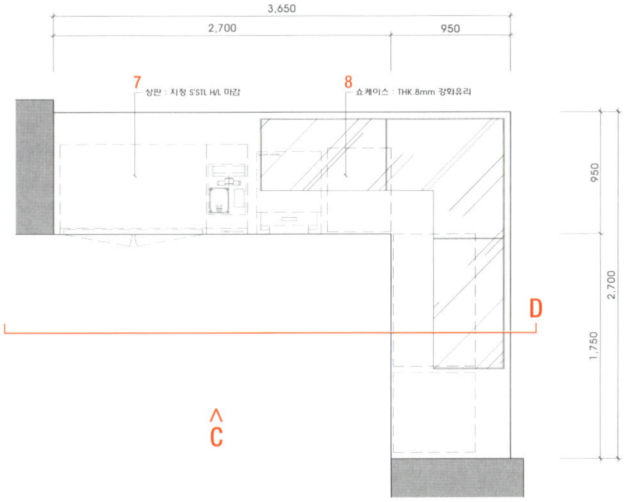

가구 평면 B / furniture top view B

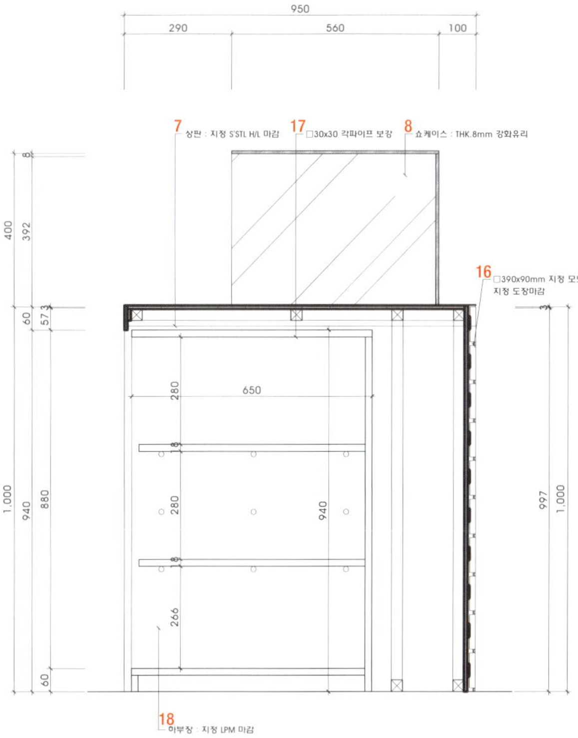

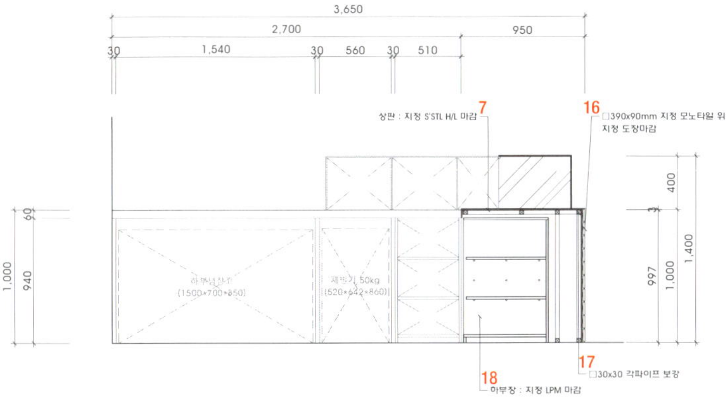

가구 단면 C / furniture section C

가구 단면 D / furniture section D

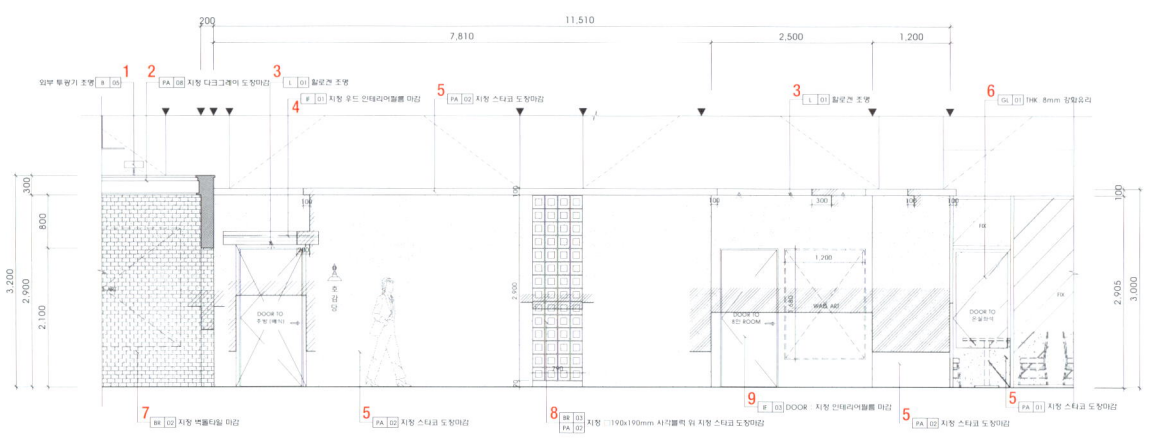

홀 입면 E1 / hall elevation E1

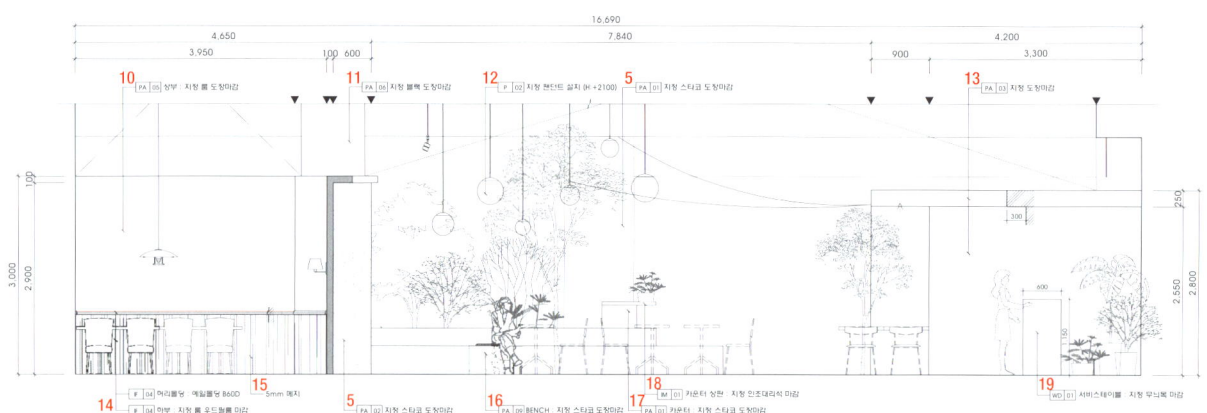

홀 입면 F / hall elevation F

1 외부 투광기 조명 2 지정 진회색 도장 3 할로겐 조명 4 지정 우드 인테리어 필름 5 지정 스타코 도장 6 T8 강화유리 7 지정 벽돌 타일 8 □190X190 블록 위 지정 스타코 도장 9 문 : 지정 인테리어 필름 10 룸 : 지정 도장 11 지정 검은색 도장 12 지정 펜던트 조명 13 지정 도장 14 몰딩 / 하부 : 지정 우드 필름 15 5mm 줄눈 16 벤치 : 지정 스타코 도장 17 카운터 : 지정 스타코 도장 18 카운터 상판 : 지정 인조대리석 19 서비스 테이블 : 지정 무늬목 20 프레임 : 지정 도장 21 상판 : T3 스테인리스 스틸 헤어라인 22 쇼케이스 : 강화유리 23 지정 벽돌 / 줄눈 : 지정 진회색 24 서비스 테이블 상판 : 지정 인조대리석

1 Floodlight 2 App. dark gray painting 3 Halogen light 4 App. wood interior film 5 App. stucco painting 6 T8 tempered glass 7 App. brick tile 8 App. stucco painting on □190X190 block 9 Door : App. interior film 10 Room : App. painting 11 App. black painting 12 App. pendant lighting 13 App. painting 14 Moulding / Lower : App. wood film 15 5mm joint reveal 16 Bench : App. stucco painting 17 Counter : App. stucco painting 18 Counter top : App. stucco painting 19 Service table : App. engineered marble 20 Frame : App. painting 21 Top : T3 stainless steel hairline 22 Showcase : Tempered glass 23 App. brick / App. dark gray joint reveal 24 Service table top : App. engineered marble

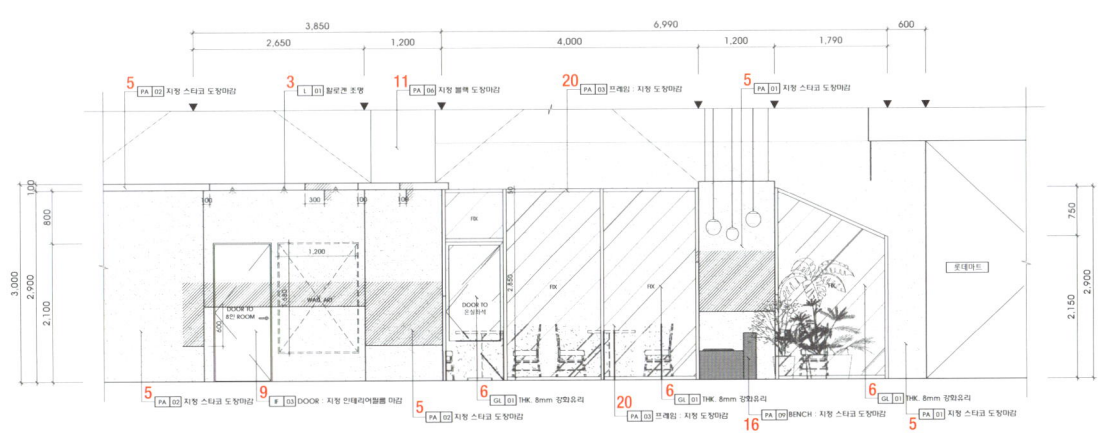

홀 입면 E2 / hall elevation E2

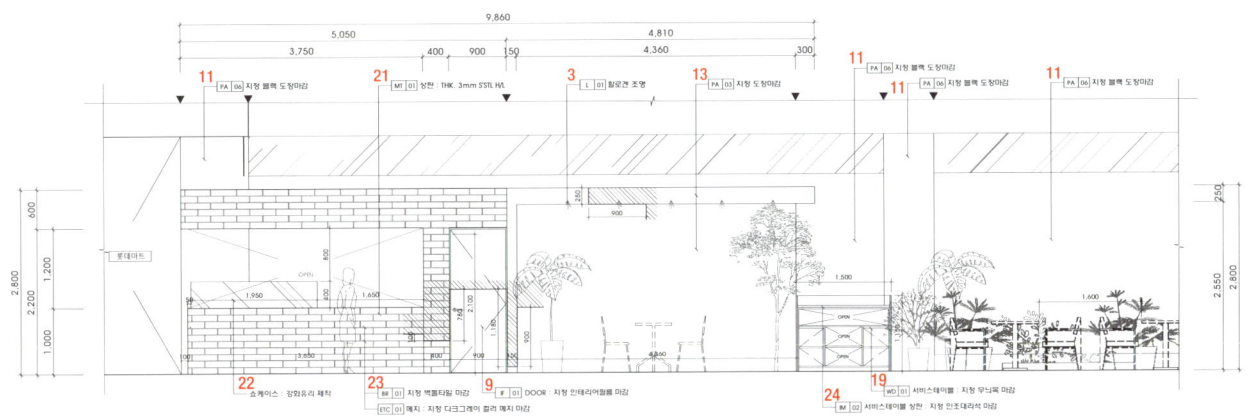

홀 입면 G / hall elevation G

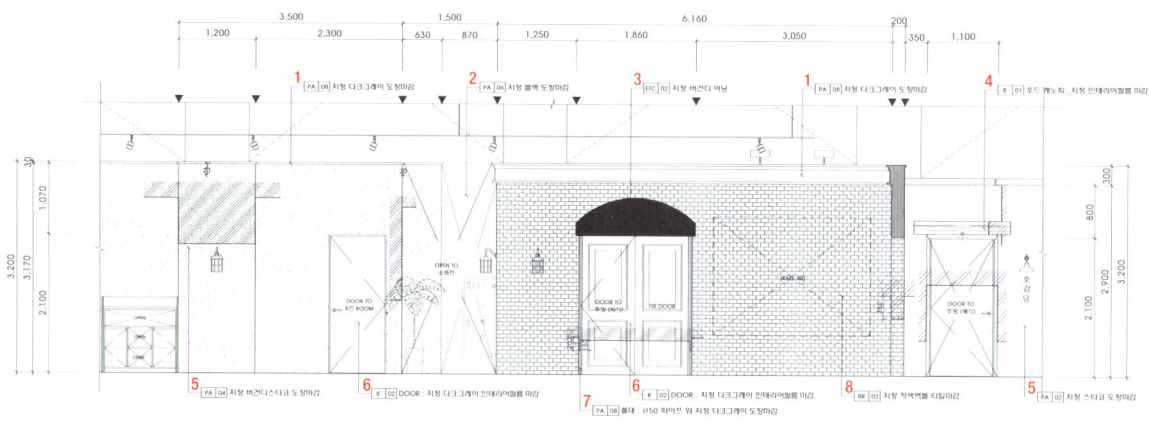

홀 입면 H / hall elevation H

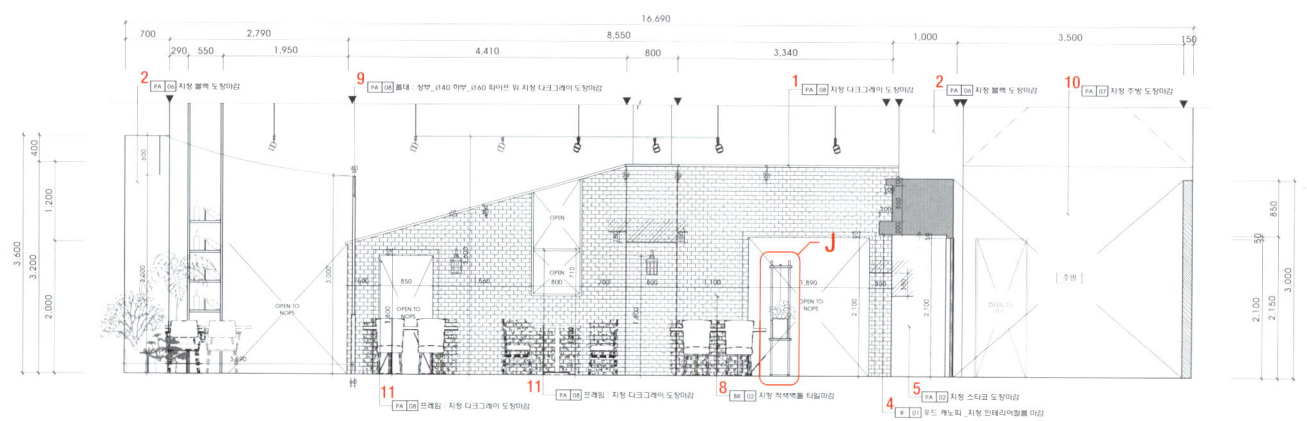

홀 입면 I / hall elevation I

1 지정 진회색 도장 2 지정 검은색 도장 3 지정 어닝 4 우드 캐노피 : 지정 인테리어 필름 5 지정 스타코 도장 6 문 : 지정 진회색 인테리어 필름 7 Ø50 파이프 위 지정 진회색 도장 8 지정 벽돌 타일 9 상부 : Ø40 / 하부 : Ø60 파이프 위 지정 진회색 도장 10 주방 : 지정 도장 11 프레임 : 지정 진회색 도장 12 지정 진회색 LPM 13 □25X25 각파이프 위 지정 갈색 도장 14 □25X25 각파이프 위 지정 갈색 도장 / 지정 패브릭 위 로고 인쇄

1 App. dark gray painting 2 App. black painting 3 App. awning 4 Wood canopy : App. interior film 5 App. stucco painting 6 Door : App. dark gray interior film 7 App. dark gray painting on Ø50 pipe 8 App. brick tile 9 Upper Ø40 / Lower : App. dark gray painting on Ø60 pipe 10 Kitchen : App. painting 11 Frame : App. dark gray painting 12 App. dark gray LPM 13 App. brown painting on □25X25 square pipe 14 App. brown painting on □25X25 square pipe / Logo printing on app. fabric

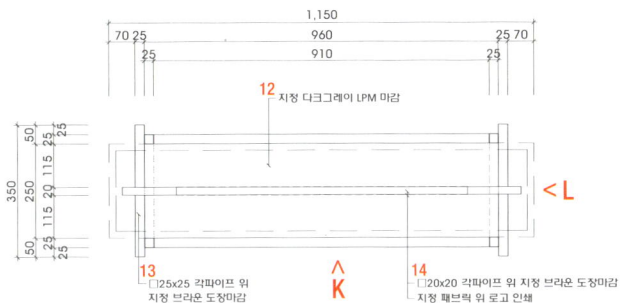

파티션 평면 J / partition top view J

파티션 정면 K / partition front view K

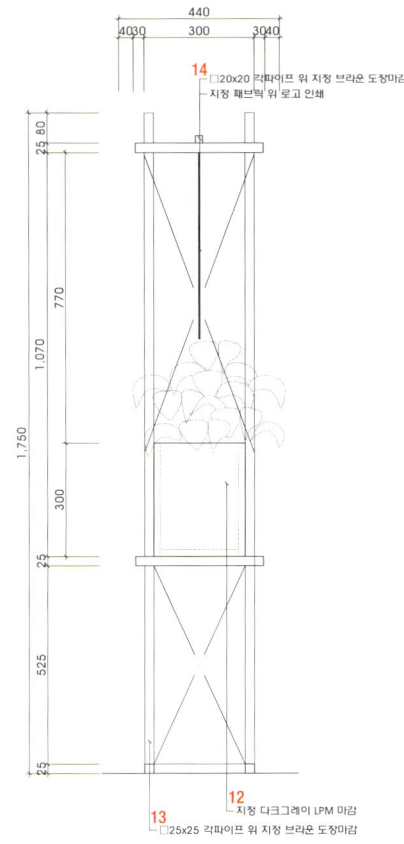

파티션 측면 L / partition side view L

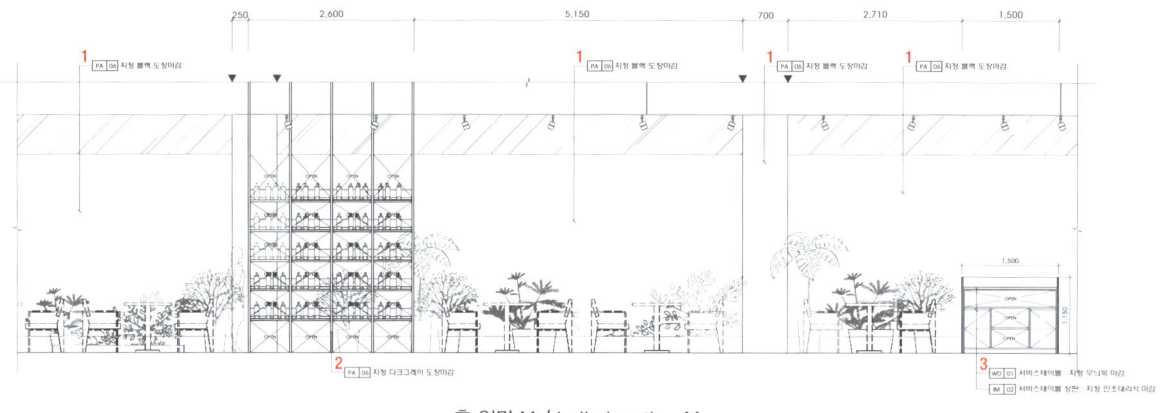

홀 입면 M / hall elevation M

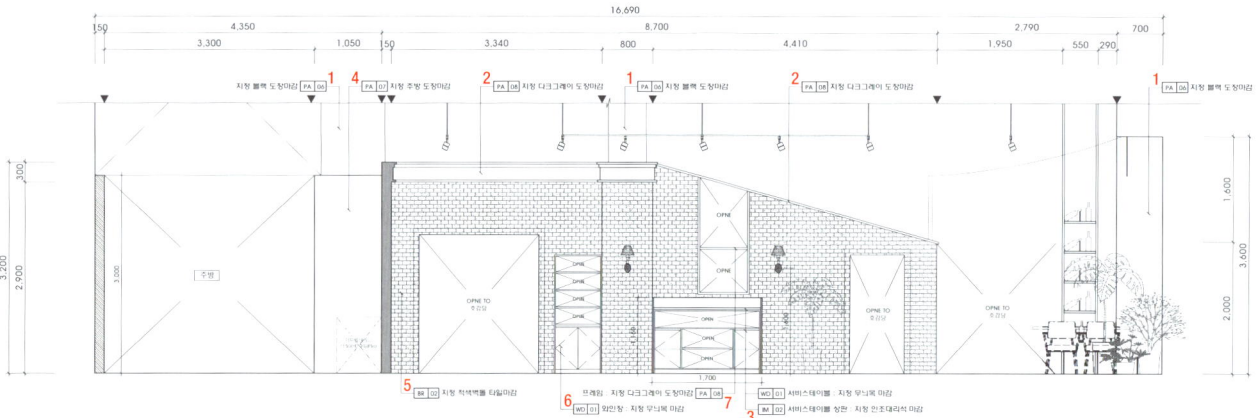

홀 입면 N / hall elevation N

1 지정 검은색 도장 2 지정 진회색 도장 3 서비스 테이블 : 지정 무늬목 / 서비스 테이블 상판 : 지정 인조대리석 4 주방 : 지정 도장 5 지정 벽돌 타일 6 와인장 : 지정 무늬목 7 프레임 : 지정 진회색 도장 8 LED 바 9 T1.6 갈바륨 위 지정 진회색 도장 10 □30X30 각파이프 위 지정 진회색 도장

1 App. black painting 2 App. dark gray painting 3 Service table : App. wood veneer / Service table top : App. engineered marble 4 Kitchen : App. painting 5 App. brick tile 6 Wine cabinet : App. wood veneer 7 Frame : App. dark gray painting 8 LED bar 9 App. dark gray painting on T1.6 galvalume 10 App. dark gray painting on □30X30 square pipe

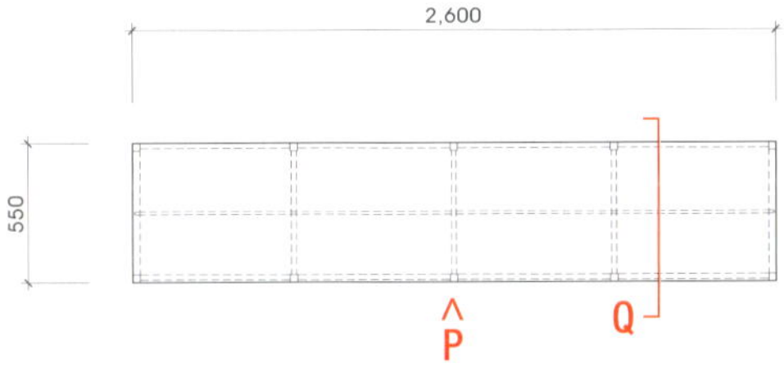

와인랙 평면 O / wine rack top view O

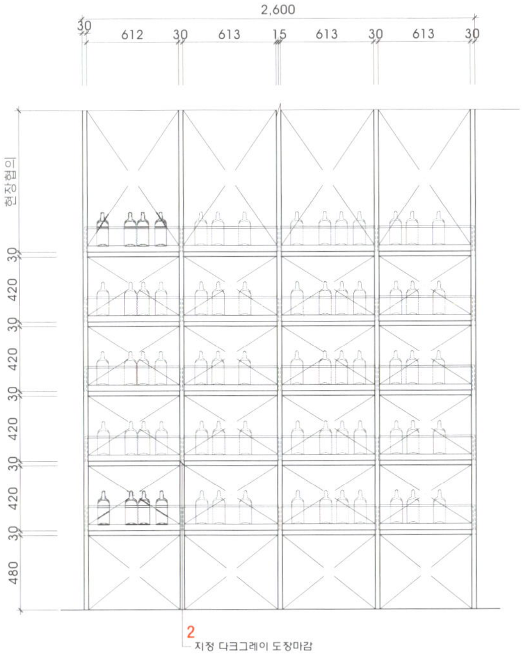

와인랙 정면 P / wine rack front view P

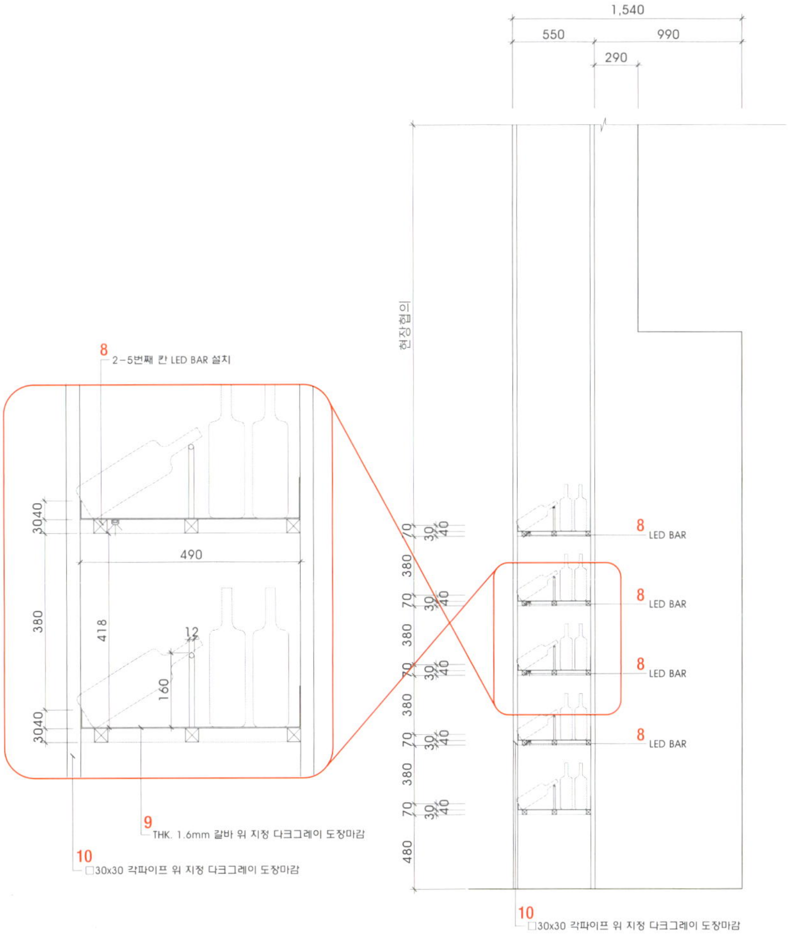

와인랙 단면 Q / wine rack section Q

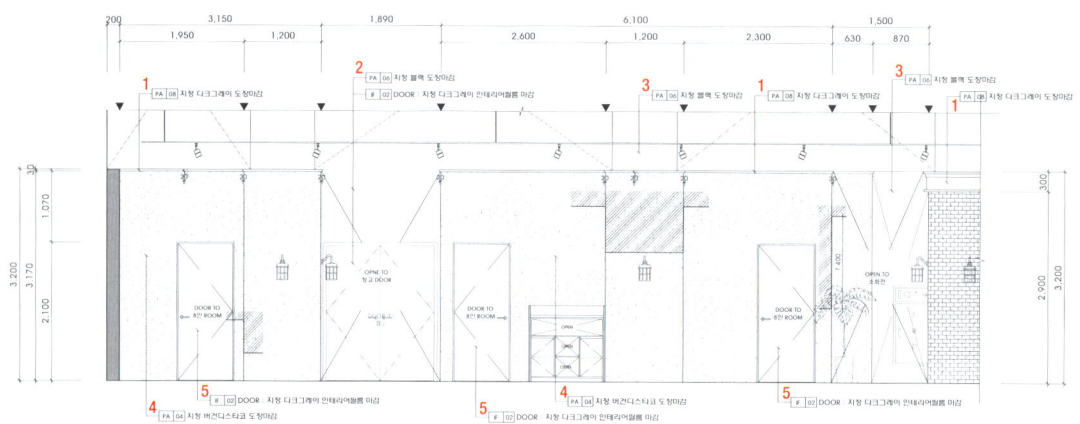

홀 입면 R / hall elevation R

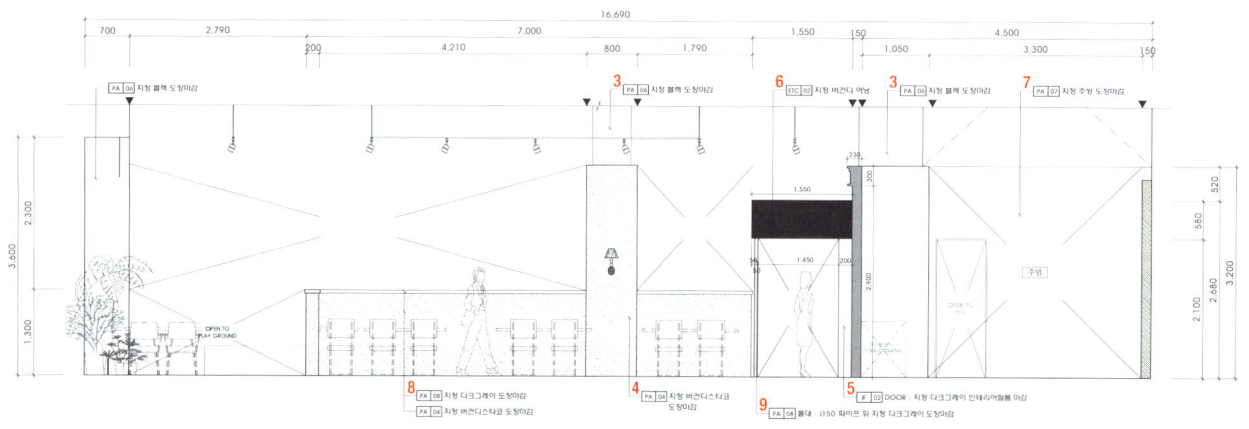

홀 입면 S / hall elevation S

1 지정 진회색 도장   2 지정 검은색 도장 / 문 : 지정 진회색 인테리어 필름   3 지정 검은색 도장   4 지정 스타코 도장   5 문 : 지정 진회색 인테리어 필름   6 지정 어닝   7 주방 : 지정 도장   8 지정 진회색 도장 / 지정 스타코 도장   9 Ø50 파이프 위 진회색 도장   10 방화문 : 지정 검은색 도장   11 서비스 테이블 : 지정 무늬목 / 서비스 테이블 상판 : 지정 인조대리석   12 지정 인조가죽   13 빔 프로젝터 벽 : 지정 도장

1 App. dark gray painting   2 App. black painting / Door : App. dark gray interior film   3 App. black painting   4 App. stucco painting   5 Door : App. dark gray interior film   6 App. awning   7 Kitchen : App. painting   8 App. dark gray painting / App. stucco painting   9 Dark gray painting on Ø50 pipe   10 Fire door : App. black painting   11 Service table : App. wood veneer / Service table top : App. engineered marble   12 App. artificial leather   13 Wall for beam projector : App. painting

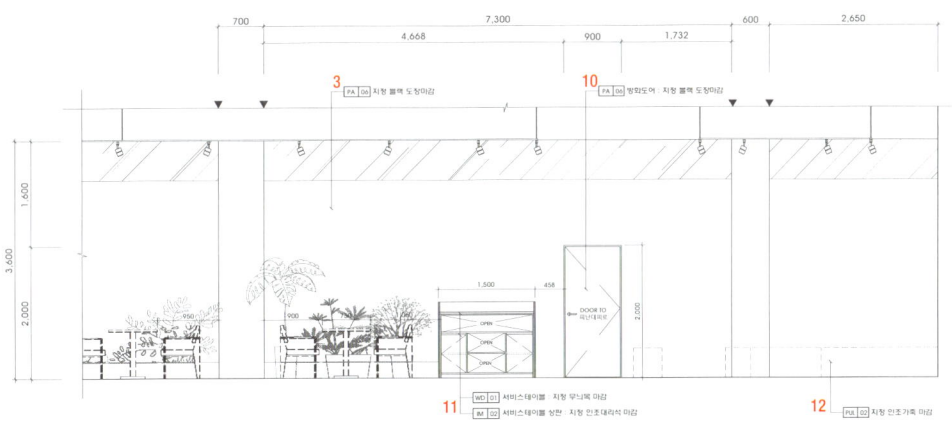

홀 입면 T / hall elevation T

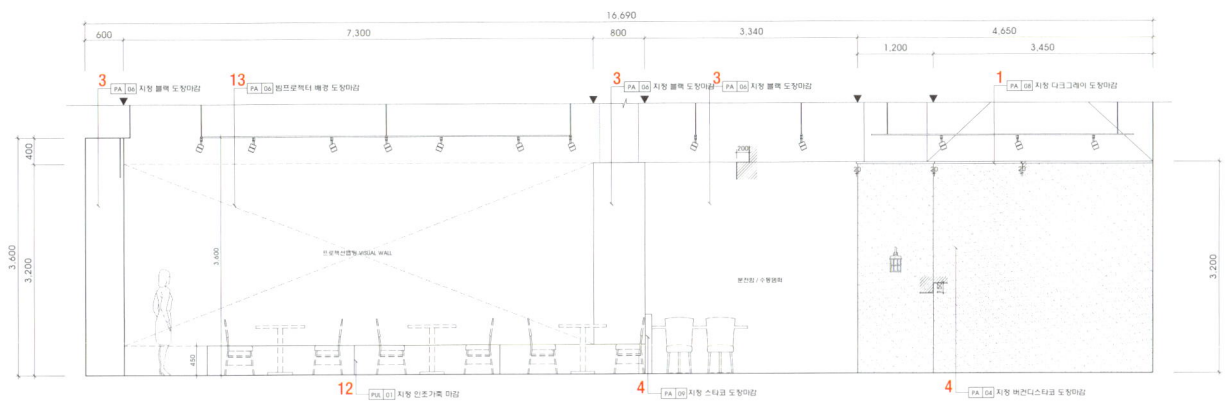

홀 입면 U / hall elevation U

HAENAM CHEONILGWAN

LABOTORY | Jinho Jung

해남천일관은 대한민국 한식당 중 100년의 역사를 자랑하는 유일한 브랜드이다. 해남천일관이 특별한 이유는 선대로부터 진심 어린 음식을 만들고, 이를 보존하며, 대를 이어 전승해 왔기 때문이다. 디자이너는 100년 동안 이곳을 특별하게 만들어온 이야기와 앞으로도 계속될 이야기, 그리고 해남천일관에서 생겨날 새로운 이야기들이 오래도록 지속되길 바라는 마음을 담아 브랜딩과 공간 디자인을 전개했다.

많은 사람들이 일상적인 한식을 평범하다고 생각한다. 그러나 해남천일관은 이 평범함을 정성스레 빚어내어 비범함으로 승화시킨다. 4대에 걸쳐 고스란히 지켜온 해남천일관의 음식은 정성이 담긴 도자기 그릇에 소중하게 담겨진다. 디자이너는 이 공간을 하나의 도자기처럼 생각하며 해남천일관의 진심이 담기도록 공간 디자인을 진행했다. 이 공간은 화려하지 않은 배경을 통해 음식이 더욱 돋보이게 하는 장소로, 해남천일관의 특별함을 느낄 수 있도록 설계되었다.

Haenam Cheonilgwan is the only brand among Korean restaurants in South Korea with a history of 100 years. The reason Haenam Cheonilgwan is special is that it has been sincerely crafting and preserving heartfelt food, passing it down through generations. Labotory has developed branding and space design with the hope that the stories that have made this place special over the past 100 years, as well as those that will continue and new ones that will emerge, will endure for a long time.

Many people consider everyday Korean cuisine to be ordinary. However, Haenam Cheonilgwan elevates this simplicity into something extraordinary through careful craftsmanship. The dishes, preserved over four generations, are thoughtfully served in ceramic bowls filled with care. We approached the design of this space as if it were a piece of pottery, ensuring that the sincere essence of Haenam Cheonilgwan is reflected in the environment. This space is designed to enhance the food against an unadorned backdrop, allowing guests to truly feel the uniqueness of Haenam Cheonilgwan.

디자인 정진호 / 라보토리
위치 서울특별시 송파구 올림픽로 300, 6층
용도 레스토랑
면적 264㎡
마감 바닥 – 타일 / 벽 – 스페셜 페인트, 세라믹, 타일 / 천장 – 비닐페인트, 스페셜 페인트
완공 2024. 5
디자인팀 유슬기, 이서정, 김유화, 조은미
사진 최용준

Location 6F, 300, Olympic-ro, Songpa-gu, Seoul
Use Restaurant
Area 264㎡
Finishing Floor - Tile / Wall - Special paint, Ceramic, Tile / Ceiling - Vinyl paint, Special paint
Completion 2024. 5
Design team Seulgi Yoo, Seojeong Lee, Youhwa Kim, Eunmi Cho
Photographer Yongjoon Choi

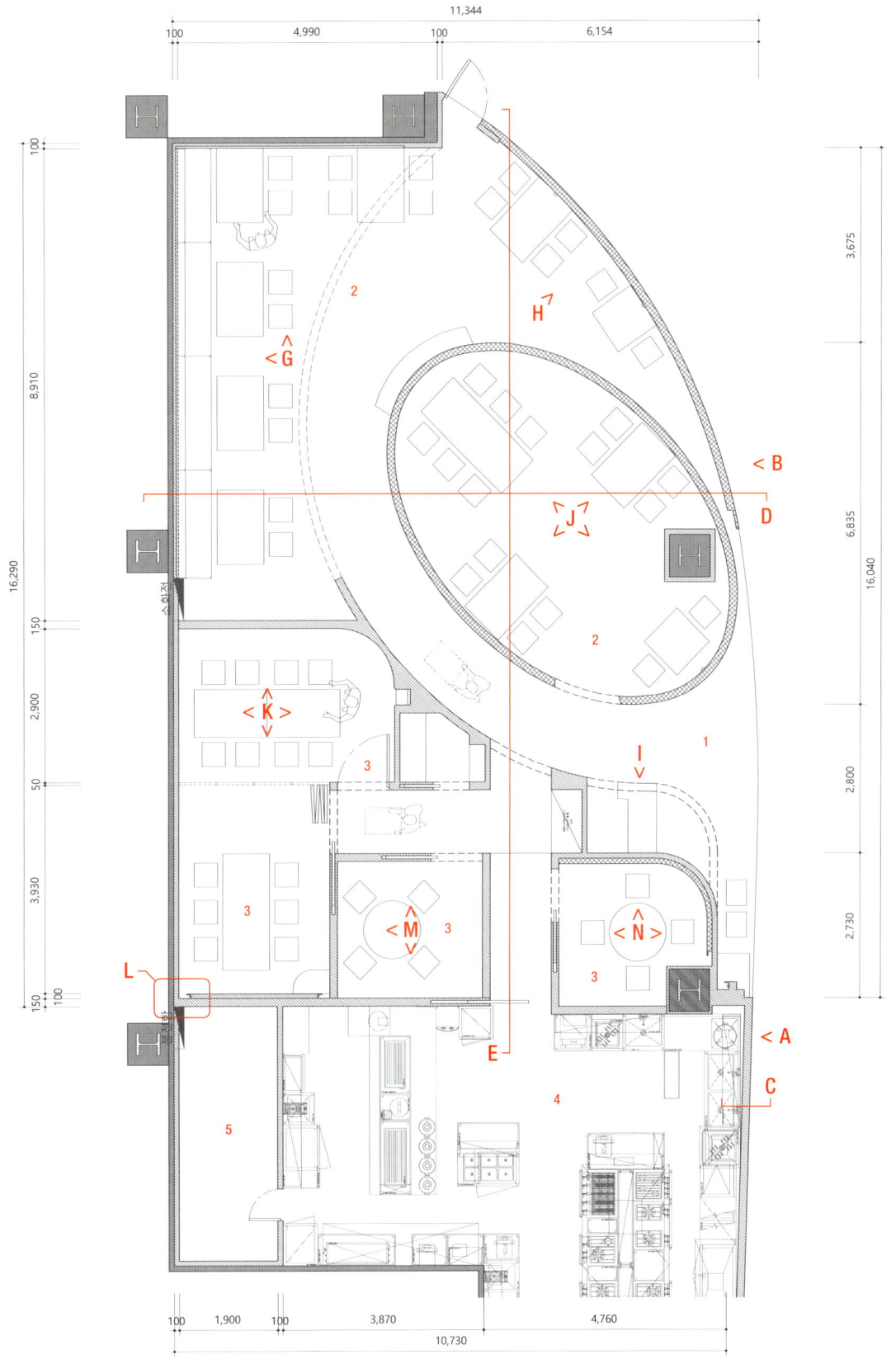

평면도 / floor plan

1 입구 2 홀 3 룸 4 주방 5 직원실

1 Entrance 2 Hall 3 Room 4 Kitchen 5 Staff room

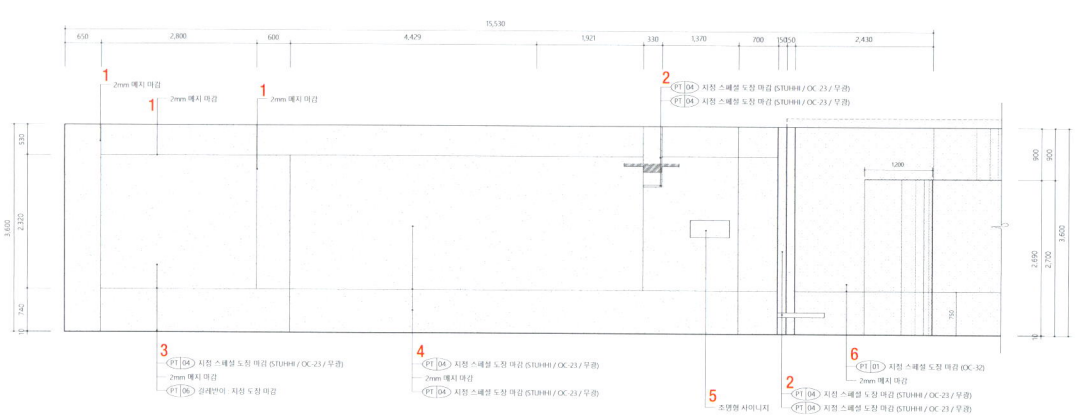

파사드 A / facade A

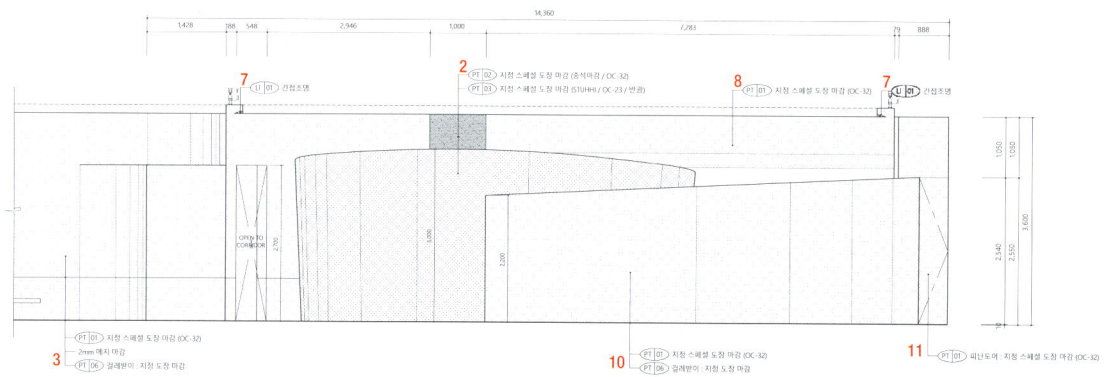

파사드 B / facade B

1 2mm 줄눈 2 지정 스페셜 페인트 / 지정 스페셜 페인트 3 지정 스페셜 페인트 / 2mm 줄눈 / 걸레받이 : 지정 도장 4 지정 스페셜 페인트 / 2mm 줄눈 / 지정 스페셜 페인트 5 조명형 사이니지 6 지정 스페셜 페인트 / 2mm 줄눈 7 간접조명 8 지정 스페셜 페인트 9 걸레받이 : 지정 도장 10 지정 스페셜 페인트 / 걸레받이 : 지정 도장 11 피난도어 : 지정 스페셜 페인트 12 T1.5 금속 위 지정 스페셜 페인트 / T2 금속 위 지정 스페셜 페인트 / ㅁ20X40X2.3T 각파이프 구조틀 13 T9.5 석고보드 2겹 위 지정 스페셜 페인트 / ㅁ100X50X2.3T 각파이프 구조틀 / 흡음재 / T9.5 석고보드 2겹 / 지정 타일 14 T1.5 금속 위 지정 스페셜 페인트 15 T9.5 석고보드 2겹 위 지정 스페셜 페인트 / ㅁ50X50X2.3T 각파이프 구조틀 / 흡음재 / T9.5 석고보드 2겹 / 지정 타일 16 걸레받이 : 지정 도장

1 2mm joint reveal 2 App. special paint / App. special paint 3 App. special paint / 2mm joint reveal / Baseboard : App. paint 4 App. special paint / 2mm joint reveal / App. special paint 5 Illuminated signage 6 App. special paint / 2mm joint reveal 7 Indirect lightning 8 App. special paint 9 Baseboard : App. paint 10 App. special paint / Baseboard : App. paint 11 Emergency exit door : App. special paint 12 App. special paint on T1.5 metal / App. special paint on T2 metal / ㅁ20X40X2.3T square pipe structure 13 App. special paint on T9.5 gypsum board 2ply / ㅁ100X50X2.3T square pipe structure / Sound absorbent / T9.5 gypsum board 2ply / App. tile 14 App. special paint on T1.5 metal 15 App. special paint on T9.5 gypsum board 2ply / ㅁ50X50X2.3T square pipe structure / Sound absorbent / T9.5 gypsum board 2ply / App. tile 16 Baseboard : App. paint

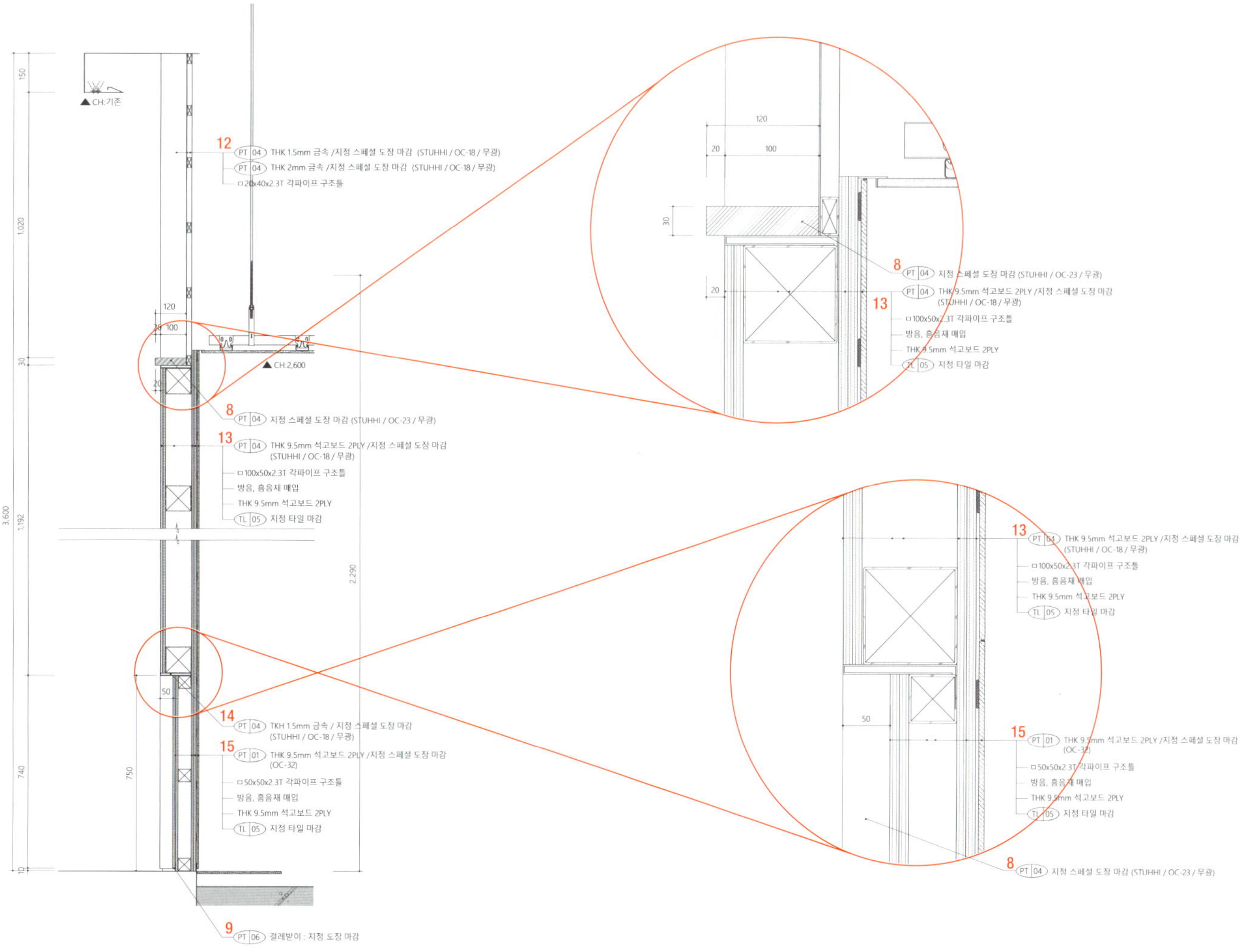

파사드 단면 상세 C / facade section detail C

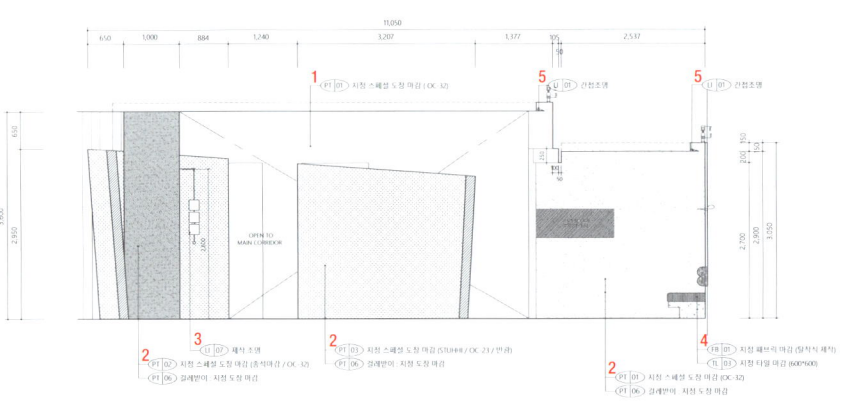

홀 단면 D / hall section D

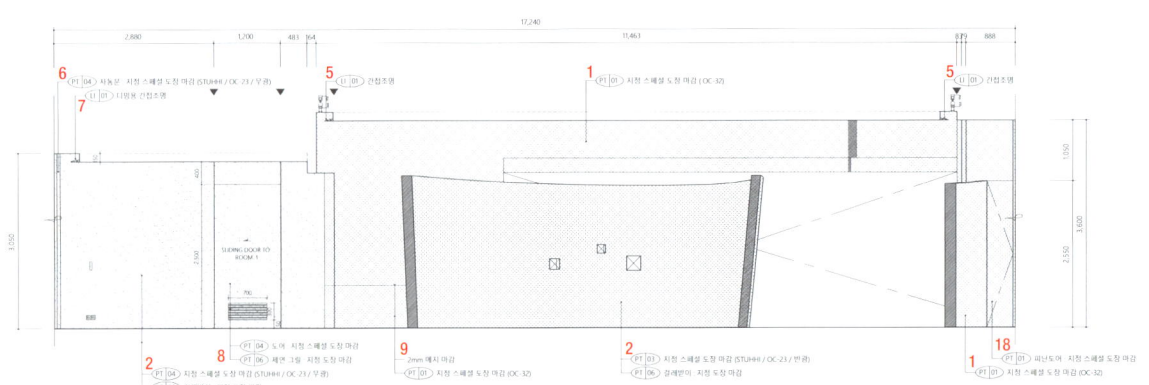

홀 단면 E / hall section E

1 지정 스페셜 페인트 2 지정 스페셜 페인트 / 걸레받이 : 지정 도장 3 제작 조명 4 지정 패브릭 / 지정 타일 5 간접조명 6 자동문 : 지정 스페셜 페인트 7 디밍용 간접조명 8 문 : 지정 스페셜 페인트 / 제연 그릴 지정 도장 9 2mm 줄눈 / 지정 스페셜 페인트 10 T9.5 석고보드 2겹 위 지정 스페셜 페인트 / C-스터드 65X45X0.8T 11 제작 도자기 타일 / 에폭시 본드 / T1.5 금속 위 지정 스페셜 페인트 / □50X50X2.3T 각파이프 구조틀 12 T1.5 금속 위 지정 스페셜 페인트 / T9.5 석고보드 2겹 위 지정 스페셜 페인트 13 지정 스페셜 페인트 / 제작 도자기 타일 14 지정 스페셜 페인트 / 소화전 : 지정 스페셜 페인트 15 지정 패브릭 / 지정 패브릭 16 지정 스페셜 페인트 / 지정 타일 17 지정 패브릭 / 지정 패브릭 / 지정 타일 18 피난도어 : 지정 스페셜 페인트 19 지정 하드웨어 20 제작 조명 21 바리솔 조명 / 지정 도장

1 App. special paint 2 App. special paint / Baseboard : App. paint 3 Custom-made light 4 App. fabric / App. tile 5 Indirect lighting 6 Automatic door : App. special paint 7 Dimmable indirect lighting 8 Door : App. special paint / Smoke exhaust grille 9 2mm joint reveal / App. special paint 10 App. special paint on T9.5 gypsum board / C-stud 65X45X08.T 11 App. porcelain tile / Epoxy glue / App. special paint on T1.5 metal / 50X50X2.3T square pipe structure 12 App. special paint on T1.5 metal / App. special paint on T9.5 gypsum board 2ply 13 App. special paint / App. porcelain tile 14 App. special paint / Fire hydrant : App. special paint 15 App. fabric / App. fabric 16 App. special paint / App. tile 17 App. fabric / App. fabric / App. tile 18 Emergency exit door : App. special paint 19 App. hardware 20 Custom-made lighting 21 Barrisol light / App. paint

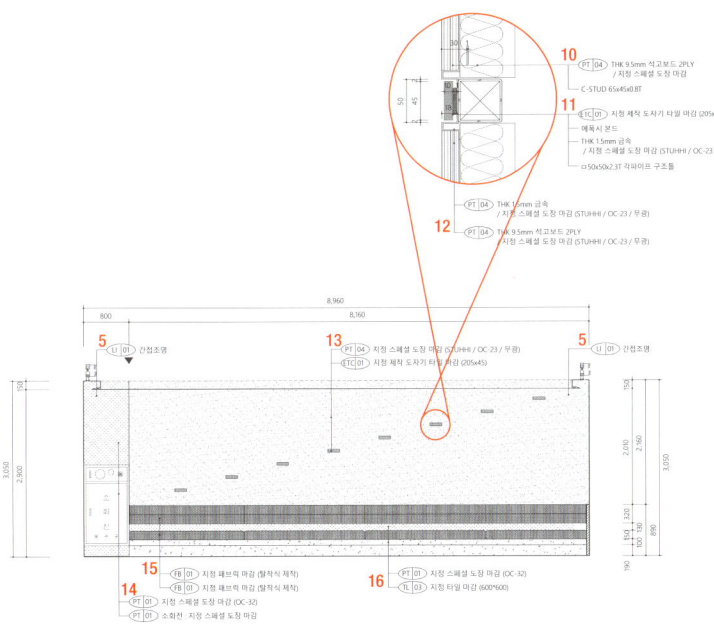

홀 입면 G1 / hall elevation G1

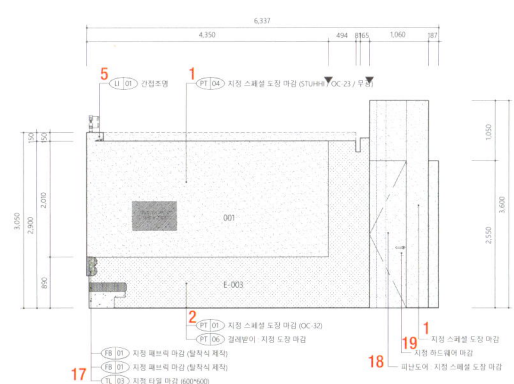

홀 입면 G2 / hall elevation G2

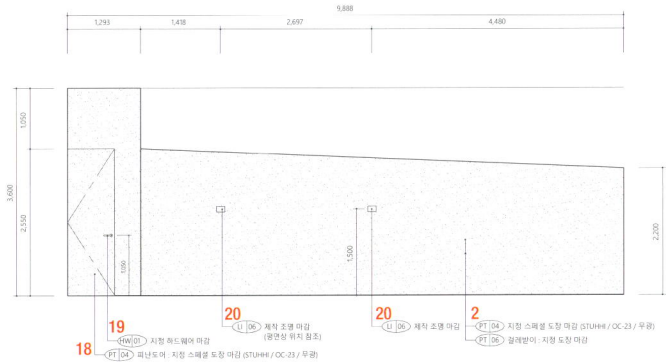

홀 입면 H / hall elevation H

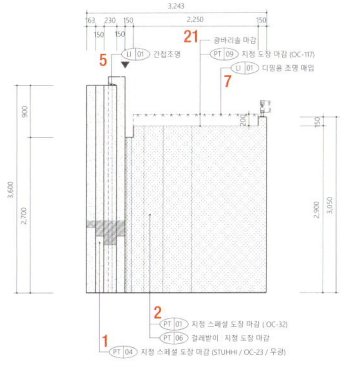

홀 입면 I / hall elevation I

1 지정 스페셜 페인트 / 걸레받이 : 지정 도장   2 지정 스페셜 페인트   3 제작 조명   4 지정 스페셜 페인트 / 지정 타일   5 2mm 줄눈 / 지정 타일   6 T9.5 석고보드 2겹 위 지정 스페셜 페인트   7 디밍용 간접조명   8 마그네틱 레일 / 제작 조명   9 간접조명   10 지정 석재   11 T2 SUS 바이브레이션   12 지정 석재 / 지정 타일   13 지정 타일   14 문 : 지정 스페셜 페인트 / 제연 그릴 : 지정 도장   15 마그네틱 레일 매입   16 문 : 지정 스페셜 페인트   17 지정 하드웨어

1 App. special paint / Baseboard : App. paint   2 App. special paint   3 Custom-made lighting   4 App. special paint / App. tile   5 2mm joint reveal / App. tile   6 App. special paint on T9.5 gypsum board 2ply   7 Dimmable indirect lighting   8 Magnetic rail / Custom-made lighting   9 Indirect lighting   10 App. stone   11 T2 SUS vibration   12 App. stone / App. tile   13 App. tile   14 Door : App. special paint / Smoke exhaust grille   15 Embedding magnetic rail   16 Door : App. special paint   17 App. hardware

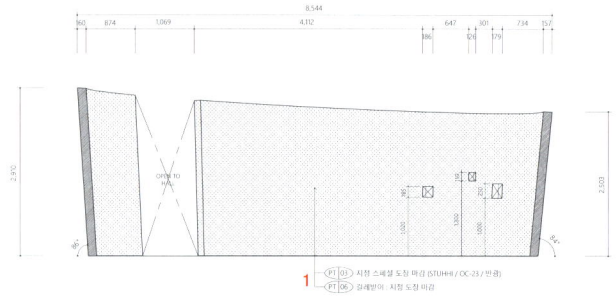

홀 입면 J1 / hall elevation J1

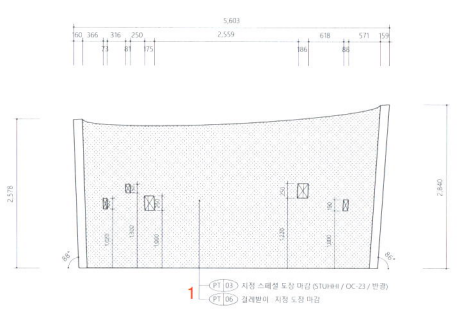

홀 입면 J2 / hall elevation J2

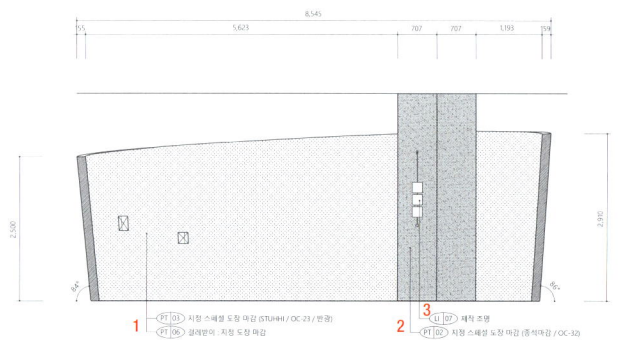

홀 입면 J3 / hall elevation J3

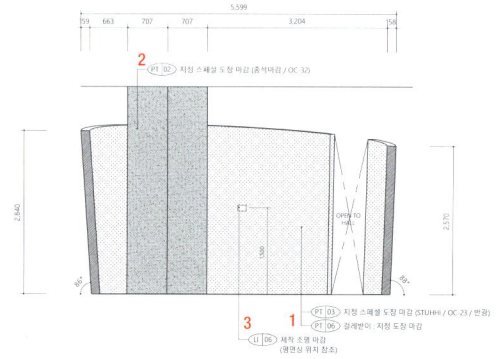

홀 입면 J4 / hall elevation J4

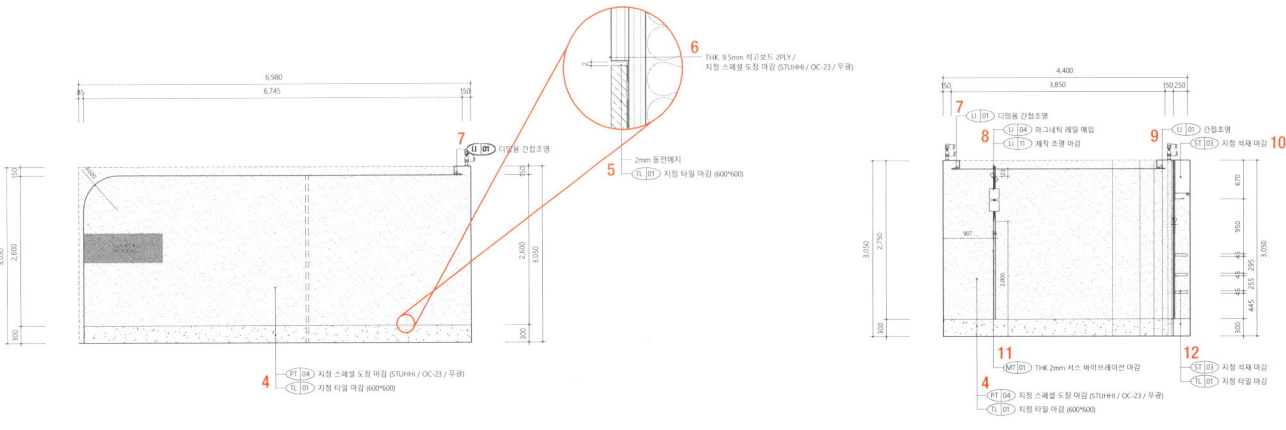

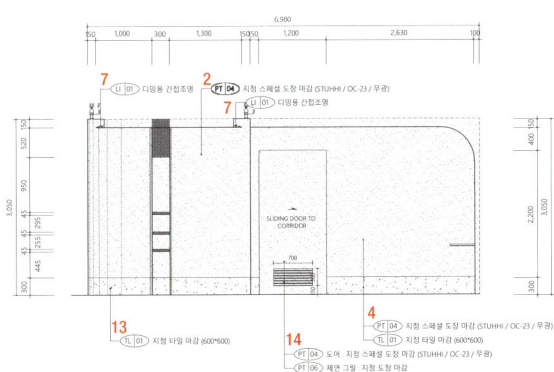

룸 입면 K1 / room elevation K1

룸 입면 K2 / room elevation K2

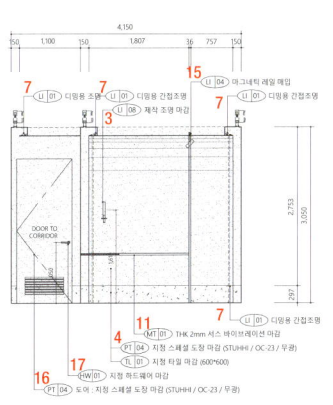

룸 입면 K3 / room elevation K3

룸 입면 K4 / room elevation K4

1 26.5X51 마그네틱 레일등 매입 / T2 SUS 바이브레이션 2 파워바 / T9.5 석고보드 2겹 위 지정 스페셜 페인트 3 마그네틱 레일등 4 T9.5 석고보드 2겹 위 지정 스페셜 페인트 / T1.5 금속 위 지정 스페셜 페인트 5 간접조명 6 T9.5 석고보드 2겹 위 지정 스페셜 페인트 7 T2 SUS 바이브레이션 / T9.5 석고보드 2겹 위 지정 스페셜 페인트 8 후크 고정 자리 9 T2 SUS 바이브레이션 10 T18 합판 위 지정 스페셜 페인트 11 바리솔 조명 / 지정 도장 12 디밍용 조명 13 지정 스페셜 페인트 / 지정 타일 14 T3 SUS 바이브레이션 / T3 SUS 바이브레이션 15 문 : 지정 스페셜 페인트 / 제연 그릴 : 지정 도장 16 지정 스페셜 페인트 / 걸레받이 : 지정 도장 17 지정 스페셜 페인트 18 문 : 지정 스페셜 페인트 / 걸레받이 : 지정 도장

1 Embedding 26.5X51 magnetic track lighting / T2 SUS vibration 2 Power bar / App. special paint on T9.5 gypsum board 2ply 3 Magnetic track lighting 4 App. special paint on T9.5 gypsum board 2ply / App. special paint on T1.5 metal 5 Indirect lighting 6 App. special paint on T9.5 gypsum board 2ply 7 T2 SUS vibration / App. special paint on T9.5 gypsum board 2ply 8 Room for joint hooks 9 T2 SUS vibration 10 App. special paint on T18 plywood 11 Barrisol lighting / App. painting 12 Dimmable indirect lighting 13 App. special paint / App. tile 14 T3 SUS vibration / T3 SUS vibration 15 Door : App. special paint / Smoke exhaust grille : App. paint 16 App. special paint / Baseboard : App. paint 17 App. special paint 18 Door : App. special paint / Baseboard : App. paint

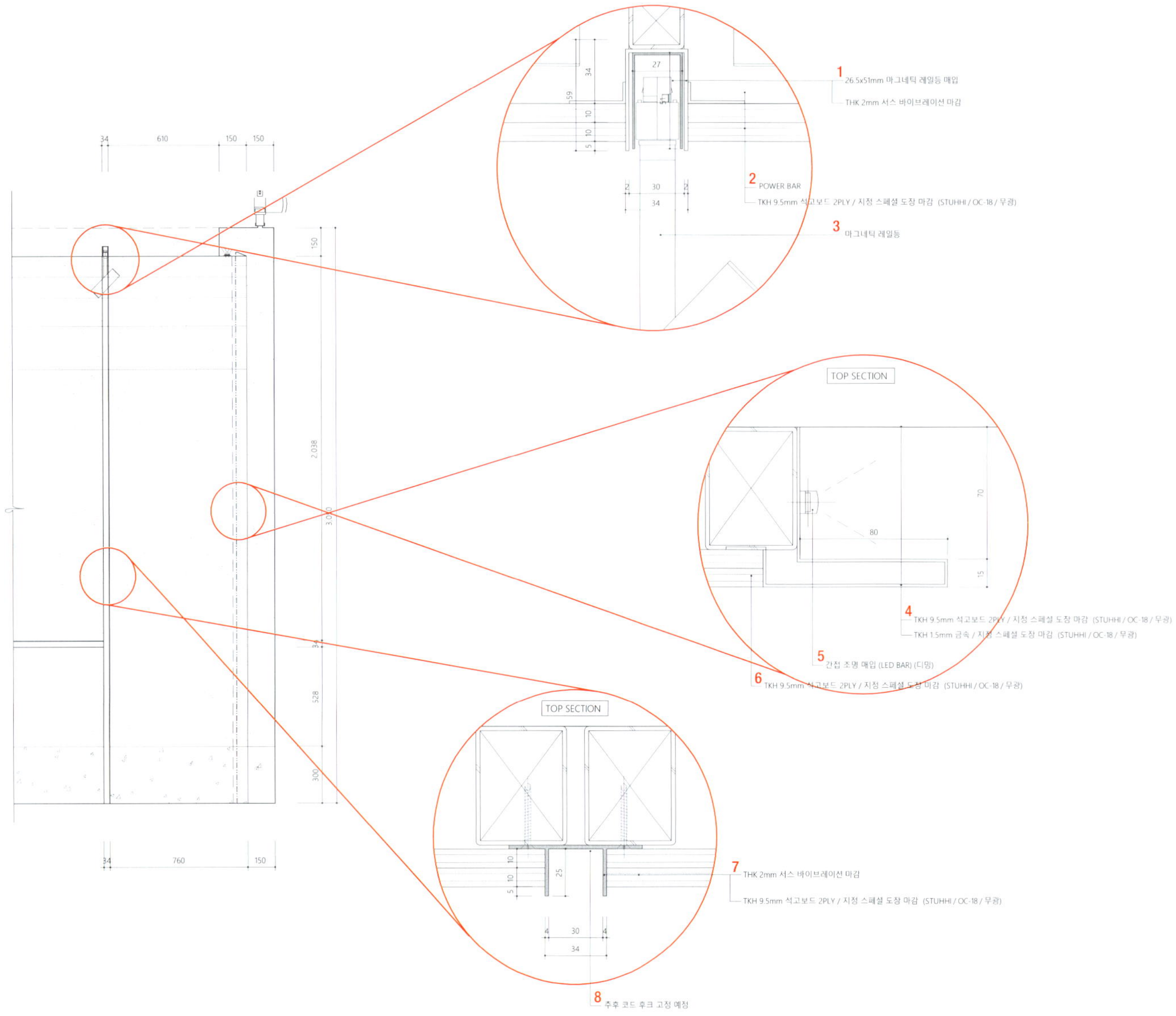

벽 단면 상세 L / wall section detail L

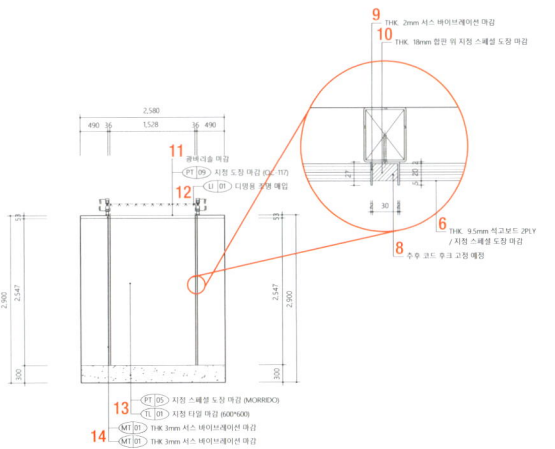

룸 입면 M1 / room elevation M1

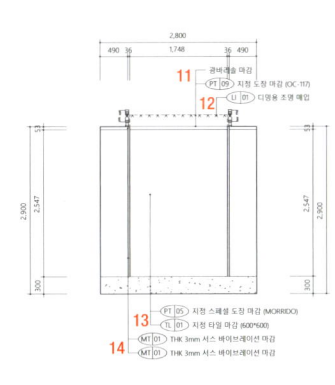

룸 입면 M2 / room elevation M2

룸 입면 M3 / room elevation M3

룸 입면 N1 / room elevation N1

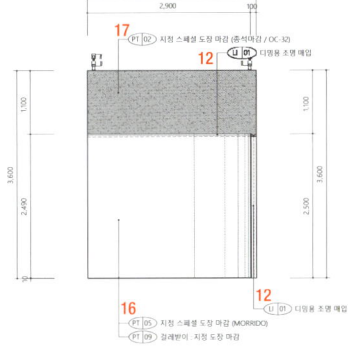

룸 입면 N2 / room elevation N2

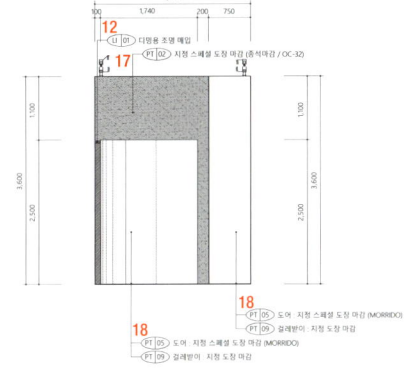

룸 입면 N3 / room elevation N3

VIKEN HUS

design by 83 | Minsuk Kim, Donghyun Nam, Chanun Park, Suhyun Jo, Hansol Jo

바이킹의 집을 콘셉트로 한 레스토랑으로 전체적인 무드도 이와 동일한 맥락으로 디자인했다. 1층에 진입함과 동시에 마주하는 입구에는 상징적인 그래픽이 손님을 맞이한다. 고풍스러운 유럽을 연상케 하는 2층에는 강렬한 스테인드글래스가 설치되어 있는데, 여기에는 직접적으로 공간 디자인의 모티브가 된 시원한 바다에 떠 있는 바이킹의 배를 묘사하고 있다. 계단을 지나 마주하는 공간은 3가지 섹션으로 구분된다. 첫 번째 영역에는 공간을 상징하는 배 모양의 테이블이 위치하는데, 이는 바이킹에서 시작됐다고 전해지는 뷔페 문화를 은유적으로 표현한 것이다. 단순 눈요기로 그치지 않고, 실제 배를 만들 때의 느낌을 최대한 살리고자 조립식으로 구조를 구성했다. 배 모양의 테이블을 지나 오른쪽 문을 지나면 바이킹들이 모여 살던 주거 형태를 현대적으로 재해석해 연출한 두 번째 영역이 나온다. 천장을 지지하는 구조적 기능의 기둥들과 그 사이사이 현대적 느낌의 금속 소재를 통해 이질적이면서도 세부적인 디테일을 표현하고자 했다. 기둥 하부에는 어린 고객을 위해 인조가죽에 솜을 넣어 안전성을 더하고, 카운터와 서비스 테이블 및 진열장, 문 등은 모두 비슷한 무드로 제작했다. 프라이빗한 세 번째 영역은 다른 두 영역의 밝은 느낌과는 차별화된 차분하면서도 고급스러운 공간이다. 이를 위해 가구 또한 다른 영역의 가구에 비해 조금 더 편안한 소재와 디자인으로 제작하여 고객들이 공간에서 쾌적하게 머물며 즐길 수 있게 했다. 전체적으로 색션별 가구디자인을 저마다 다르게 디자인하고 오브제 및 소품 또한 콘셉트와 조화되도록 적절히 배치했다. 벽체는 유럽 미장으로 마감하고, 바닥은 다소 밋밋하게 보이지 않게 서로 다른 사이즈와 패턴의 타일을 활용하여 공간을 더욱더 밀도 있게 구성했다.

The retaurant was conceptualized as Viking's home, and the overall mood was designed to reflect the same theme. Upon entering the 1st floor, guests are welcomed by an iconic visual display. On the 2nd floor, a stained-glass window with a vintage European aesthetic displays a Viking vessel sailing in the chilly ocean straightforwardly. The area beyond the stairs is separated into three distinct sections. The first area showcases a ship-shaped table, which symbolizes space. The table serves as a metaphor for the buffet culture that allegedly traces back to the Vikings. They didn't just want it to be an eye-catcher, so they wanted it to be prefabricated to intesify the sensation of constructing an actual ship. Past the shipshaped table and through the door on the right lies the second area - a contemporary interpretation of the Vikings' clustered dwellings. Columns providing the foundational support for the ceiling and contemporary metallic materials between them express disparate yet intricate details. The bottom of the columns is decorated with imitation leather cushions for the safety of younger customers. The counters, service tables, display cases, and doors are made in a similar style. The private third area is calm and luxurious space that is different from the bright other two areas. To achieve this goal, the furniture is crafted with more comfortable materials and designs than that of other areas, allowing customers to relax and enjoy the space without discomfort. In general, each section has a distinct furniture design, and objects and accessories are arranged to complement the overall concept. The walls were finished with European plaster, while the floor was tiled with a variety of sizes and patterns to add dimension and prevent a flat appearance to the space.

디자인 김민석, 남동현, 박찬연 / 디자인바이팔삼
위치 부산광역시 동구 중앙대로196번길 10, 2층
용도 레스토랑
면적 260m²
마감 바닥 – 타일 / 벽 – 스페셜 페인트, 스테인리스 스틸, 낙엽송 합판, 우드 필름, 앤틱거울 / 천장 – 스페셜 페인트
완공 2023. 10
디자인 조수현, 조한솔
시공팀 남동현, 조한솔
가구 배지헌 / TT
사진 김동규

Location 2F, 10, Jungang-daero 196beon-gil, Dong-gu, Busan
Use Restaurant
Area 260m²
Finishing Floor - Tile / Wall - Special paint, Stainless steel, Larch plywood, Wood film, Antique mirror / Ceiling - Special paint
Completion 2023. 10
Design team Suhyun Jo, Hansol Jo
Construction team Donghyun Nam, Hansol Jo
Furniture Jiheon Bae / TT
Photographer Donggyu Kim

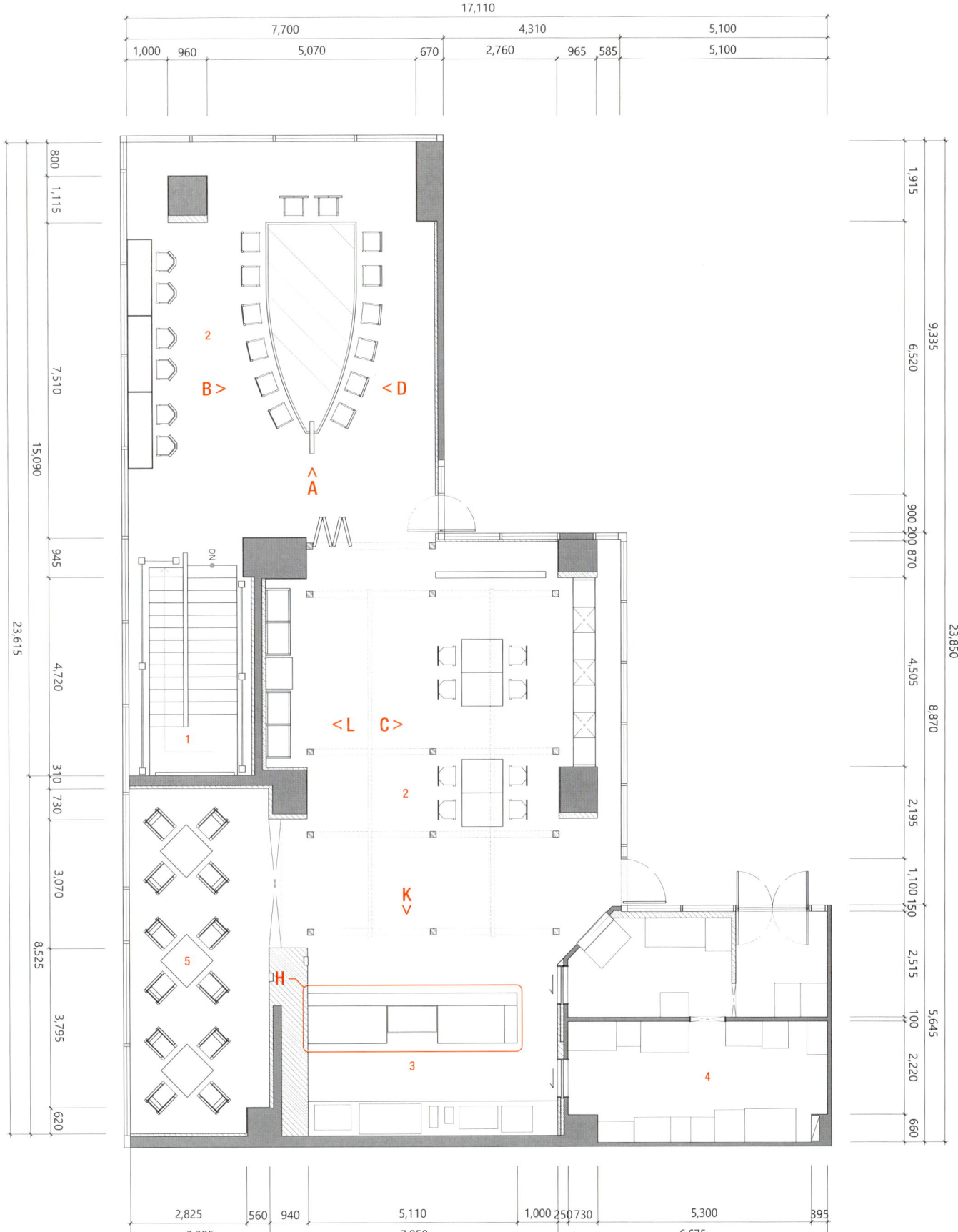

평면도 / floor plan

1 계단실 2 홀 3 카운터 4 주방 5 프라이빗 룸

1 Stairs 2 Hall 3 Counter 4 Kitchen 5 Private room

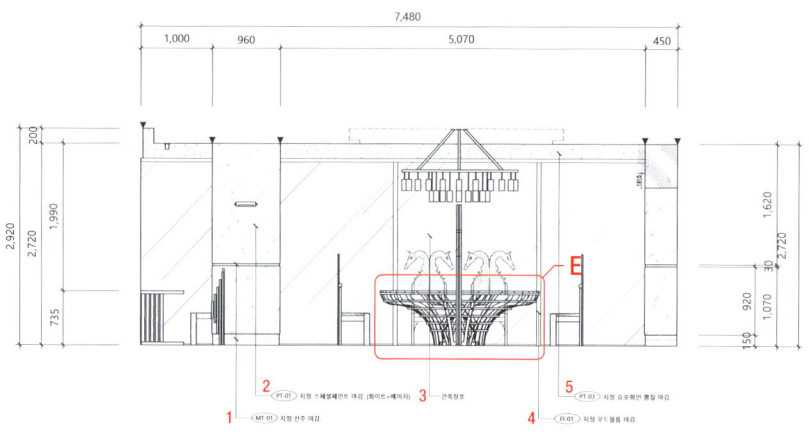

홀 입면 A / hall elevation A

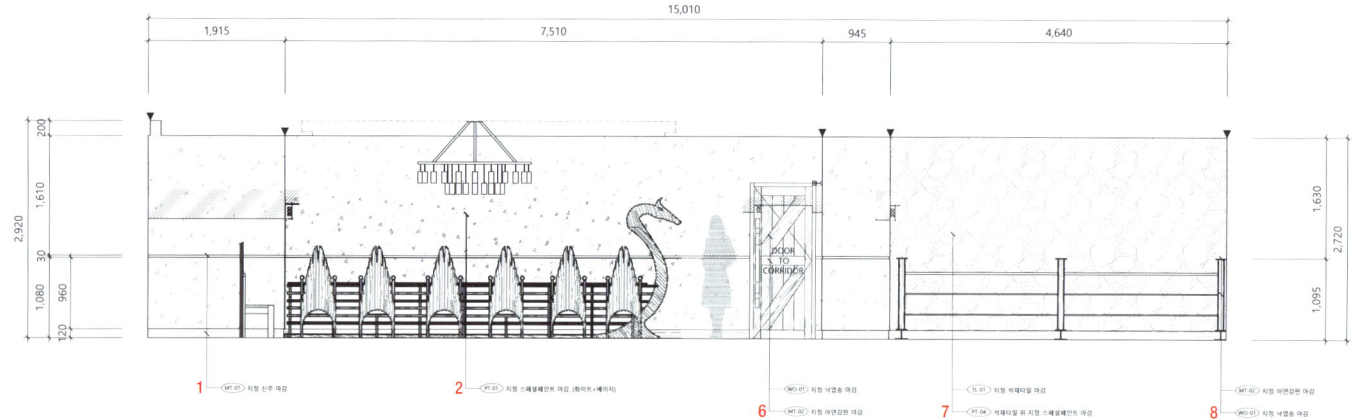

홀 입면 B / hall elevation B

1 지정 황동 2 지정 스페셜 페인트 3 건축창호 4 지정 우드 필름 5 지정 슈퍼화인 뿜칠 6 지정 낙엽송 / 지정 아연강판 7 지정 석재 타일 / 석재 타일 위 지정 스페셜 페인트 8 지정 아연강판 / 지정 낙엽송 9 지정 아연강판 / 지정 낙엽송 / 지정 가죽 10 지정 낙엽송 11 지정 아연강판 12 지정 스페셜 페인트 / 지정 아연강판 13 지정 철근 / 철근 위 지정 실버 래커

1 App. brass 2 App. special paint 3 Architecture windows 4 App. wood film 5 App. super fine spray 6 App. Japanese larch / App.hot rolled steel plate 7 App. stone tile / App. special paint on stone tile 8 App. galvanized steel sheet / App. Japanese larch 9 App. galvanized steel sheet / App. Japanese larch / App. leather 10 App. Japanese larch 11 App. galvanized steel sheet 12 App. special paint / App. galvanized steel sheet 13 App. reinforcing steel bar / App. silver lacquer on reinforcing steel bar

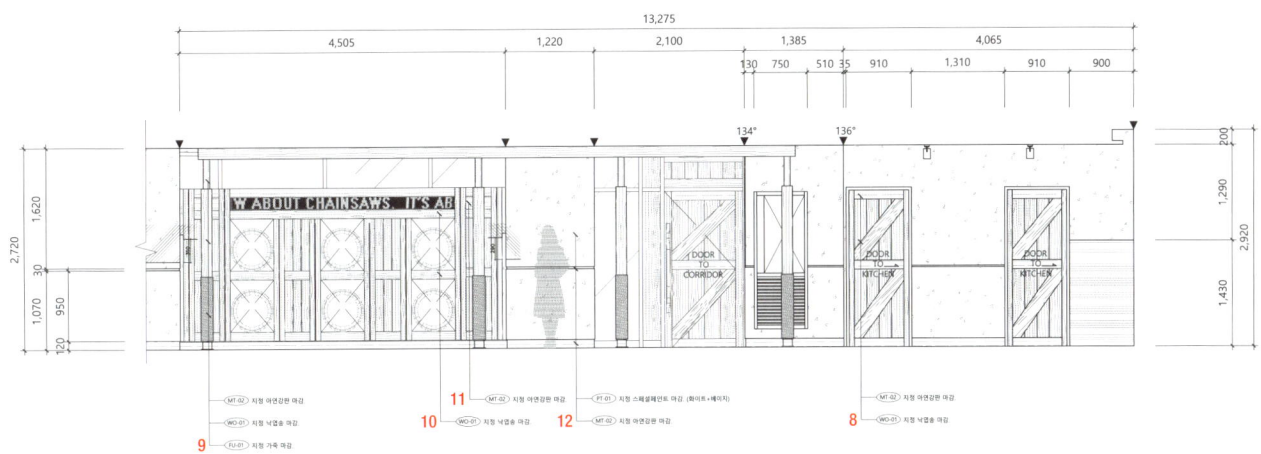

홀 입면 C / hall elevation C

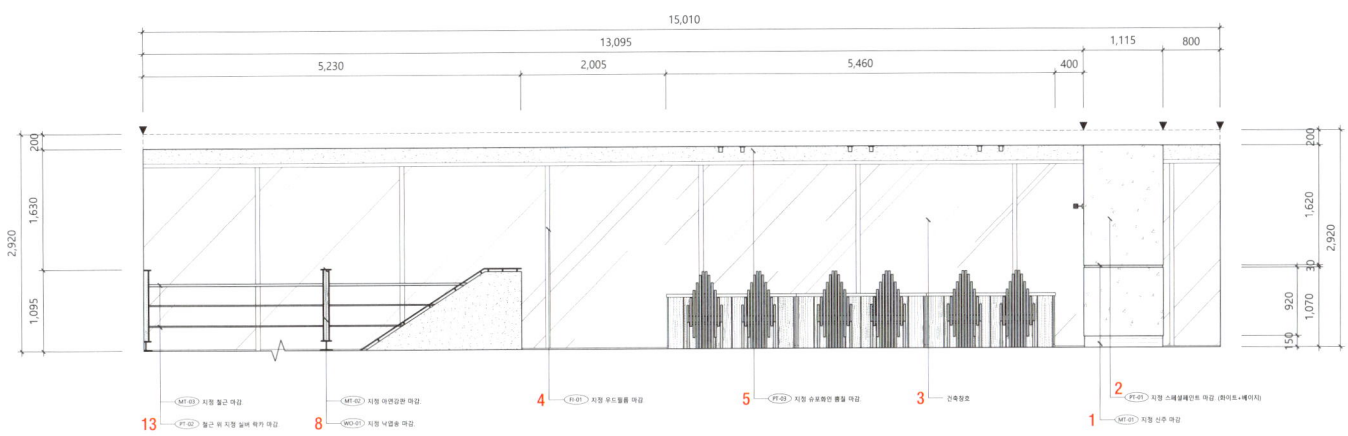

홀 입면 D / hall elevation D

1 지정 투명 강화유리 2 지정 애쉬 원목 3 지정 유리 4 지정 아연강판 5 지정 낙엽송 6 T1.2 아연강판 위 은분 마감 7 T1.2 아연강판 절곡

1 App. clear tempered glass 2 App. ash wood 3 App. glass 4 App. galvanized steel sheet 5 App. Japanese larch 6 Metallic powder on T1.2 galvanized steel sheet 7 T1.2 galvanized steel sheet bending

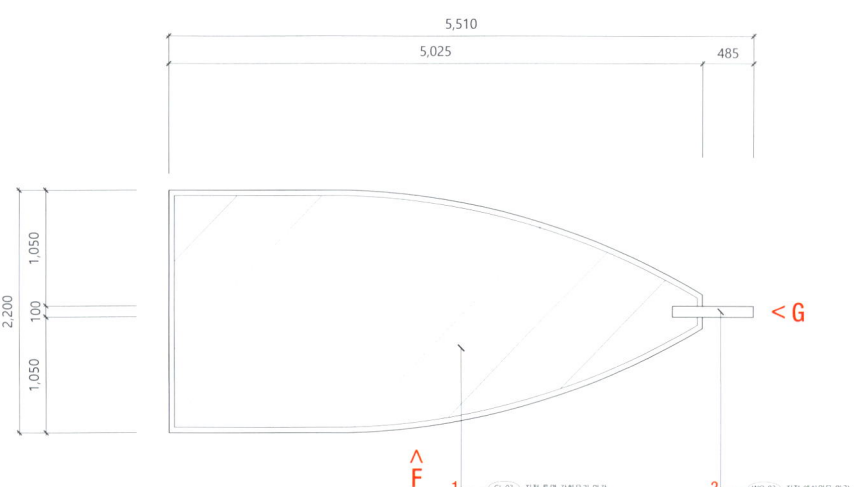

테이블 평면 E / table top view E

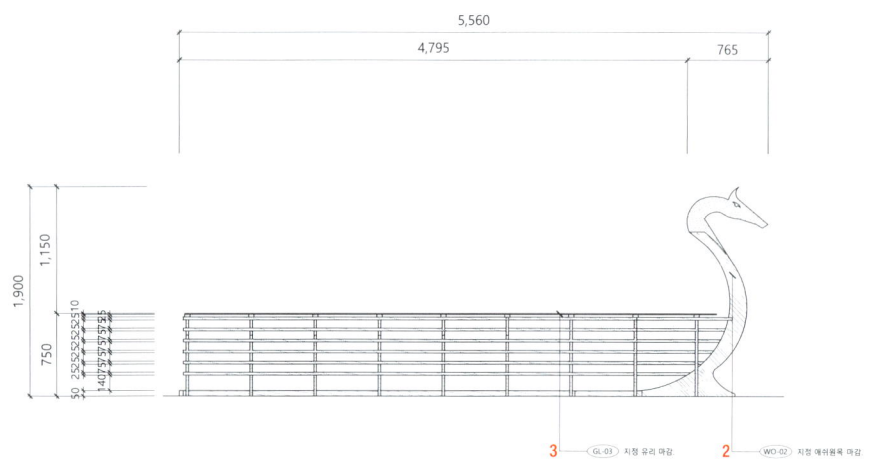

테이블 정면 F / table front view F

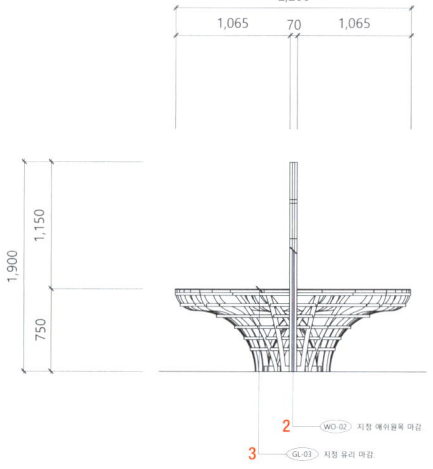

테이블 측면 G / table side view G

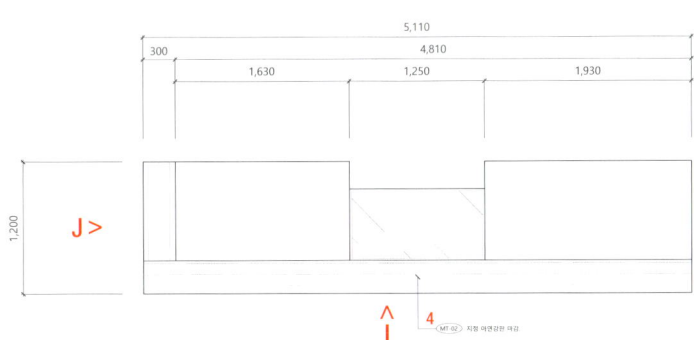

카운터 평면 H / counter top view H

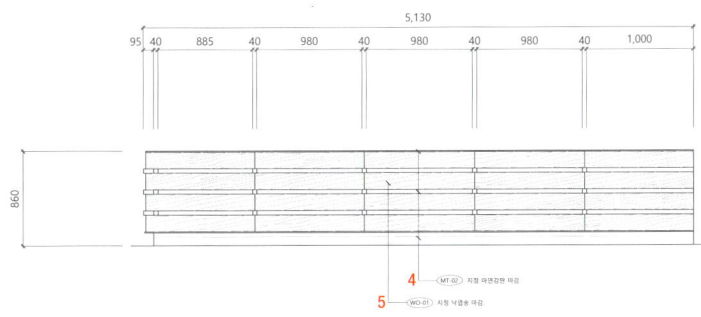

카운터 정면 I / counter front view I

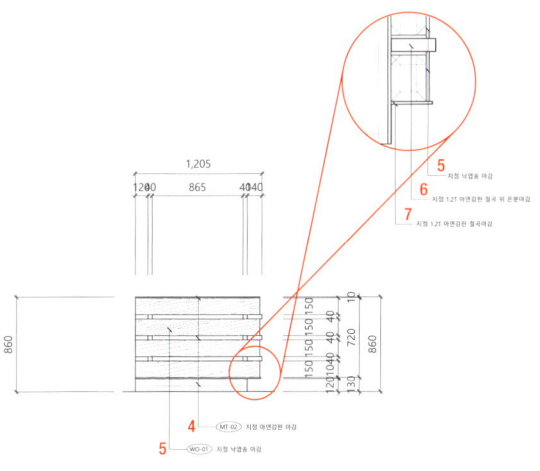

카운터 측면 J / counter side view J

1 지정 아연강판 2 지정 스페셜 페인트 3 지정 낙엽송 / 지정 가죽 / 지정 아연강판 4 지정 황동 5 지정 석재 타일 / 석재 타일 위 지정 스페셜 페인트 6 지정 낙엽송 7 T1.2 아연강판 절곡 8 지정 가죽

1 App. galvanized steel sheet 2 App. special paint 3 App. Japanese larch / App. leather / App. galvanized steel sheet 4 App. brass 5 App. stone tile / App. special paint on stone tile 6 App. Japanese larch 7 T1.2 galvanized steel sheet bending 8 App. leather

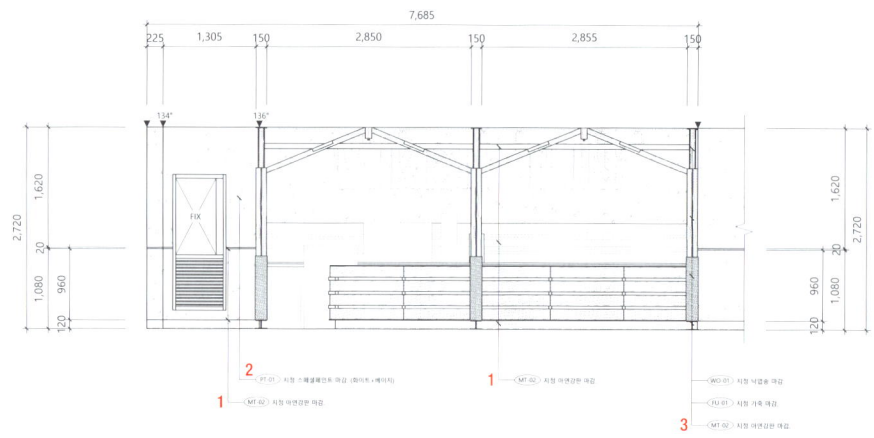

홀 입면 K / hall elevation K

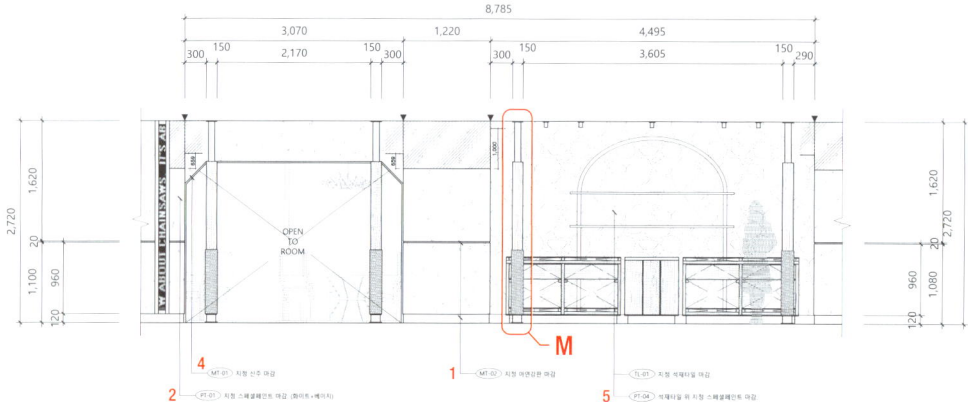

홀 입면 L / hall elevation L

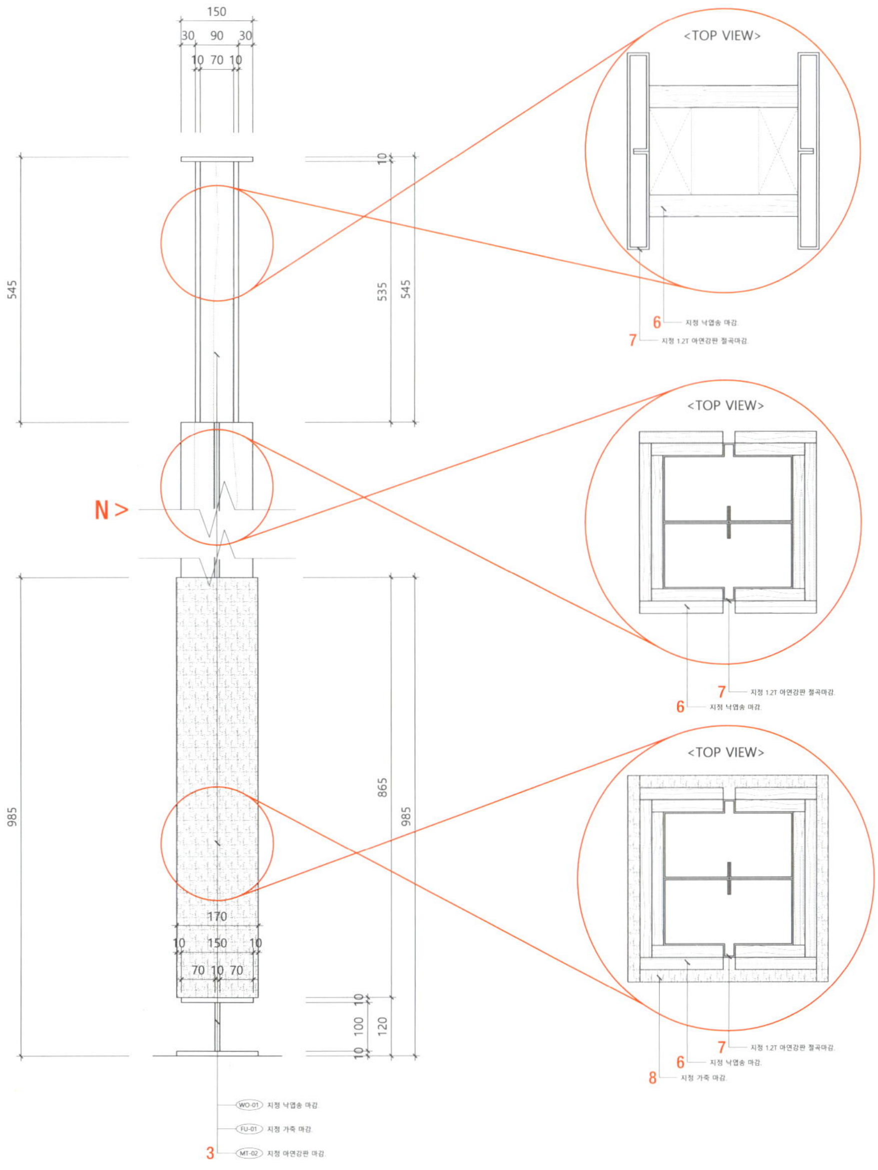

기둥 정면 M / column front view M

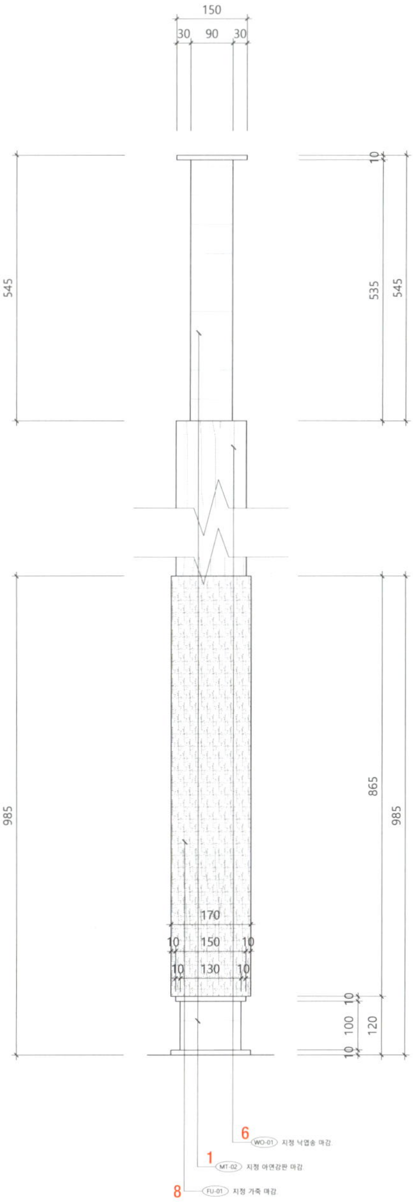

기둥 측면 N / column side view N

VARIEGATA GROTTA

PROJECT MARK | Jaehong Son, Jeeseung Yang, Jihyung Song

바리에가타 명동점은 (구)한국은행 소공별관에 위치한다. 이러한 역사적이고 금융업에서 특별한 의미가 있는 건물 지하에 레스토랑이 있다는 것은 이용객으로 하여금 특별하고 우연한 발견을 하는 경험을 준다. 공간 콘셉트는 '그로타'이다. '그로타'는 이탈리아어로 해식동굴을 뜻하는데, 파도가 만든 뭍 아래 특별함이 숨겨져 있는 동굴과 같이 이 사이트에서 특별함과 유사함을 느껴 콘셉트로 사용되었다. 분위기는 전반적으로 어두운 조도에 사람의 시선과 움직임이 머무는 곳에만 국부 조명을 설치하여 한 줄기 빛이 동굴에 들어왔을 때의 강렬함을 은유했다. 재료 마감은 벽은 바위 질감의 스페셜 페인트를 사용함과 동시에 천장은 그것과 톤온톤이 되는 베이지색 페인트를 사용하고, 검은색 무늬목과 구로철을 사용한 가구들을 배치하였다. 입구에 들어서면 마주하는 대형 테이블과 라인 펜던트는 해식동굴에 들어온 오래된 배를 상상하며 디자인했고, 러프하고 아날로그한 디테일을 추가하였다. 이러한 디자인 요소들은 바리에가타의 독특한 정체성과 분위기를 형성하며 깊이 있는 경험을 제공한다. '그로타' 콘셉트를 통해 역사와 현대의 융합을 시각적으로 표현한 '바리에가타 명동점'은 방문객들에게 단순한 식사 이상의 의미를 전달한다.

Variegata Myeongdong Branch is located in the annex building of the former Bank of Korea. As the space is situated in the basement of this historically significant building to the financial sector, visitors can experience an extraordinary and unexpected discovery. The spatial design concept is 'Grotta,' which means sea cave in Italian. This concept was proposed because visitors can experience similar uniqueness found in incredible caves created by waves beneath the terrain. The overall atmosphere maintains dim lighting, while spotlights are installed where people's eyes and movements linger. This arrangement expresses the intense feeling of light streaming into a cave. The walls are finished with special paint that creates a rocky texture, while the ceiling in a tone-on-tone relationship with the walls is finished in beige paint. Furniture pieces made of black wood veneer and black steel sheet are placed throughout the space. The large table and linear pendant lighting that welcome visitors at the entrance are designed to evoke an image of an old ship entering a sea cave, with added details that convey a rough and classical feel. These design elements complete Variegata's unique identity and atmosphere, delivering a profound experience. Designed based on the concept of 'Grotta' to create a visual aesthetic that combines history and modernity, Variegata Myeongdong offers visitors an experience that goes beyond the conventional concept of dining.

디자인 손재홍, 양지승, 송지형 / 프로젝트 마크
위치 서울특별시 중구 남대문로 551, 지하 1층
용도 레스토랑
면적 200m²
마감 타일, 무늬목, 스페셜 페인트, 열연강판
완공 2024. 9
디자인팀 김어진, 팽종인
사진 김한빛

Location B1, 551, Namdaemun-ro, Jung-gu, Seoul
Use Restaurant
Area 200m²
Finishing Tile, Wood veneer, Special paint, Hot rolled steel sheet
Completion 2024. 9
Photographer Hanbit Kim

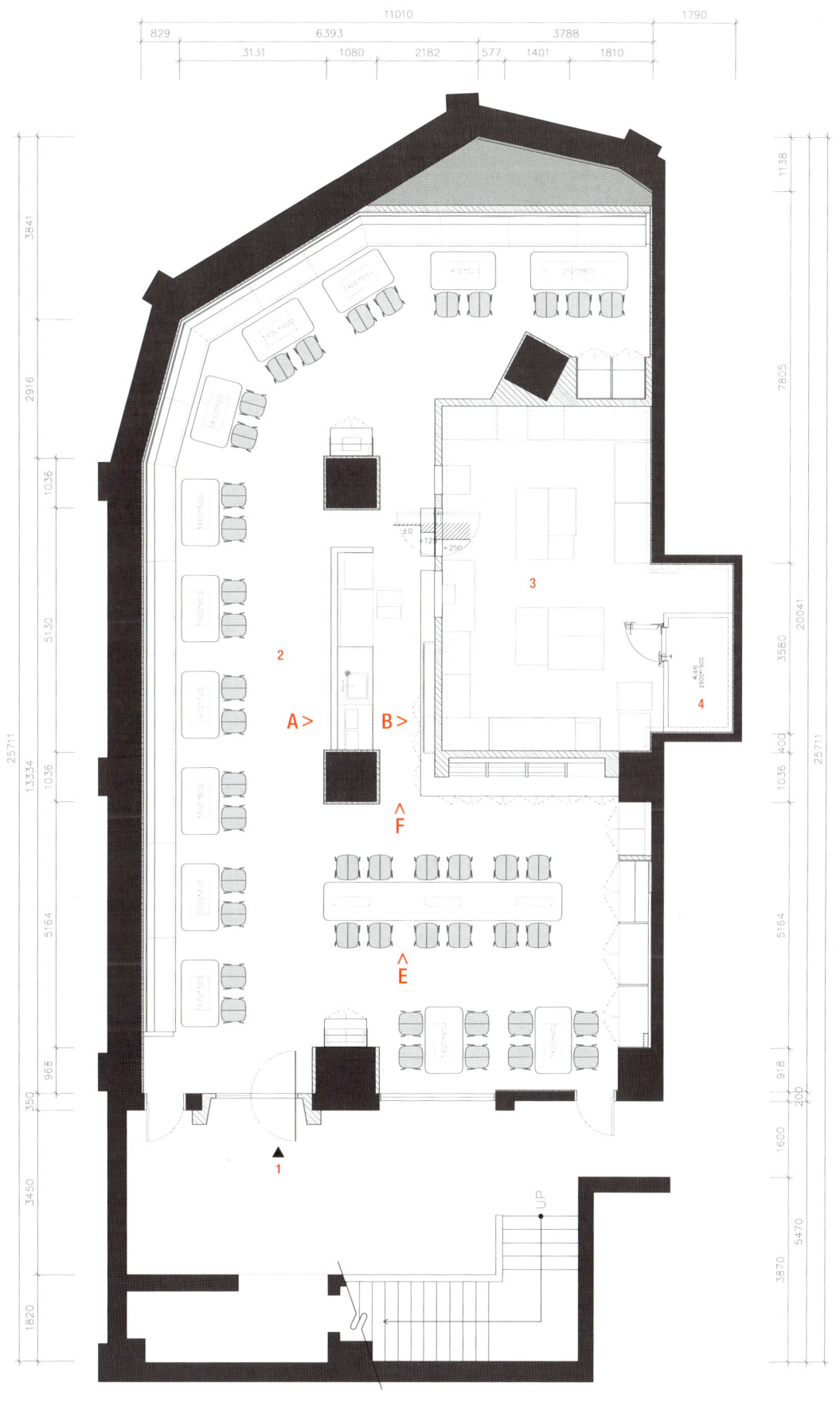

평면도 / floor plan

1 입구 2 홀 3 주방 4 워크인 냉장고

1 Entrance 2 Hall 3 Kitchen 4 Walk-in fridge

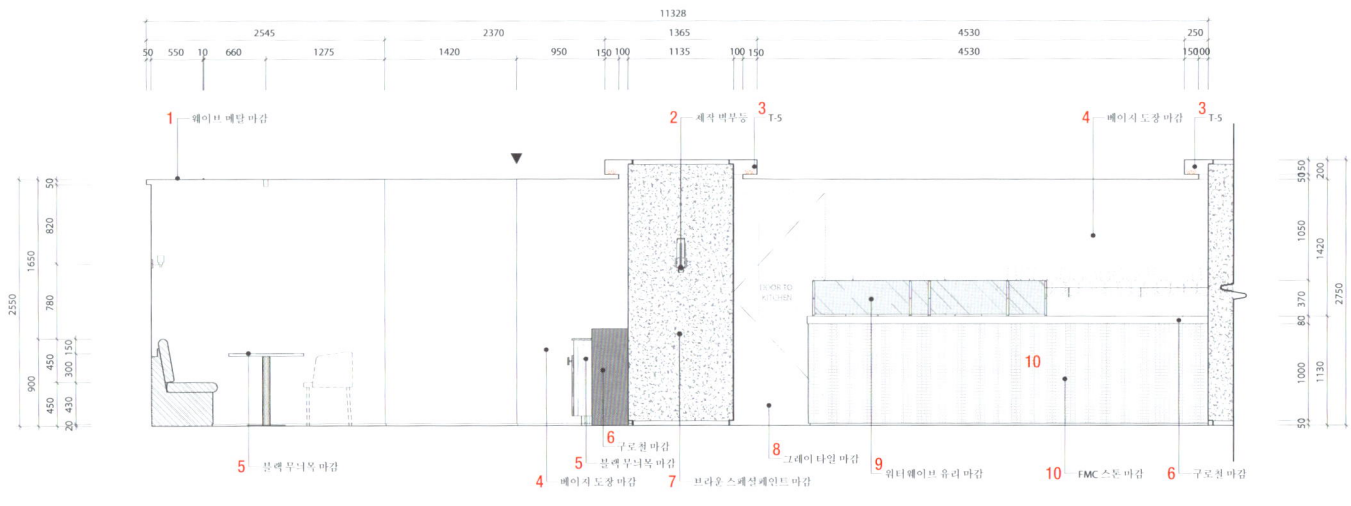

홀 입면 A1 / hall elevation A1

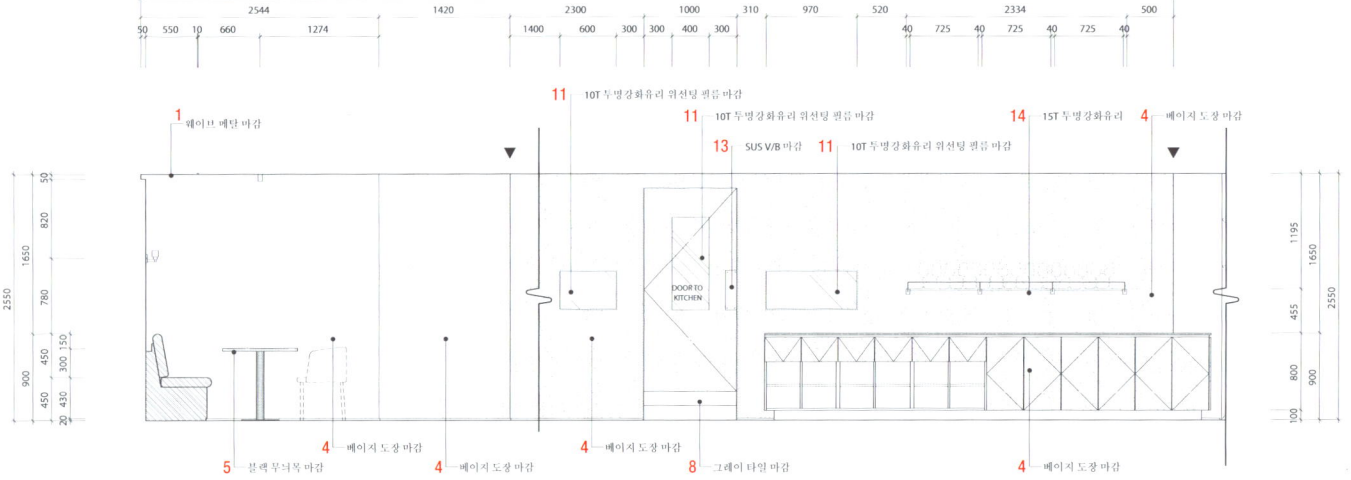

홀 입면 B1 / hall elevation B1

1 웨이브 메탈 2 제작 벽부등 3 지정 조명 4 베이지 도장 5 블랙 무늬목 6 열연강 7 브라운 스페셜 페인트 8 회색 타일 9 워터웨이브 유리 10 FMC 스톤 11 T10 투명 강화유리 위 선팅 필름 12 제작 펜던트 조명 13 SUS 바이브레이션 14 T15 투명 강화유리 15 합판 보강 및 앵커 고정 16 3mm 와이어

1 Wave metal 2 Custom-made wall lamp 3 App. painting 4 Beige painting 5 Black wood veneer 6 Hot rolled steel 7 Brown special paint 8 Gray tile 9 Water wave textured glass 10 FMC stone 11 Tint film on T10 clear tempered glass 12 Custom-made pendant lighting 13 SUS vibration 14 T15 clear tempered glass 15 Plywood reinforcement, Jointing anchor 16 3mm wire

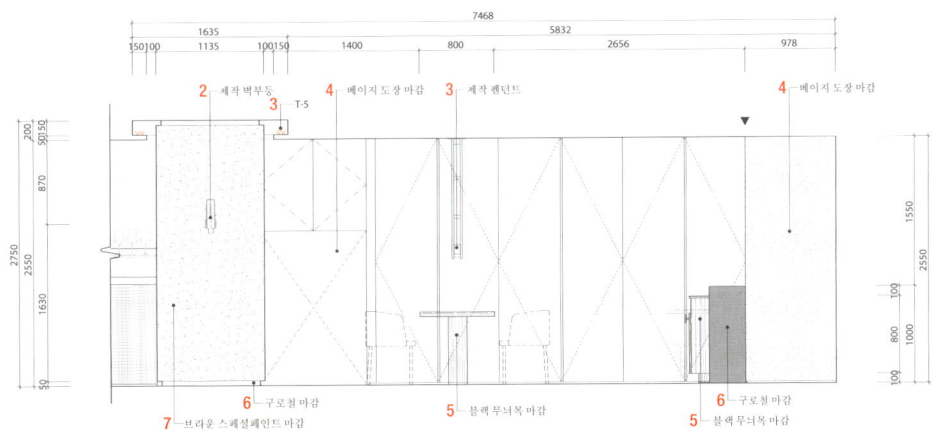

홀 입면 A2 / hall elevation A2

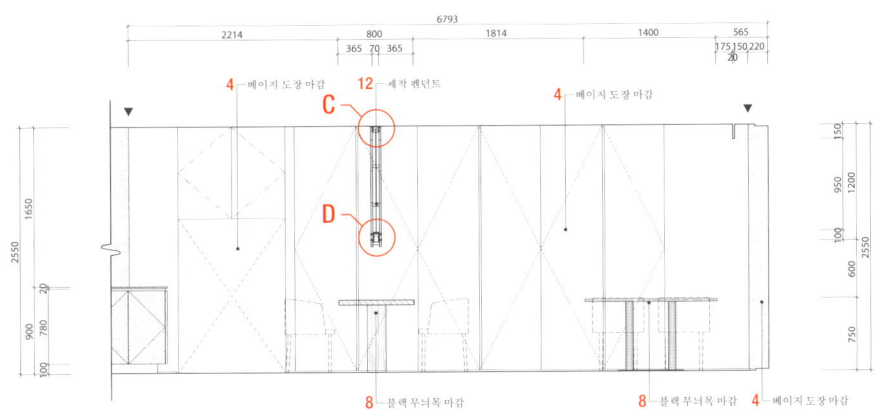

홀 입면 B2 / hall elevation B2

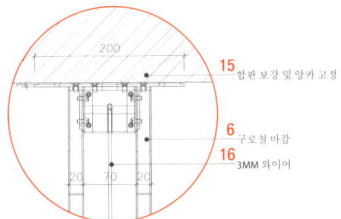

조명 상세 C / lighting detail C

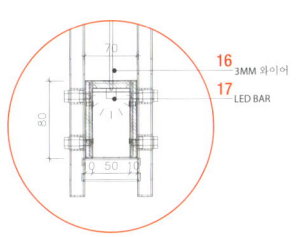

조명 상세 D / lighting detail D

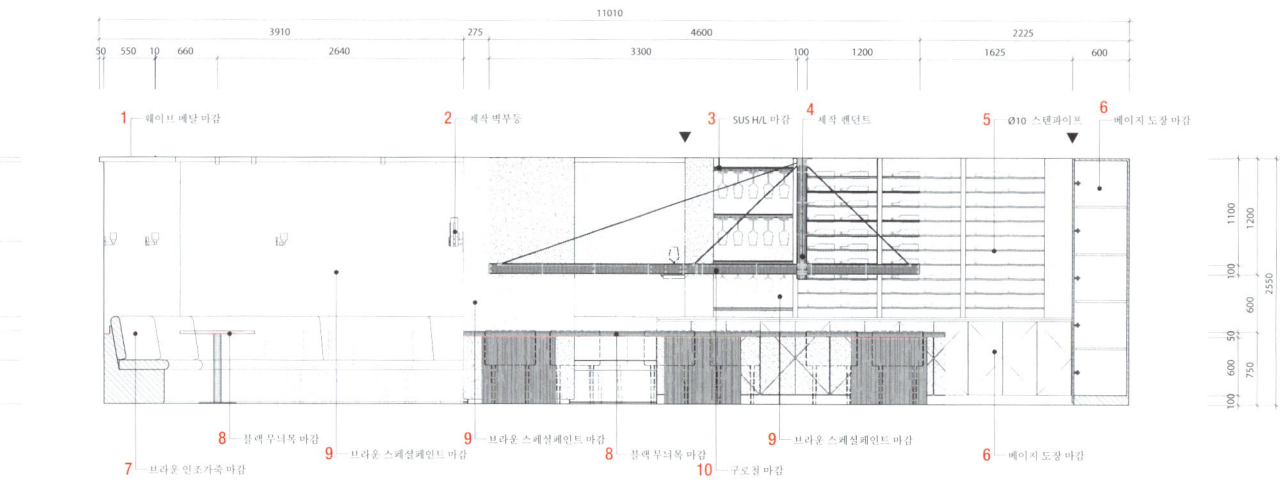

홀 입면 E / hall elevation E

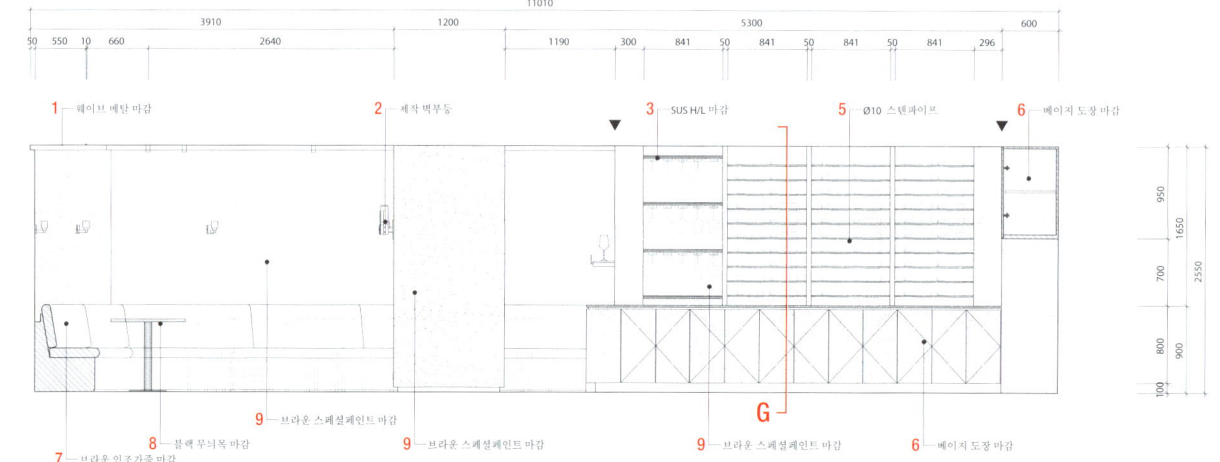

홀 입면 F / hall elevation F

1 웨이브 메탈 2 제작 벽부등 3 SUS 헤어라인 4 제작 펜던트 조명 5 Ø10 스테인리스 스틸 파이프 6 베이지 도장 7 브라운 인조가죽 8 블랙 무늬목 9 브라운 스페셜 페인트 10 열연강 11 T10 투명 강화유리 12 SUS 실버 헤어라인 13 와인랙 고정 14 도장 15 M-바 16 600X600 알루미늄 평패널 17 혹두기 도장 18 T9 석고보드 2겹 19 T9 타일 + 시멘트 20 T12.5 방화 석고보드 21 T9 석고보드, 합판 22 50X30 금속 구조 23 시멘트 벽돌

1 Wave metal 2 Custom-made wall lamp 3 SUS hairline 4 Custom-made pendant lighting 5 Ø10 stainless steel pipe 6 Beige painting 7 Brown artificial leather 8 Black wood veneer 9 Brown special paint 10 Hot rolled steel 11 T10 clear tempered glass 12 SUS silver hairline 13 Wine rack joint 14 Painting 15 M-bar 16 600X500 aluminum flat panetl 17 Textured painting 18 T9 gypsum board 2ply 19 T9 tile + Cement 20 T12.5 fireproof gypsum board 21 T9 gypsum board, Plywood 22 50X30 metal structure 23 Cement brick

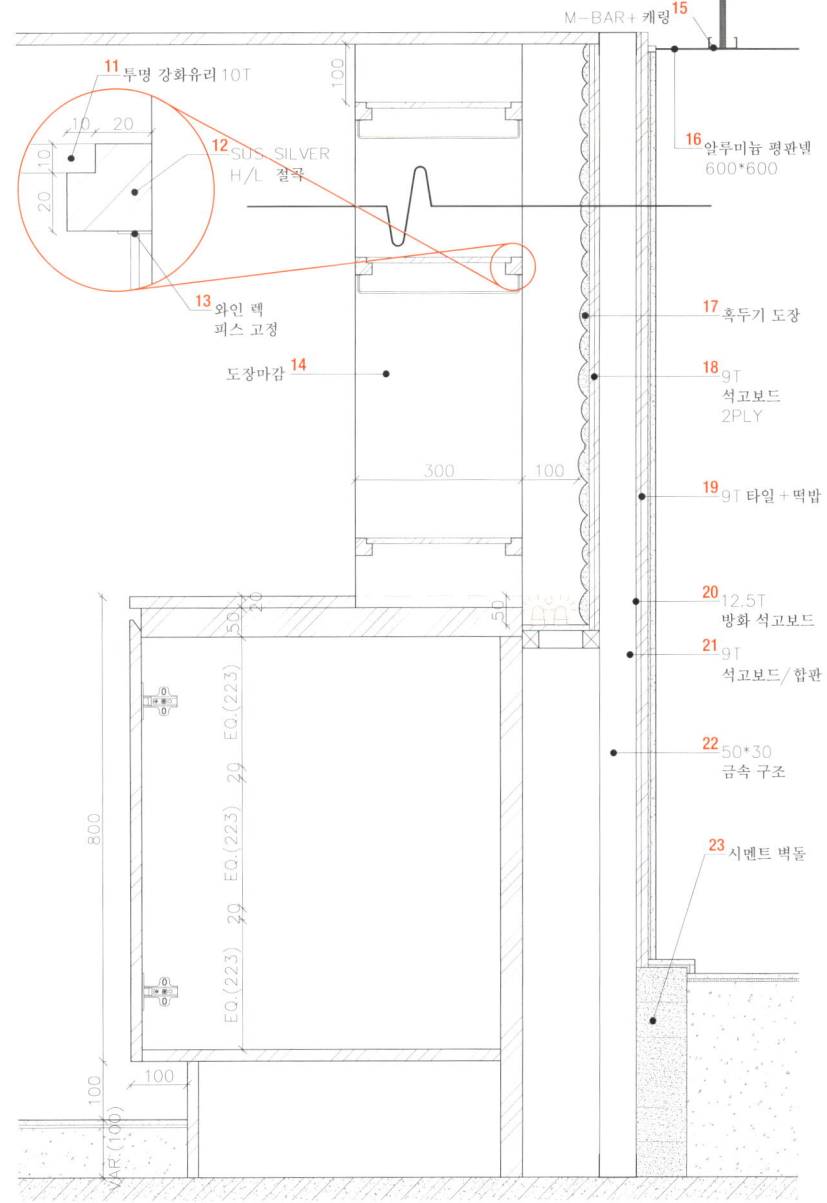

와인 랙 단면 상세 G / wine rack section detail G

2025
ANNUA(L)
CAFE
INTER(I)
DETAI(L)

104	**SETT**	
	세트	
118	**KNOTTED GIMPO**	
	노티드 김포	
126	**INC COFFEE**	
	인크커피	
134	**SECOND ONE LAKE ASAN**	
	세컨드원 레이크 아산	
148	**OHVENU HANNAM**	
	오베뉴 한남	
154	**LES MAINS DORÉES**	
	레망도레	
164	**AISO SOUND**	
	아이소 사운드	
180	**HORONG**	
	호롱	
188	**FILLMATE**	
	필메이트	
206	**SOOSOO COFFEE**	
	수수커피	

SETT

Design studio maoom | Minkyu Choi

"질서의 미학"은 사물이나 현상이 규칙적이고 조화롭게 배열될 때 느껴지는 아름다움이다. SETT는 존중의 태도를 바탕으로 감각적인 질서가 존재할 수 있도록, 조화로운 규칙과 명확한 구조(오브제-자리)를 통해 시각적 흐름을 감각적인 미적 경험으로 이끌어낸다. SETT는 고유한 규칙과 질서를 기반으로 다양한 변주를 선보이는 공간이다. 잘 세팅된 메뉴와 F&B를 통해 수준 높은 맛과 정돈된 서비스를 제공하며, 공간을 통해 더 나은 취향과 서로 존중하는 방식을 제안하는 철학이 깃들어 있는 브랜드다. 디자이너는 모든 순간이 존중과 배려를 바탕으로, 높은 품질의 서비스와 다채로운 경험을 제공하는 방법을 모색했다. 묵묵히 그 자리를 지키며 배려의 마음을 담아 준비된 모든 순간들이 공존의 가치를 실현한다. 공간은 자연과 사람 사이의 관계를 이어주며, 다투지 않고 자연에 순응하는 태도로 머무는 시간을 존중한다. 이러한 접근을 바탕으로 디자이너는 섬세한 배려와 세련된 존중이 감각적인 질서를 이루며, 휴식의 새로운 가치를 창출하는 공간을 이야기하고자 했다. 머무르는 장소에 따라 경험하는 시간이 공간에 대한 감각으로 확장되며, 배려의 태도는 시각적 질서로 구현된다. 불필요한 요소를 배제하고, 건축적 흐름에 따라 자리마다의 의미를 두고자 했다. 계절의 소리를 가까이에서 느끼며, 자연과 교감하는 깊은 경험으로까지 확장되는 섬세한 배려와 존중을 통해 휴식의 새로운 가치를 만들어 가고자 한다.

The 'aesthetics of order' means the beauty found in systematically and harmoniously arranged objects or phenomena. SETT uses harmonious orders and intuitive structures to transform visual flow into a sensory and aesthetic experience, with a goal of establishing a sensory order that promotes respectful interaction. The spatial narrative of SETT presents diverse variations grounded in unique rules and order. The brand's well-crafted menus and systems provide high-quality cuisine and organized services, while the spatial design promotes sophisticated sensibility and mutual respect. The designer meticulously studied methods to deliver high-quality services and diverse experiences, rooted in respect and care. Individual spatial elements, reflecting the attitude of mutual respect, humbly stay in their position to showcase the beauty of coexistence. The space reconciles the relationship between nature and people and values the time for surrendering to nature. In this way, the designer wanted to create a space where delicate care and refined respect establish a sensory order, redefining the meaning of rest. The experience of visitors expands into spatial perception depending on where they stay, and the attitude of care is represented in the form of visual order. Unnecessary elements are eliminated, and each spot carries its own meaning that corresponds to the architectural narrative. The space of subtle care and respect, where visitors can listen to the sounds of seasons and interact with nature, creates a new meaning of rest.

디자인 최민규 / 디자인스튜디오 마음
위치 경기도 용인시 수지구 고기로 437, 1~3층
용도 베이커리 카페
면적 1층 - 689㎡ / 2층 - 538㎡ / 3층 - 641㎡
마감 바닥 - 테라조, 카펫, 원목마루, 타일 / 벽 - 스페셜 페인트, 대리석, 마이크로시멘트, 페인트 / 천장 - 페인트
완공 2023. 12
디자인팀 김연종, 이정환, 김재년, 박현석, 김종호
사진 김동규

Location 1-3F, 437, Gogi-ro, Suji-gu, Yongin-si, Gyeonggi-do
Use Bakery cafe
Area 1F - 689m² / 2F - 538m² / 3F - 641m²
Finishing Floor - Terazzo, Carpet, Wood flooring, Tile / Wall - Special paint, Marble, Microcement, Paint / Ceiling - Paint
Completion 2023. 12
Photographer Donggyu Kim

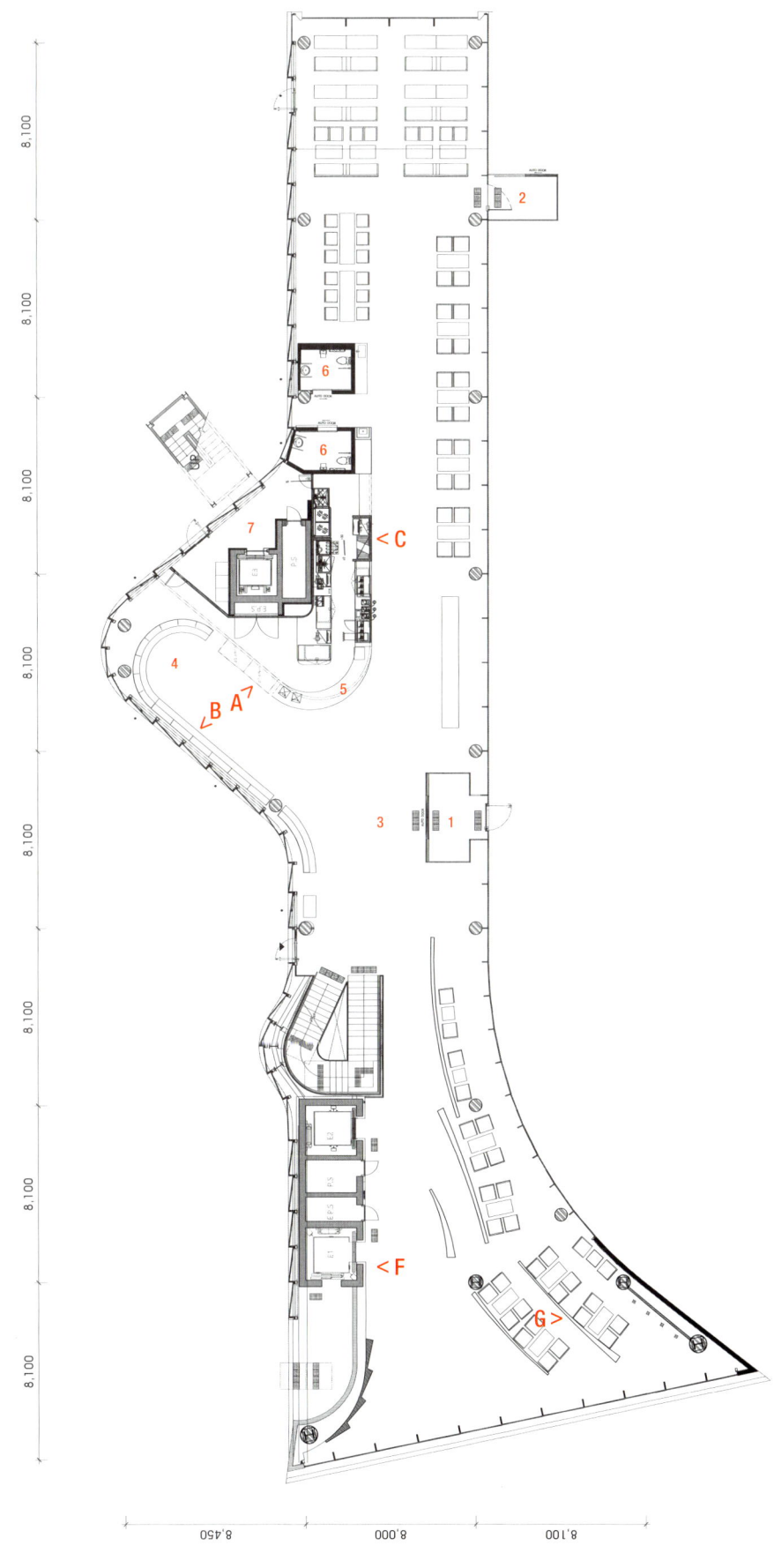

1층 평면도 / 1st floor plan

1 주출입구 2 부출입구 3 홀 4 베이커리 디스플레이 존 5 카운터 바 6 화장실 7 창고

1 Main entrance 2 Sub entrance 3 Hall 4 Bakery display zone 5 Counter bar 6 Restroom 7 Storage

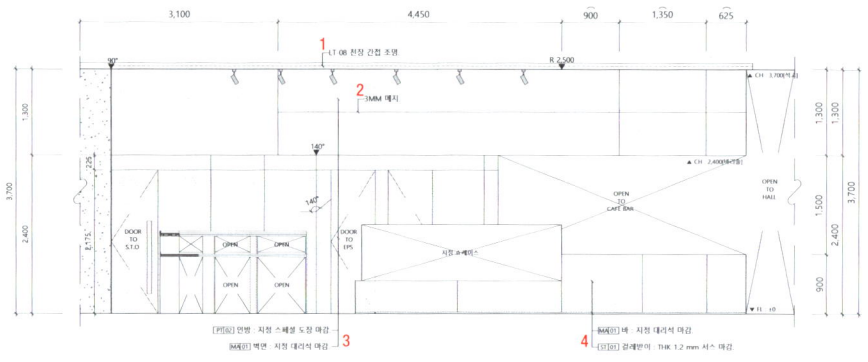

1층 베이커리 존 입면 A / 1F bakery zone elevation A

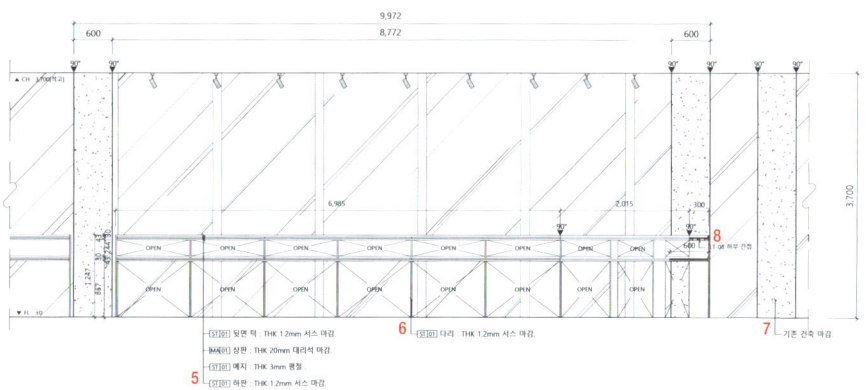

1층 베이커리 존 입면 B / 1F bakery zone elevation B

1 천장 간접조명 2 3mm 줄눈 3 인방 : 지정 스페셜 페인트 / 벽 : 지정 대리석 4 바 : 지정 대리석 / 걸레받이 : T1.2 SUS 5 T1.2 SUS / 상판 : T20 대리석 / 줄눈 : T3 평철 / 하판 : T1.2 SUS 6 다리 : T1.2 SUS 7 기존 건축마감 8 하부 간접조명 9 인방 : 지정 스페셜 페인트 / 바 : 지정 대리석 10 프레임 : T1.2 SUS 바이브레이션 / 줄눈 : T3 평철 11 지정 마이크로시멘트 / 선반 : T1.2 SUS / 하부장 : T1.2 SUS 12 간접조명 / 음수대 13 줄눈 14 T10 SUS 파이프 15 지정 대리석 / 지정 타일 본드 / T9 합판 16 지정 타일 / 지정 타일 본드 / T9.5 석고보드 2겹 / C-65X45X0.8T@450 17 T3 평철 줄눈 18 내부 석고보드 보강, T1.2 SUS 바이브레이션 19 Ø10 스테인리스 스틸 파이프 20 지정 간접조명 21 상부 선반 : T1.2 SUS 절곡 / 상판 : T1.2 SUS 절곡 / 하부장 : PB장 위 SUS

1 Ceiling indirect lighting 2 3mm joint reveal 3 Lintel : App. special paint / Wall : App. marble 4 Bar : App. marble / Baseboard : T1.2 SUS 5 T1.2 SUS / Top : T20 marble / Joint reveal : T3 flat steel / Lower plate : T1.2 SUS 6 Legs : T1.2 SUS 7 Existing finishing 8 Lower indirect lighting 9 Lintel : App. special paint / Bar : App. marble 10 Frame : T1.2 SUS vibration / Joint reveal : T3 flat steel 11 App. microcement / Shelf : T1.2 SUS / Lower cabinet : T1.2 SUS 12 Indirect lighting / Drinking fountain 13 T3 flat steel joint reveal 14 T10 SUS spipe 15 App. marble / App. tile adhesive / T9 plywood 16 App. tile / App. tile adhesive / T9.5 gypsum board 2ply / C-65X45X0.8T@450 17 T3 flat steel joint reveal 18 Gypsum board reinforcement, T1.2 SUS vibration 19 Ø10 stainless steel pipe 20 App. indirect lighting 21 Upper shelf : T1.2 SUS bending / Top : T1.2 SUS bending / Lower cabinet : SUS on PB cabinet

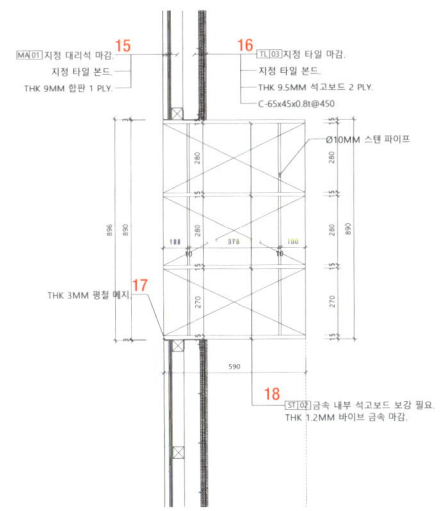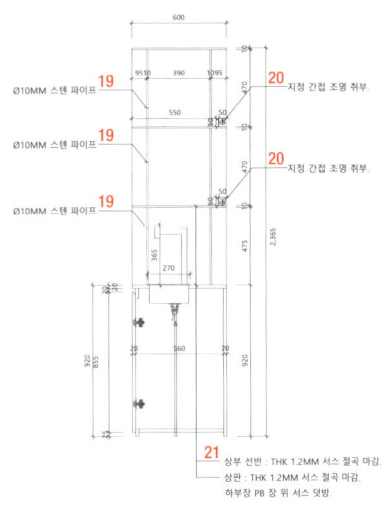

선반 단면 상세 D / shelf section detail D

선반 단면 상세 E / shelf section detail E

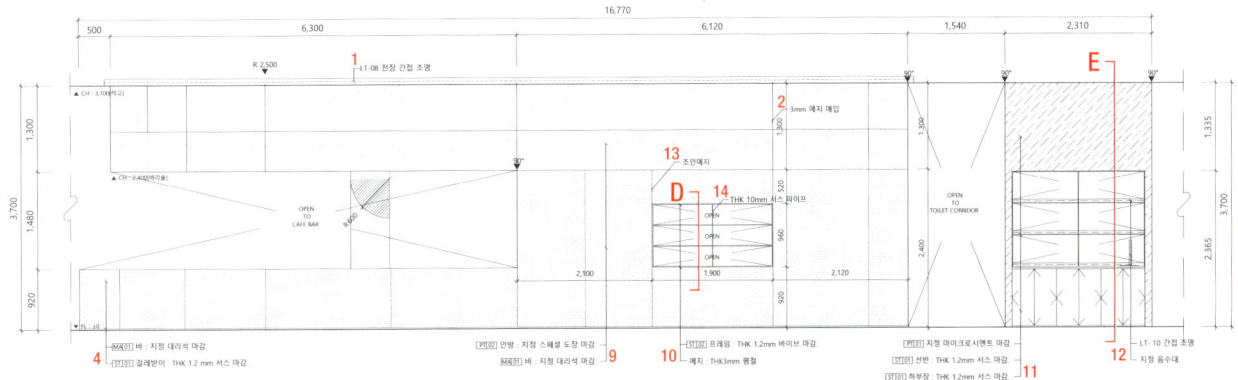

1층 홀 입면 C / 1F hall elevation C

107

1 기존 난간 2 지정 도장 3 지정 마이크로시멘트 4 문 : 지정 마이크로시멘트 5 상부 선반 : T1.2 SUS 절곡 / T9.5 석고보드 2겹 위 지정 도장 6 스탠드 조명형 사인 7 지정 대리석 위 음각 사인

1 Existing balustrade 2 App. painting 3 App. microcement 4 Door : App. microcement 5 Upper shelf : T1.2 SUS bending / App. painting on T9.5 gypsum board 2ply 6 Stand tyle illuminated sign 7 Engraving app. marble

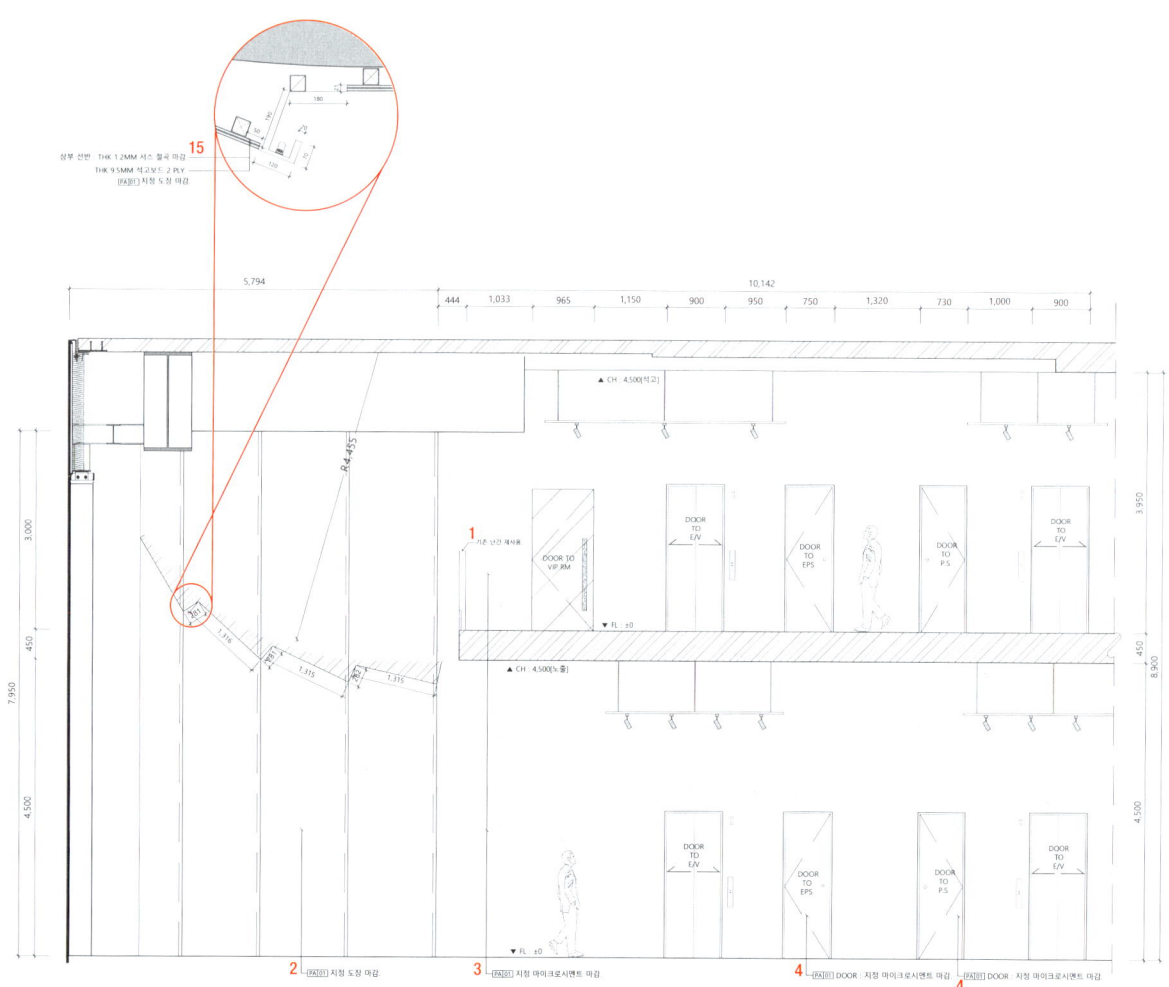

1층 홀 입면 F / 1F hall elevation F

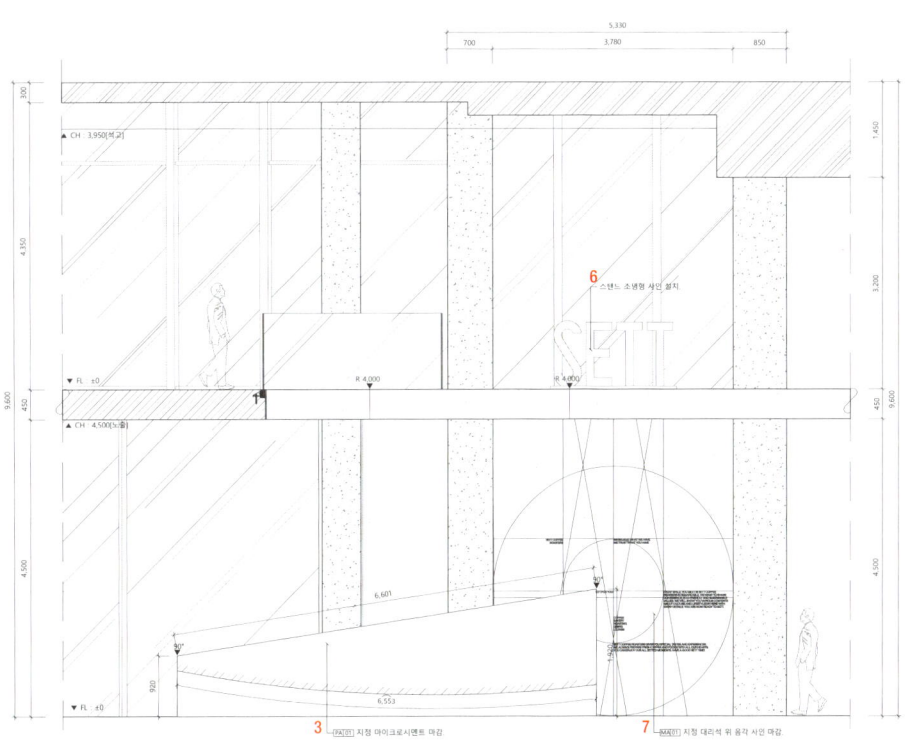

1층 홀 입면 G / 1F hall elevation G

2층 평면도 / 2nd floor plan

1 엘리베이터 홀 2 라운지 3 리턴 존 4 스페셜 라운지 5 로스팅 룸 6 화장실 7 VIP 룸

1 Elevator hall 2 Lounge 3 Return zone 4 Special lounge 5 Roasting room 6 Restroom 7 VIP room

1 지정 스페셜 페인트   2 가로 바 : T3 평철 위지정 소부도장   3 지정 스페셜 페인트 / 세로 바 : T3 평철 위 지정 소부도장

1 App. special paint   2 Horizontal bar : App. baked finish on T3 flat steel   3 App. special paint / Vertical bar : App. baked finish on T3 flat steel

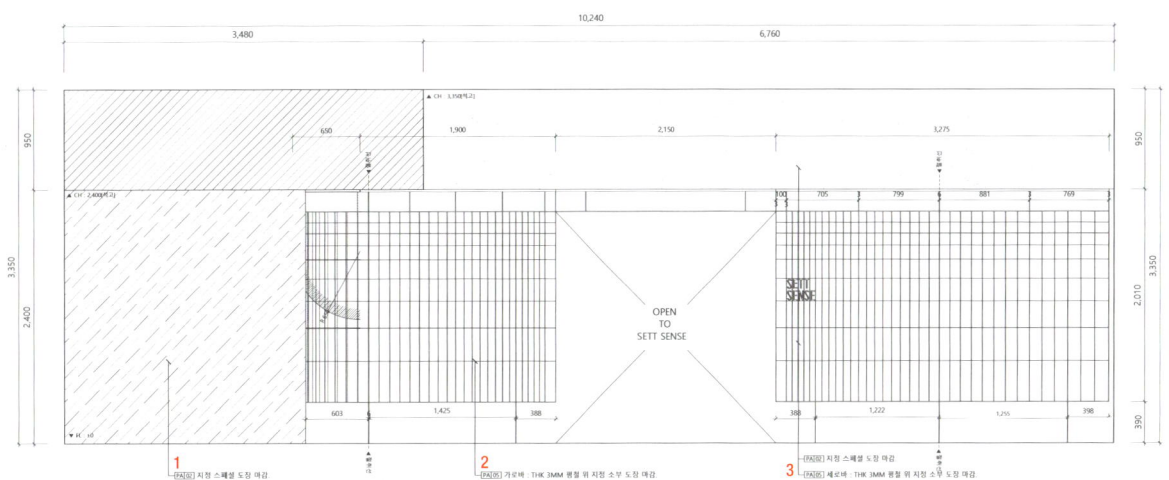

2층 라운지 입면 H / 2F lounge elevation H

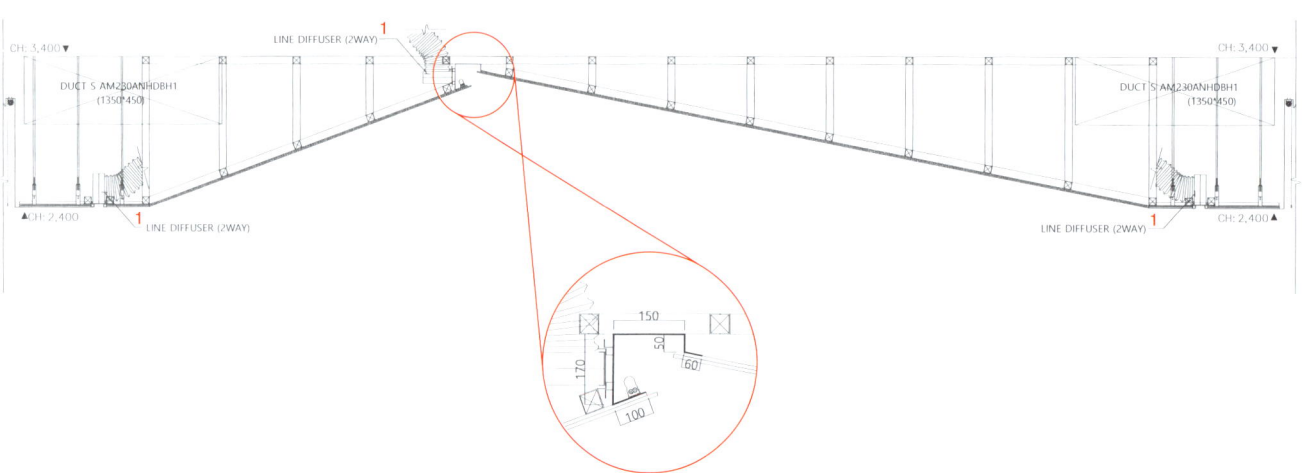

천장 상세 J / ceiling detail J

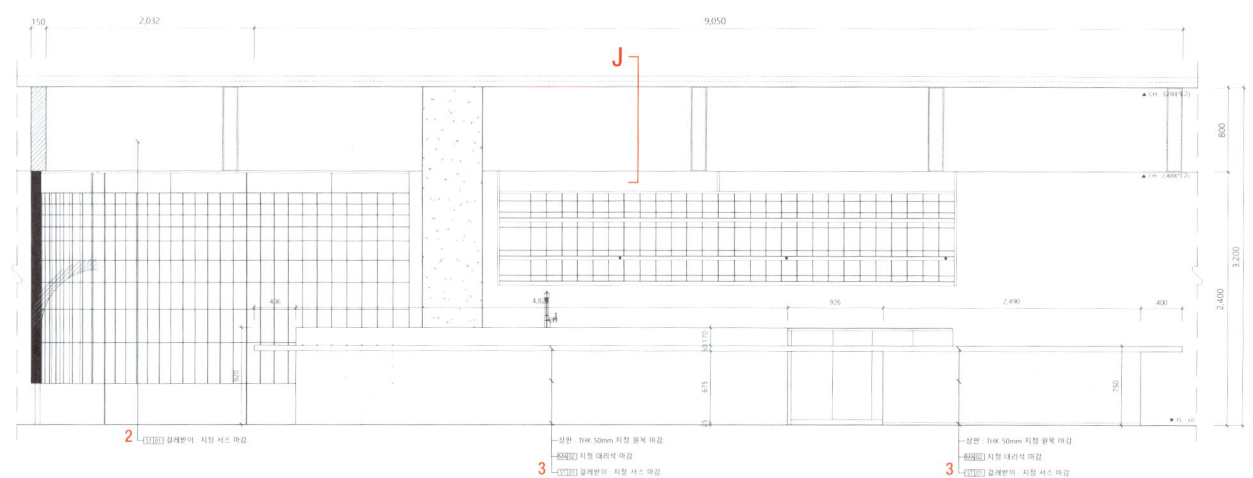

2층 스페셜 라운지 입면 I / 2F special lounge elevation I

1 라인디퓨저 2 걸레받이 : 지정 SUS 3 상판 : T50 지정 원목 / 지정 대리석 / 걸레받이 : 지정 SUS 4 T12 지정 유리 5 지정 은경 / 지정 수전 6 상부장 : T1.2 SUS / PB장 위 T1.2 SUS 헤어라인 7 상부장 : T1.2 SUS 8 간접조명 9 문 : 지정 마이크로시멘트 10 T1.2 철판 위 지정 도장 / T9.5 석고보드 2겹 11 □50X50X1.6T 철제 파이프 12 T3 철판 13 지정 조명

1 Line diffuser 2 Baseboard : App. SUS 3 Top : T50 app. solid wood / App. marble / Baseboard : App. SUS 4 T12 app. glass 5 App. mirror / App. faucet 6 Upper cabinet : T1.2 SUS / T1.2 SUS hairline on PB cabinet 7 Upper cabinet : T1.2 SUS 8 Indirect lighting 9 Door : App. microcement 10 App. painting on T1.2 steel plate / T9.5 gypsum board 2ply 11 □50X50X1.6T steel pipe 12 T3 steel plate 13 App. lighting

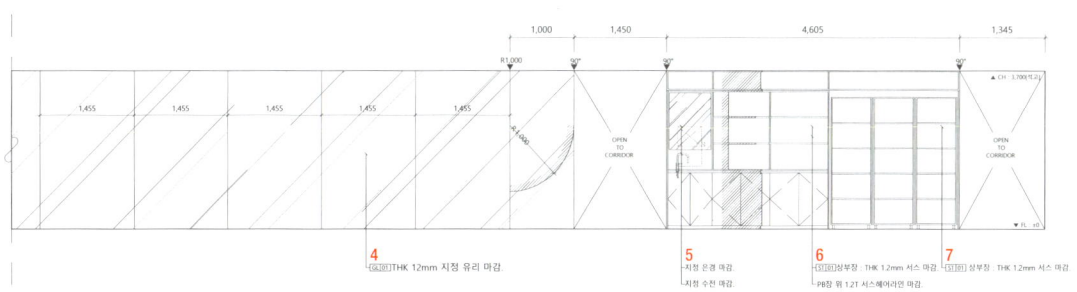

2층 라운지 입면 K / 2F lounge elevation K

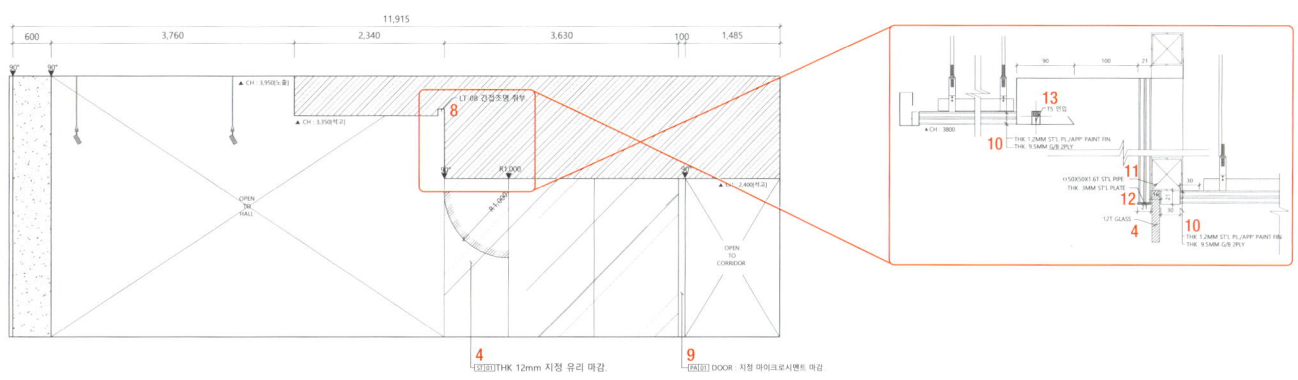

2층 라운지 입면 L / 2F lounge elevation L

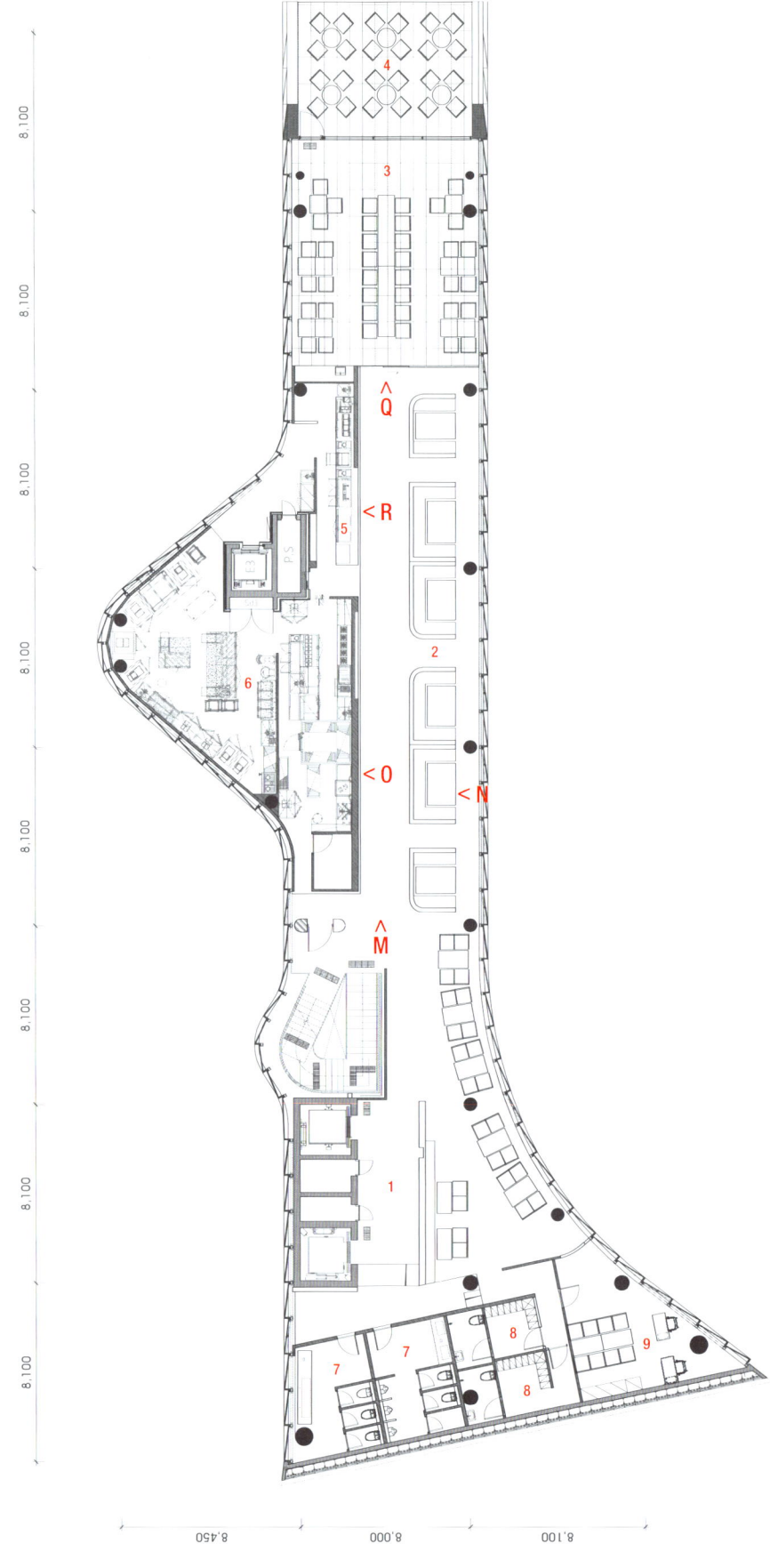

3층 평면도 / 3rd floor plan

1 엘리베이터 홀 2 라운지 3 파티룸 4 테라스 5 카운터 6 주방 7 화장실 8 라커룸 9 사무실

1 Elevator hall 2 Lounge 3 Party room 4 Terrace 5 Counter 6 Kitchen 7 Restroom 8 Locker room 9 Office

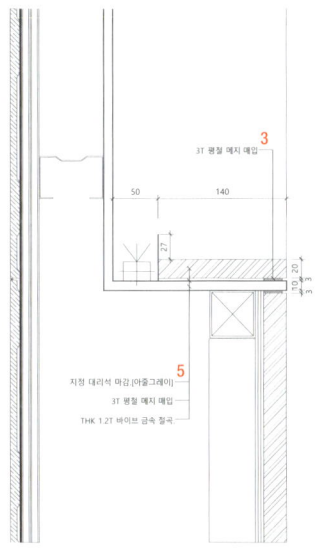

선반 단면 상세 P / shelf section detail P

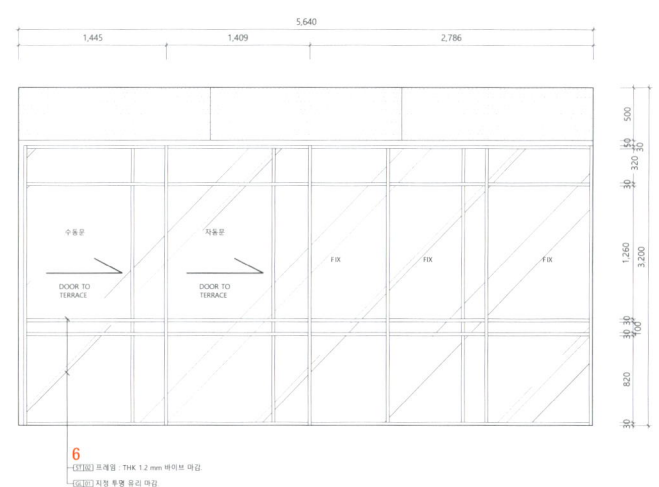

3층 라운지 입면 Q / 3F lounge elevation Q

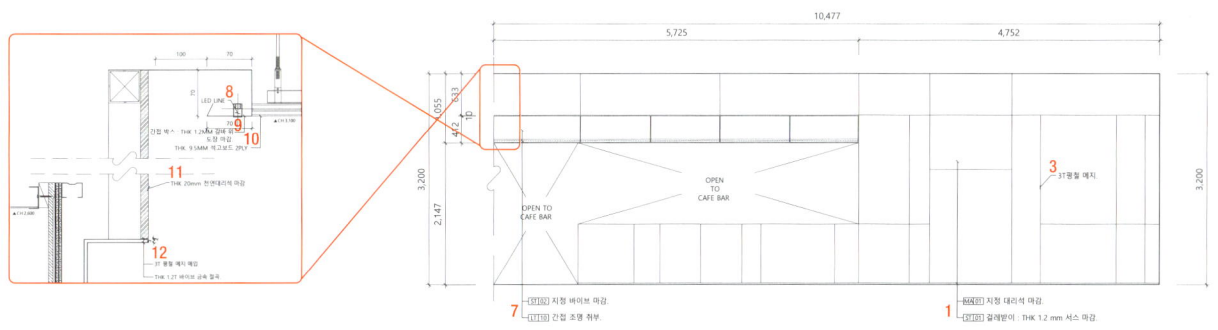

3층 라운지 입면 R / 3F lounge elevation R

KNOTTED GIMPO

SUBTEXT | Jangeun Won, Eunyoung Kim

어릴 때 놀았던 놀이터는 많은 아티스트들에게 영감을 준다. 즐거움이라는 건 단지 기구를 타는 것 외에도 오감의 모든 감각이 작동되기 때문이다. 어린 시절 놀이터에서 놀던 움직임의 감각 중 신체적인 반응을 가장 빠르게 인지하는 촉감과 시각을 공간에 표현하였다.

단순히 도넛의 맛을 전달하는 것뿐 아니라 포장을 받을 때의 기분 좋은 즐거움을 고객들에게 선사한다는 브랜드 헤리티지를 가지고 있는 노티드를 통해 우리의 삶에서 생겨나는 기분 좋은 감각을 공간 안에서 표현하고자 하였다. 도넛을 고르기 위해 머무르는 공간에서 무엇을 먹을지 고민하며 서있을 때의 행복한 표정과 손짓을 담아내고자 천장에 반사체를 사용하여 행복한 순간을 담아낼 수 있도록 하였으며, 앉아있는 의자뿐만 아니라 우리가 공간을 가장 크게 인지하는 벽체를 폭신한 소재로 마감하여 벽에 기대어 서서 공간을 바라보았을 때 폭신함이 유쾌한 즐거움을 언겨주도록 하였다. 또한 시각적 컬러에서 오는 즐거움을 표현하기 위해 브랜드 컬러 중 노란색을 포인트 색상으로 사용하였으며, 포토 존에는 한송준 작가와 함께 금속 미러 소재의 풍선으로 스마일 오브제로 재탄생 시켰다.

Playgrounds from childhood often become a source of inspiration for many artists. This is because play is not just about using playground equipment, but an activity that stimulates all the senses. The design objective was to express touches and visual experiences that were vividly remembered from playing in childhood playgrounds. Knotted's brand motto is to not only provide delicious donuts, but also the pleasant feeling of receiving donuts in a package. To embody this, the designer sought to translate the enjoyable sensations encountered in our daily lives into the spatial experience. At the donut selection area, ceiling-mounted mirrors are installed to capture the happy expressions and gestures of customers as they choose donuts to eat. Beyond just the seating, even the walls - one of the most prominent elements in the space - are finished with soft, plush materials. This allows customers leaning against the walls or simply observing the space to feel a delightful sense of coziness. In addition, to express the joy of seeing vibrant colors, yellow from the brand colors is used as a highlighter color, and metallic mirror balloon sculptures are placed as photo zone objects in collaboration with artist Han Song-jun.

디자인 원장은, 김은영 / 서브텍스트
위치 서울특별시 강서구 하늘길 77
용도 카페
면적 136.39m²
마감 바닥 - 타일 / 벽 - 인조가죽, 비닐페인트, 페인트 / 천장 - 크로뮴 페인트, 비닐페인트
완공 2023. 12
디자인팀 이병철, 김송이
사진 홍기웅

Location 77, Haneul-gil, Gangseo-gu, Seoul
Use Cafe
Area 136.39m²
Finishing Floor - Tile / Wall - Artificial leather, Vinyl paint, Paint / Ceiling - Chromium Paint, Vinyl paint
Completion 2023. 12
Design team Byoungcheol Lee, Songii Kim
Photographer Kiwoong Hong

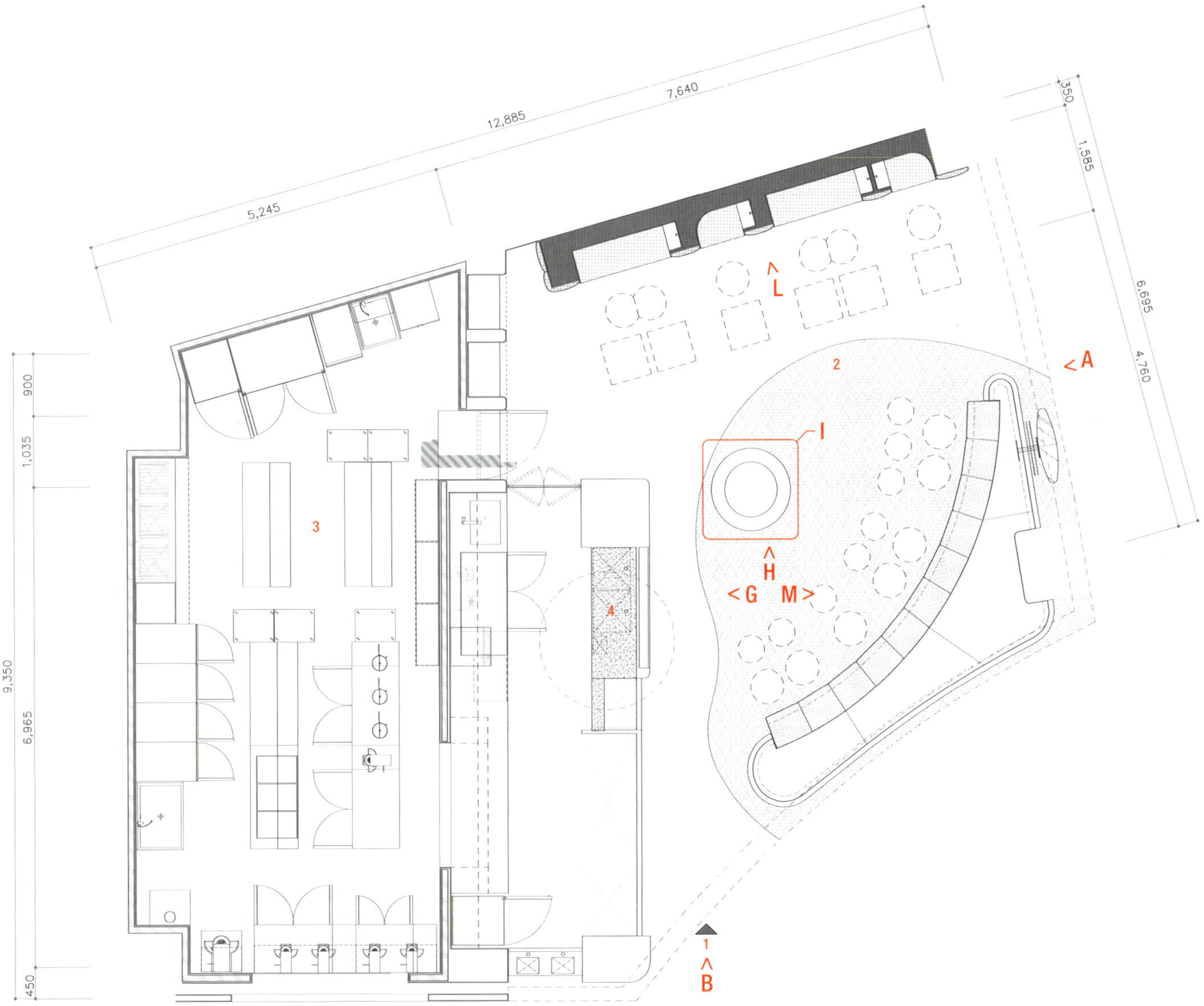

평면도 / floor plan

1 입구 2 홀 3 주방 4 카운터

1 Entrance 2 Hall 3 Kitchen 4 Counter

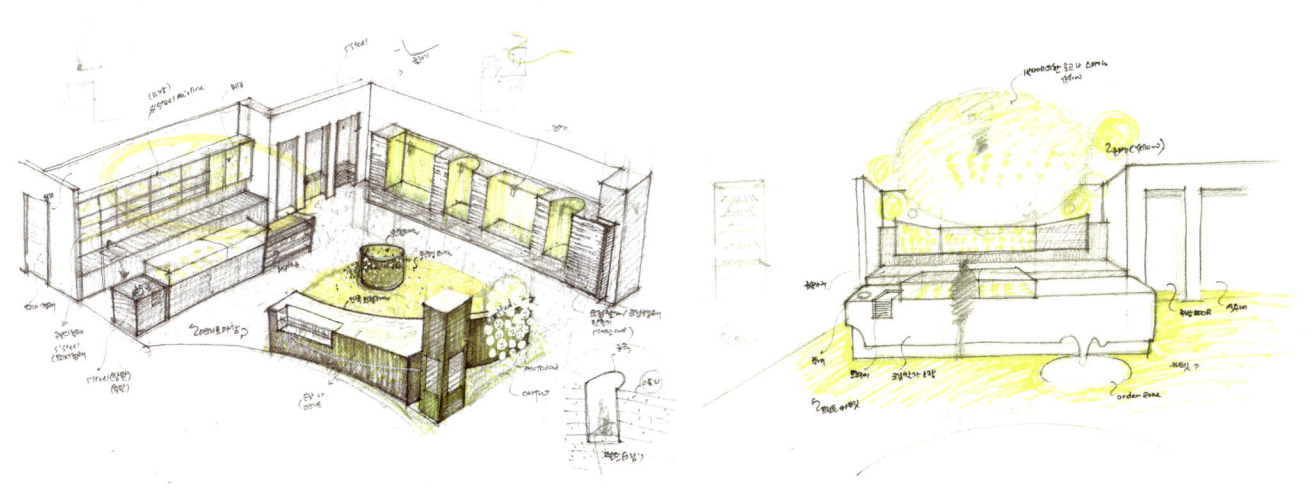

스케치 / sketch

1 채널 로고 사인　2 10mm 금속 재료분리대 : T1.2 스테인리스 스틸 헤어라인　3 지정 흰색 도장　4 제작 벤치 : 지정 패브릭 / 걸레받이 : 지정 노란색 래커 도장　5 기둥 : 지정 모자이크 타일　6 지정 네온 아크릴 사인 / 거울 오브제, 거치대 : 0.6T 스테인리스 스틸　7 방화셔터 라인　8 T10 투명 강화유리　9 지정 모자이크 타일　10 지정 도장　11 걸레받이 : 지정 스테인리스 스틸　12 카운터 : 지정 흰색 래커 도장　13 제작 벤치 : 지정 패브릭　14 8T SUS 300X300　15 거치대 : Ø800 T1.6 SUS 헤어라인 원형 레이저 및 타공 / 받침대 : Ø800 T1.6 SUS 원형 레이저　16 L자 포스트 : □60X60 T1.6 SUS 절곡　17 3T SUS 310X80 / 310X80 철재　18 오브제　19 Ø1,100 오브제　20 Ø800 오브제 거치대　21 베이스판 : 8T SUS 300X300　22 T0.6 스테인리스 스틸 폴리싱　23 Ø800 T1.6 SUS 원형 레이저　24 Ø800 T1.6 SUS 헤어라인 원형 레이저 및 타공　25 150X130 철재　26 3T SUS 310X80　27 가구 라인　28 Ø12 타공　29 Ø6 타공　30 목틀

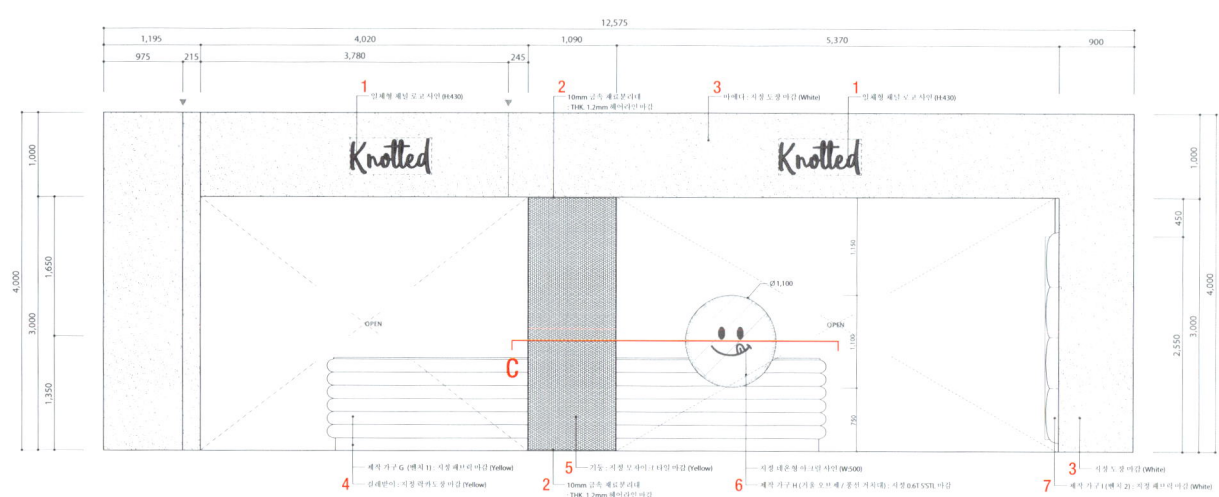

파사드 A / facade A

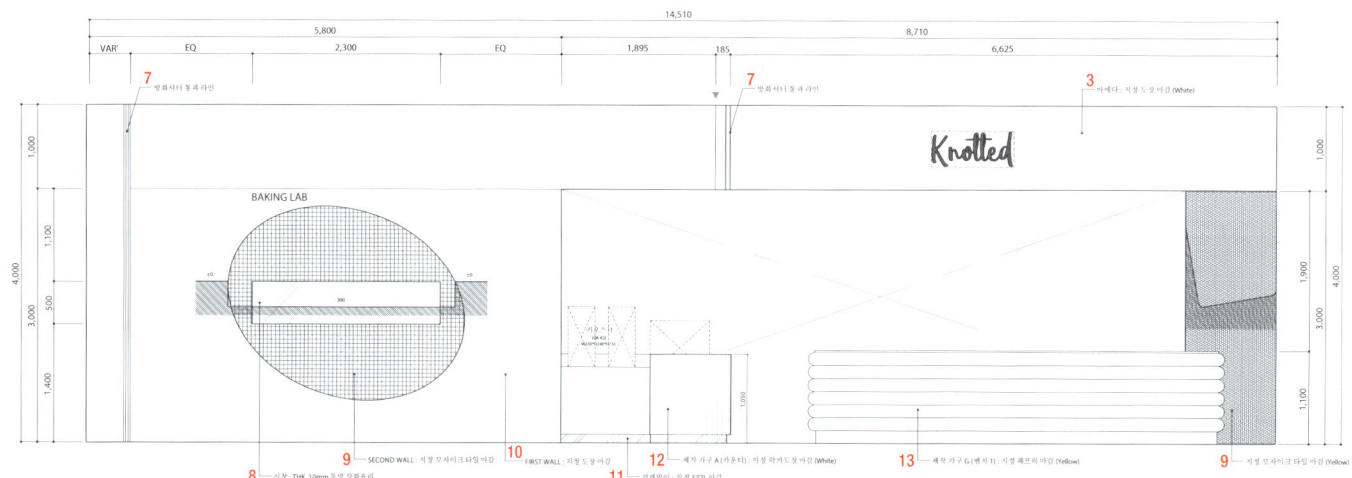

파사드 B / facade B

1 Channel logo sign 2 10mm metal material separator : T1.2 stainless steel hairline 3 App. white painting 4 Custom-made bench : App. fabric / Baseboard : App. yellow lacquer painting 5 Column : App. mosaic tile 6 App. neon acrylic sign / Mirror objet, Support : 0.6T stainless steel 7 Fire shutter line 8 T10 clear tempered glass 9 App. mosaic tile 10 App. painting 11 Baseboard : App. stainless steel 12 Counter : App. white lacquer painting 13 Custom-made bench : App. fabric 14 8T SUS 300X300 15 Holder : Ø800 T1.6 SUS laser into round and perforation / Support : Ø800 T1.6 SUS circle laser cut 16 L-post : □60X60 T1.6 SUS bending 17 3T SUS 310X80 / 310X80 steel 18 Objet 19 Ø1,100 objet 20 Ø800 objet support 21 Base plate : 8T SUS 300X300 22 T0.6 stainless steel polishing 23 Ø800 T1.6 SUS circle laser cut 24 Ø800 T1.6 SUS laser into round and perforation 25 150X130 steel 26 3T SUS 310X80 27 Furniture line 28 Ø12 perforation 29 Ø6 perforation 30 Wooden frame

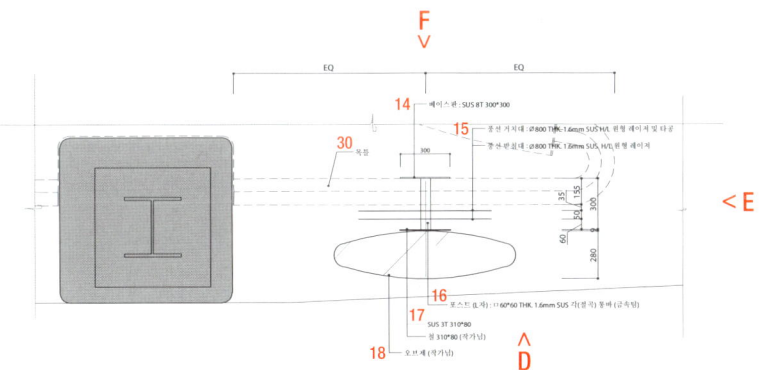

거울 오브제 평단면 C / mirror objet top section C

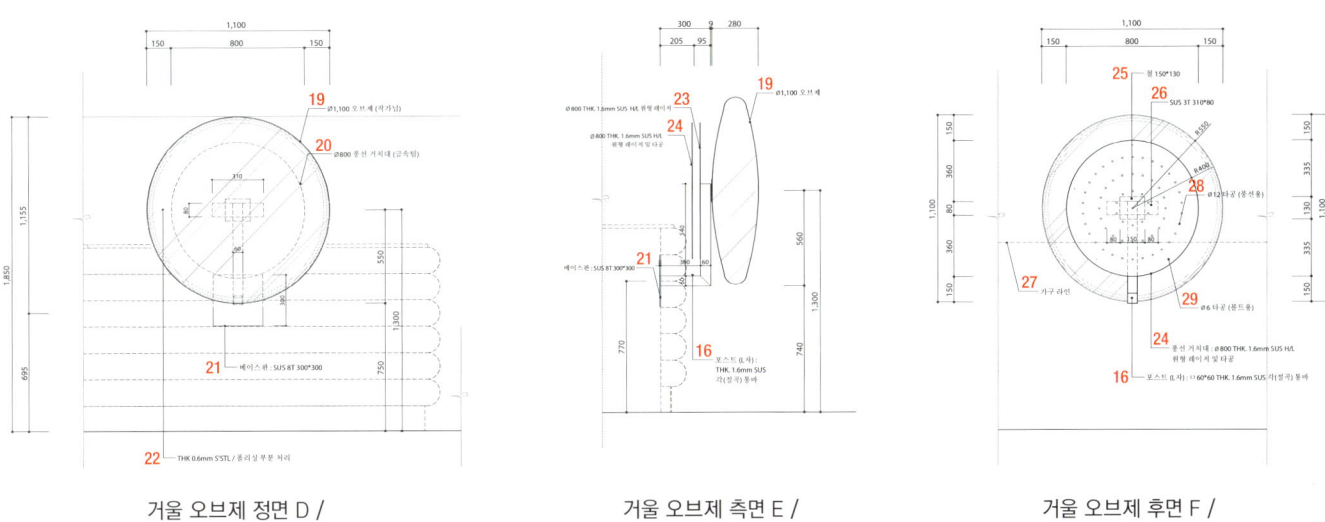

거울 오브제 정면 D /
mirror objet front view D

거울 오브제 측면 E /
mirror objet side view E

거울 오브제 후면 F /
mirror objet rear view F

1 벽 : 지정 흰색 도장 2 상부창 : T8 투명 강화유리 3 벤치 : 지정 노란색 도장 / 시트 : 지정 패브릭 4 걸레받이 : 스테인리스 스틸 헤어라인 5 카운터 : 지정 흰색 래커 도장 6 T6 지정 모자이크 타일 7 문 : 스테인리스 스틸 헤어라인 / 걸레받이 : 스테인리스 스틸 헤어라인 8 벽 : 스테인리스 스틸 헤어라인 / 서비스 스테이션 : 지정 흰색 래커 도장 / 걸레받이 : 스테인리스 스틸 헤어라인 9 퇴식구 : 스테인리스 스틸 헤어라인 / 벽 : 스테인리스 스틸 헤어라인 / 걸레받이 : 스테인리스 스틸 헤어라인 10 벽 : 지정 흰색 수성 도장 / 벽 : 지정 모자이크 타일 11 천장 오브제 : T1.6 갈바륨 위 은경막 도장 12 기둥 : 지정 흰색 수성 도장 13 벽 : 스테인리스 스틸 헤어라인 14 쇼케이스 : 스테인리스 스틸 헤어라인 15 여닫이문 : 지정 흰색 래커 도장 16 선반 : 지정 노란색 래커 도장 17 벤치 : 지정 패브릭 18 벽 : 지정 원형 모자이크 타일 19 기둥 라인 20 LED 라인 간접조명 21 노란색 지정 도장 / 노란색 지정 도장 / 노란색 지정 도장 22 걸레받이 라인 / 고정형 문 : 지정 노란색 도장 23 문 : 지정 노란색 도장 24 노란색 지정 도장 / 노란색 지정 도장 25 문 : 지정 노란색 도장 / 걸레받이 : 지정 스테인리스 스틸 헤어라인 26 LED 바 27 MDF 위 지정 노란색 도장 28 문 손잡이 29 바닥 마감라인

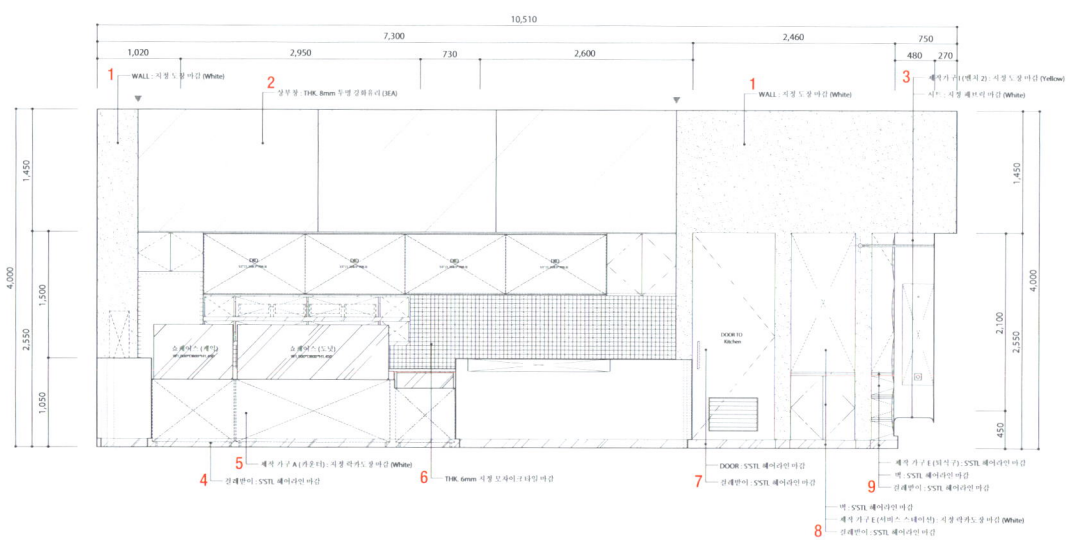

홀 입면 G / hall elevation G

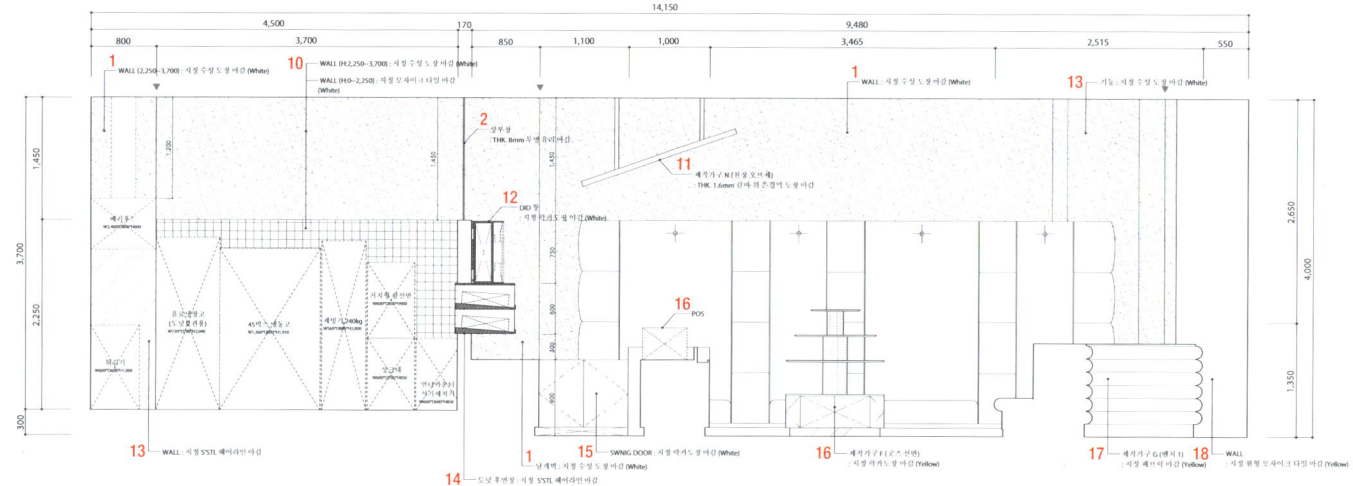

홀 입면 H / hall elevation H

1 Wall : App. white painting   2 Upper windows : T8 clear tempered glass   3 Bench : App. yellow painting / Sheet : App. fabric   4 Baseboard : Stainless steel hairline   5 Counter : App. white lacquer painting   6 T6 app. mosaic tile   7 Door : Stainless steel hairline / Baseboard : Stainless steel hairline   8 Wall : Stainless steel hairline / Service station : App. white lacquer painting / Baseboard : Stainless steel hairline   9 Return table : Stainless steel hairline / Wall : Stainless steel hairline / Baseboard : Stainless steel hairline   10 Wall : App. white water paint / Wall : App. mosaic tile   11 Ceiling objet : App. chrome painting   12 Column : App. white water paint   13 Wall : Stainless steel hairline   14 Showcase : Stainless steel hairline   15 Swing door : App. white lacquer painting   16 Shelf : App. yellow lacquer painting   17 Bench : App. fabric   18 Wall : App. round mosaic tile   19 Column finishing line   20 LED strip indirect lighting   21 App. yellow painting / App. yellow painting / App. yellow painting   22 Baseboard line / Fixed door : App. yellow painting   23 Door : App. yellow painting   24 App. yellow painting / App. yellow painting   25 Door : App. yellow painting / Baseboard : App. stainless steel hairline   26 LED bar   27 App. yellow painting on MDF   28 Door handle   29 Floor finishing line

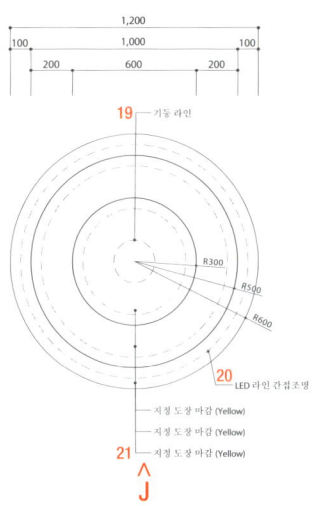

선반 평면 I / shelf top view I

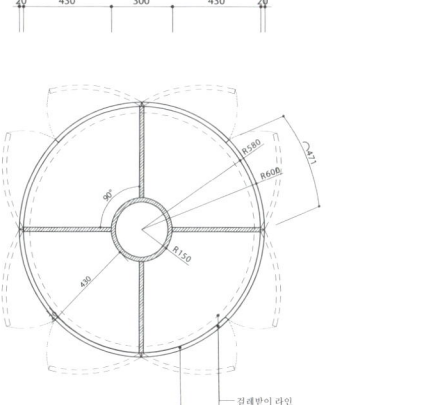

선반 단면 K / shelf section K

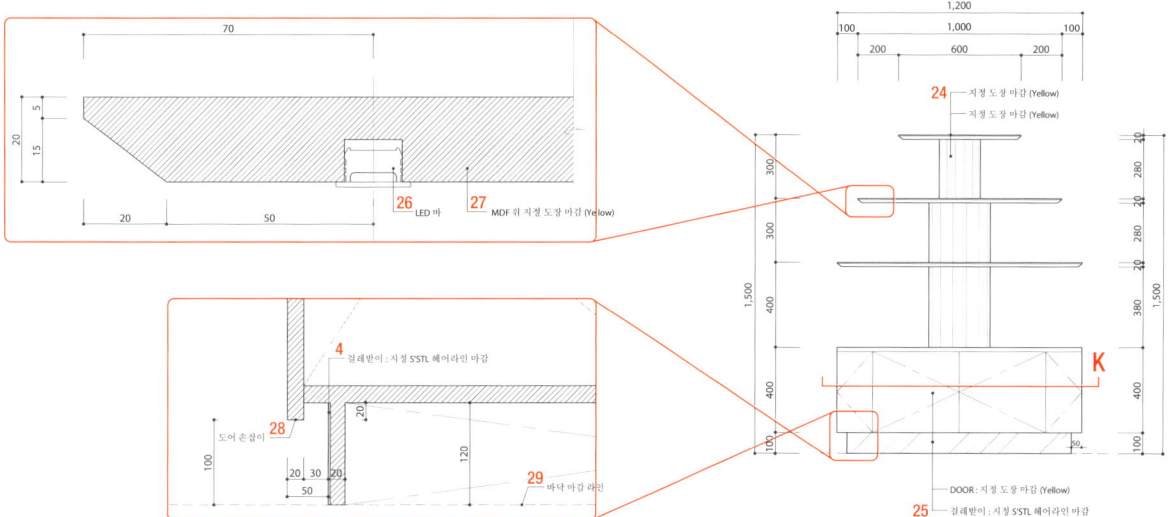

선반 정면 J / shelf front view J

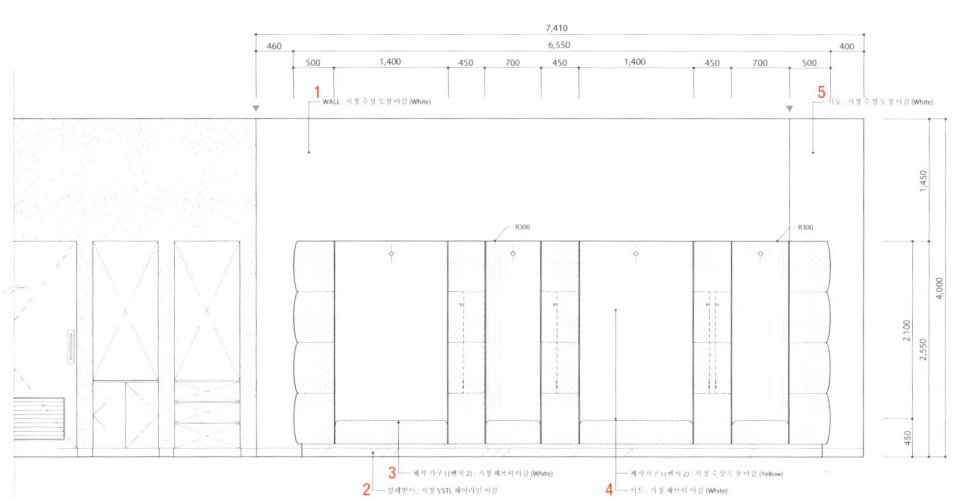

홀 입면 L / hall elevation L

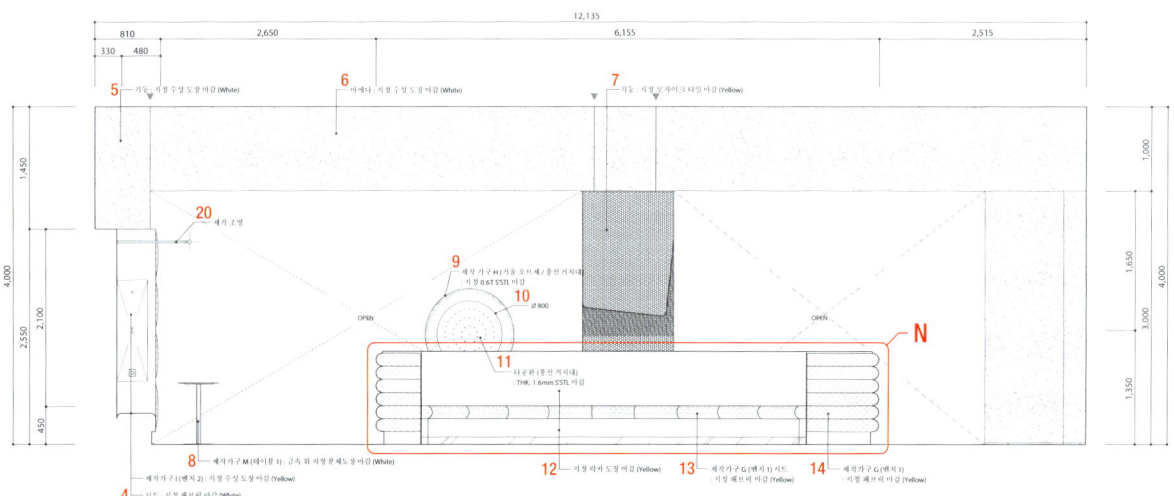

홀 입면 M / hall elevation M

1 벽 : 지정 흰색 수성 도장 2 걸레받이 : 지정 스테인리스 스틸 헤어라인 3 벤치 : 지정 흰색 패브릭 4 벤치 : 지정 노란색 수성 도장 / 시트 : 지정 패브릭 5 기둥 : 지정 흰색 수성 도장 6 지정 흰색 수성 도장 7 기둥 : 지정 노란색 모자이크 타일 8 제작 가구 : 금속 위 지정 흰색 분체 도장 9 지정 0.6T 스테인리스 스틸 10 Ø80 11 타공판 : T1.6 스테인리스 스틸 12 지정 노란색 래커 도장 13 벤치 시트 : 지정 노란색 패브릭 14 벤치 : 지정 노란색 패브릭 15 지정 노란색 래커도장 / 지정 노란색 인조가죽 16 지정 노란색 래커 도장 / 걸레받이 : 스테인리스 스틸 헤어라인 17 지정 노란색 인조가죽 18 MDF 위 지정 노란색 래커 도장 / 지정 노란색 인조가죽 19 풍선 오브제 설치 20 제작 조명

1 Wall : App. white water paint 2 Baseboard : App. stainless steel hairline 3 Bench : App. white fabric 4 Bench : App. yellow water paint / Sheet : App. fabric 5 Column : App. white water paint 6 App. white water paint 7 Column : App. yellow mosaic tile 8 Custom-made furniture : App. white powder coating on metal 9 0.6T app. stainless steel 10 Ø80 11 Perforated metal sheet : T1.6 stainless steel 12 App. yellow lacquer painting 13 Bench seat : App. yellow fabric 14 Bench : App. yellow fabric 15 App. yellow lacquer painting / App. yellow artificial leather 16 App. yellow lacquer painting / Baseboard : Stainless steel hairline 17 App. yellow artificial leather 18 App. yellow lacuqer painting on MDF / App. yellow artificial leather 19 Balloon objet installation 20 Custom-made lighting

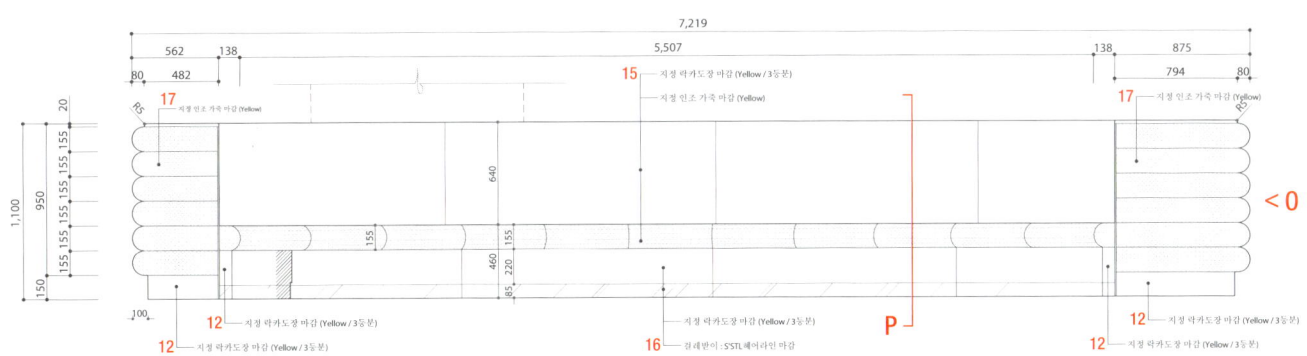

벤치 정면 N / bench front view N

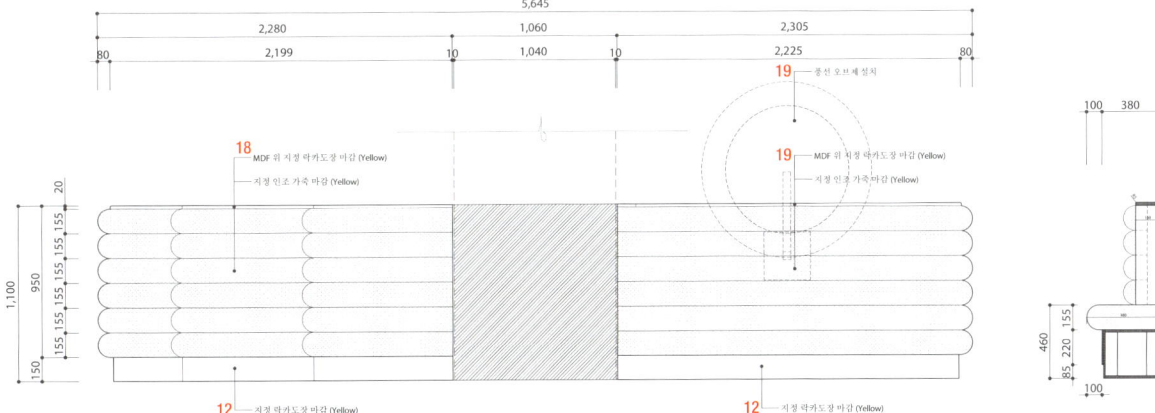

벤치 측면 O / bench side view O

벤치 단면 P / bench section P

125

INC COFFEE

HEDURBAN STUDIO | Bumgyu Kim

커피의 본질을 추구하는 인크커피에서 한 걸음 더 나아간 "INClassic". 시대를 넘나들며 높이 평가받는 "클래식"에 담긴 의미처럼 브랜드의 본질을 알리며 프리미엄 커피를 소개하는 교감의 장소이자 감성을 자극하는 예술작품과 같은 공간을 구현하고자 했다. 코너에 위치해 다각도로의 접근과 노출이 용이하다는 특징을 살려 내외부의 경계를 자연스럽게 연결하는 파사드를 포인트로 환대의 공간을 만들고, 인크커피의 시그니처 컬러인 다크 그레이를 베이스로 헤링본 우드플로어, 벽면과 천장을 감싸는 클래식한 몰딩과 조명을 더한 새로운 프리미엄 공간을 디자인했다. 각기 다른 입면의 카운터를 레이어드하고 복도를 향해 오픈된 벤치 좌석을 통해 공간 속의 또 다른 공간으로 입장하는 듯한 분위기를 연출함과 동시에 개방감과 프라이빗함을 느낄 수 있도록 유도했다. 또한 자칫 구조적 제약으로 치부될 수 있는 기둥을 카운터 바와 연결하고, 클래식 몰딩 디테일을 더해 공간의 구심점 역할을 하며 콘셉트와 조화롭게 어우러질 수 있도록 디자인했다. 천장과 벽면을 따라 전개되며 브랜드의 아이덴티티를 직관적으로 보여주는 미디어아트도 시선을 사로잡는 중요 포인트 중 하나다. 모던과 세미 클래식 감성을 동시에 느낄 수 있는 이곳에서 방문객들은 프리미엄 커피를 경험하며 인크커피의 핵심가치인 본질에 대해 깊이 이해하는 시간을 갖게 된다.

"INClassic" is a step beyond for INC Coffee, which seeks the essence of coffee. It aims to create a space that serves as a place for interaction, introducing premium coffee while also stimulating emotions like an art piece. Taking advantage of its easy access and exposure to various angles, we created a space of hospitality with a façade that naturally connects the internal and external boundaries and designed a new premium space using herringbone wood flooring, classical molding and lightings surrounding walls and ceilings, based on Dark Gray, Inc Coffee's signature color. Layered counters with different façades and open bench seating towards the corridor create an atmosphere as if entering another space within the space, while inducing a sense of openness and privacy simultaneously. Additionally, we connected the columns, which could be perceived as structural constraints, to the counter bar, integrating them seamlessly into the space and allowing them to serve as focal points with classic molding details that harmonize with the overall concept. Media art flowing along the ceiling and walls, showcasing the brand's identity intuitively, is also a captivating point. Here, where visitors can experience premium coffee and feel both modern and semi-classic sensibilities simultaneously, they will have the opportunity to deeply understand the essence of INC Coffee's core values.

디자인 김범규 / 헤드어반스튜디오
위치 경기도 수원시 장안구 수성로 175, 4층
용도 카페
면적 200m²
마감 바닥 – 우드 플로링 / 벽 – 페인트 / 천장 – 페인트
완공 2024. 1
디자인팀 백수연, 서예린
사진제공 인크커피

Location 4F, 175, Suseong-ro, Suwon-si, Gyeonggi-do
Use Cafe
Area 200m²
Finishing Floor - Wood flooring / Wall - Paint / Ceiling - Paint
Completion 2024. 1
Photos offer INC COFFEE

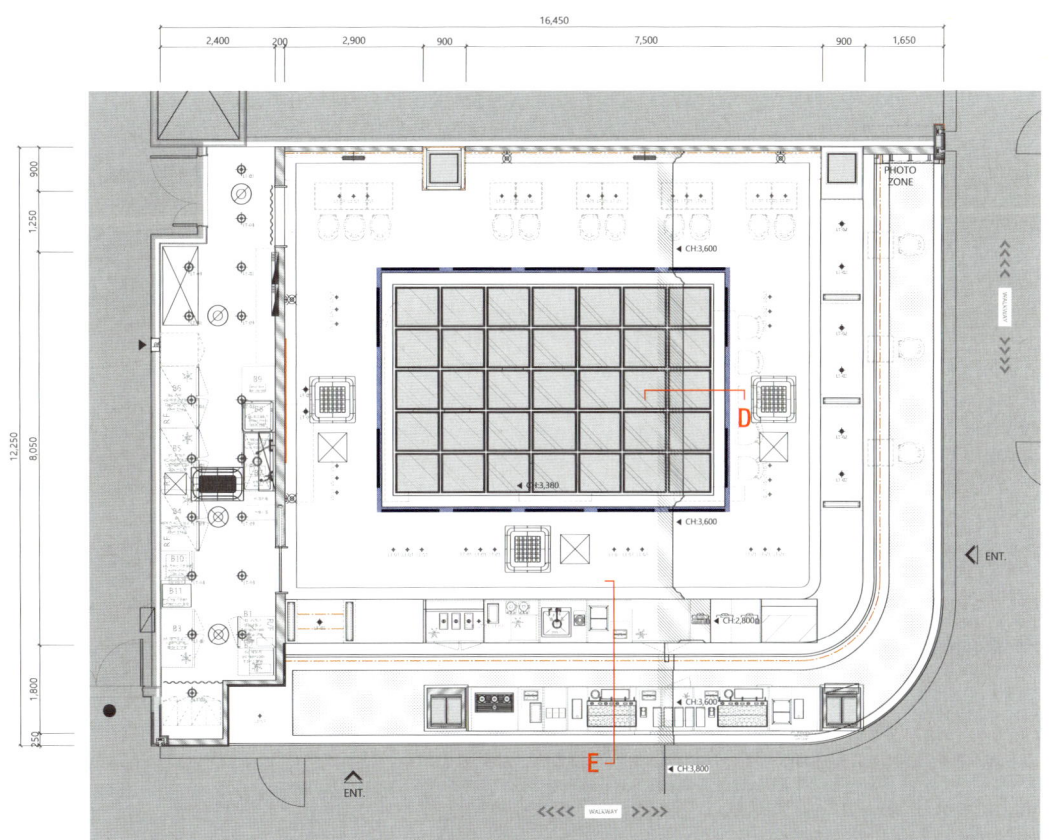

천장도 / ceiling plan

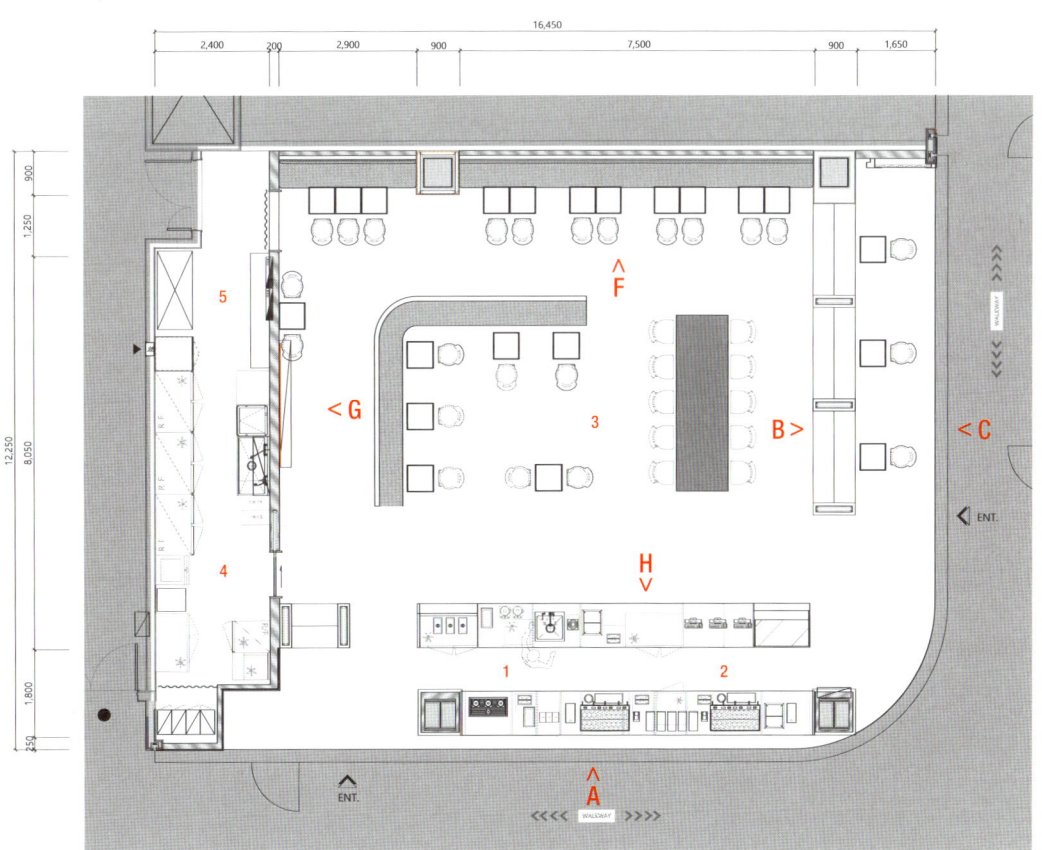

평면도 / floor plan

1 카페 바 2 카운터 3 홀 4 창고 5 직원실

1 Cafe bar 2 Counter 3 Hall 4 Storage 5 Staff room

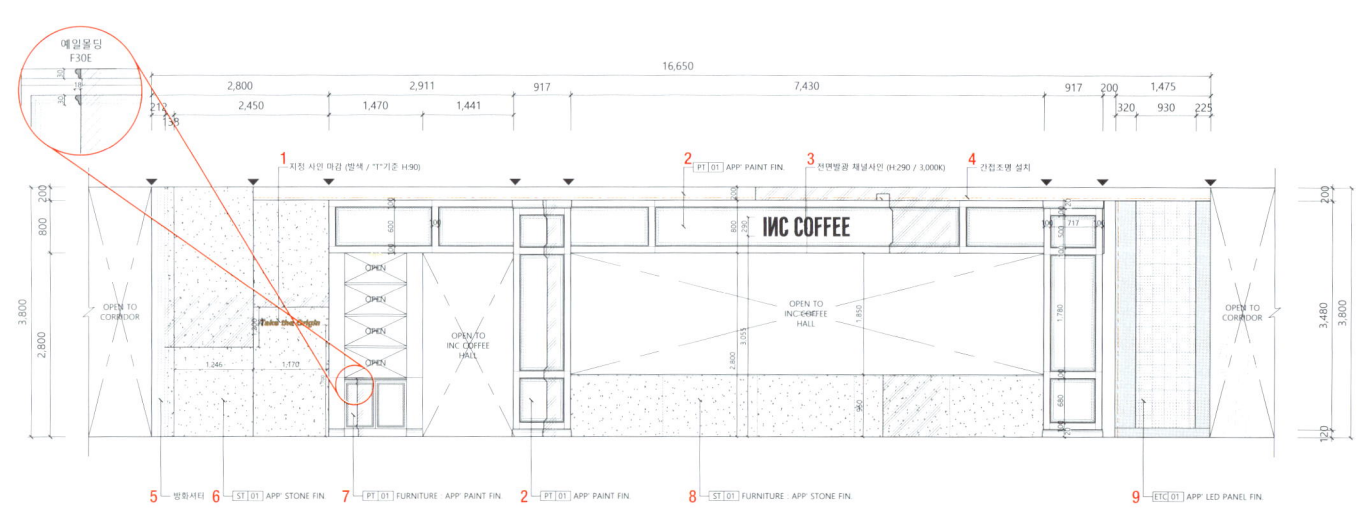

홀 입면 A / hall elevation A

1 지정 사인 2 지정 도장 3 전면 발광 채널 사인 4 간접조명 설치 5 방화 셔터 6 지정 석재 7 가구 : 지정 도장 8 가구 : 지정 석재 9 지정 LED 패널 10 지정 패브릭 / 지정 도장 11 기둥 : 지정 도장 / 가구 : 지정 석재 12 디마이징 피어 13 T9 MDF 14 지정 금속

1 App. sign 2 App. paint 3 Front fluorescent channel sign 4 Indirect lighting installation 5 Fire shutter 6 App. stone 7 Furniture : App. paint 8 Furniture : App. stone 9 App. LED panel 10 App. fabric / App. paint 11 Column : App. paint / Furniture : App. stone 12 Demising pier 13 T9 MDF 14 App. metal

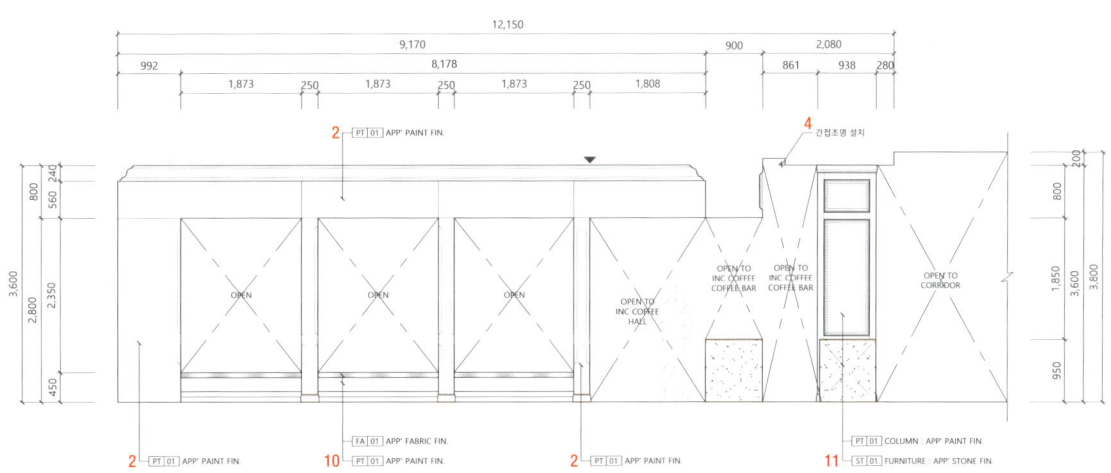

홀 입면 B / hall elevation B

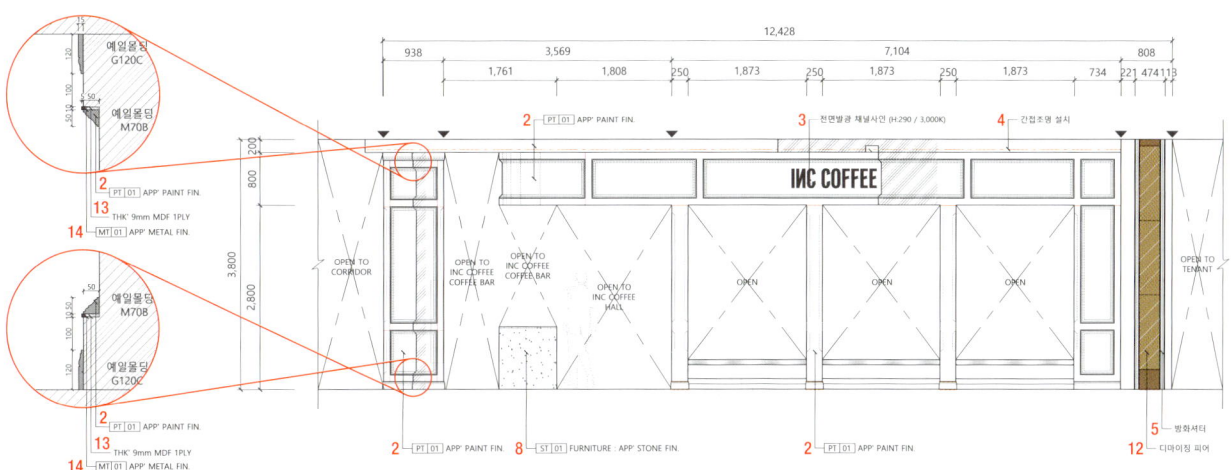

홀 입면 C / hall elevation C

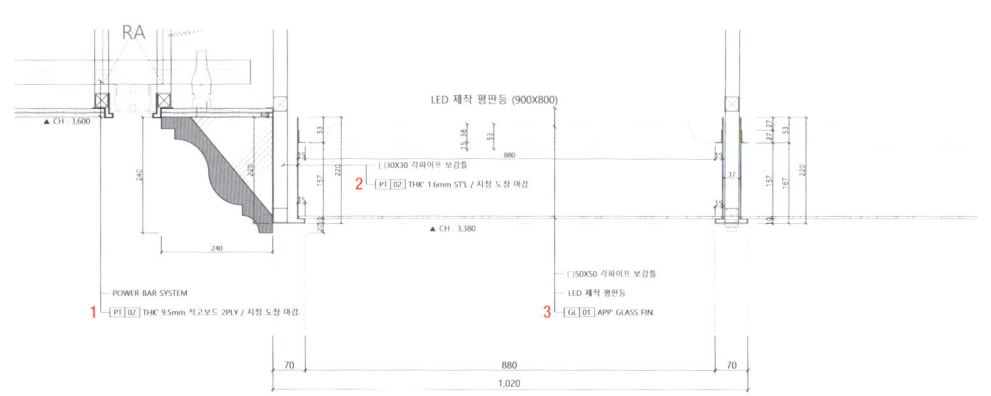

천장 단면 상세 D / ceiling section detail D

천장 단면 상세 E / ceiling section detail E

1 파워바 시스템 / T9.5 석고보드 2겹 위 지정 도장 2 □30X30 각파이프 보강틀 / T1.6 철판 위 지정 도장 3 □50X50 각파이프 보강틀 / LED 제작 평판등 / 지정 유리 4 간접조명 설치 5 □30X30 각파이프 보강틀 / 지정 도장 6 □50X50 각파이프 보강틀, LED 미디어 패널 7 지정 도장 8 □50X50 각파이프 보강틀 / T9 합판 2겹 / T9 MDF 2겹 / 지정 금속 9 T9 합판 2겹, 가구 마감 10 지정 사인 11 가구 : 간접조명 설치 / 가구 : 지정 패브릭 / 가구 : 지정 도장 12 지정 LED 패널

1 Power-bar system / App. paint on T9.5 gypsum board 2ply 2 □30X30 square pipe reinforcing frame / App. paint on T1.6 steel plate 3 □50X50 square pipe reinforcing frame / LED flat panel lighting / App. glass 4 Indirect lighting installation 5 □30X30 square pipe reinforcing frame / App. paint 6 □50X50 square pipe reinforcing frame, LED media panel 7 App. paint 8 □50X50 square pipe reinforcing frame / T9 plywood 2ply / T9 MDF 2ply / App. metal 9 T9 plywood 2ply, Furniture finishing 10 App. sign 11 Furniture : Indirect lighting installation / Furniture : App. fabric / Furniture : App. paint 12 App. LED panel

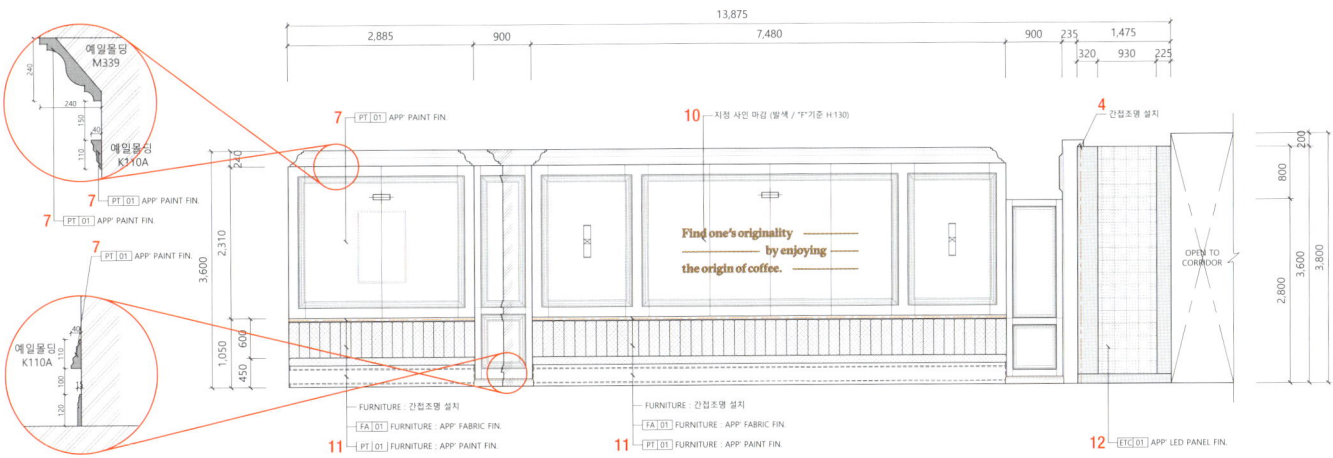

홀 입면 F / hall elevation F

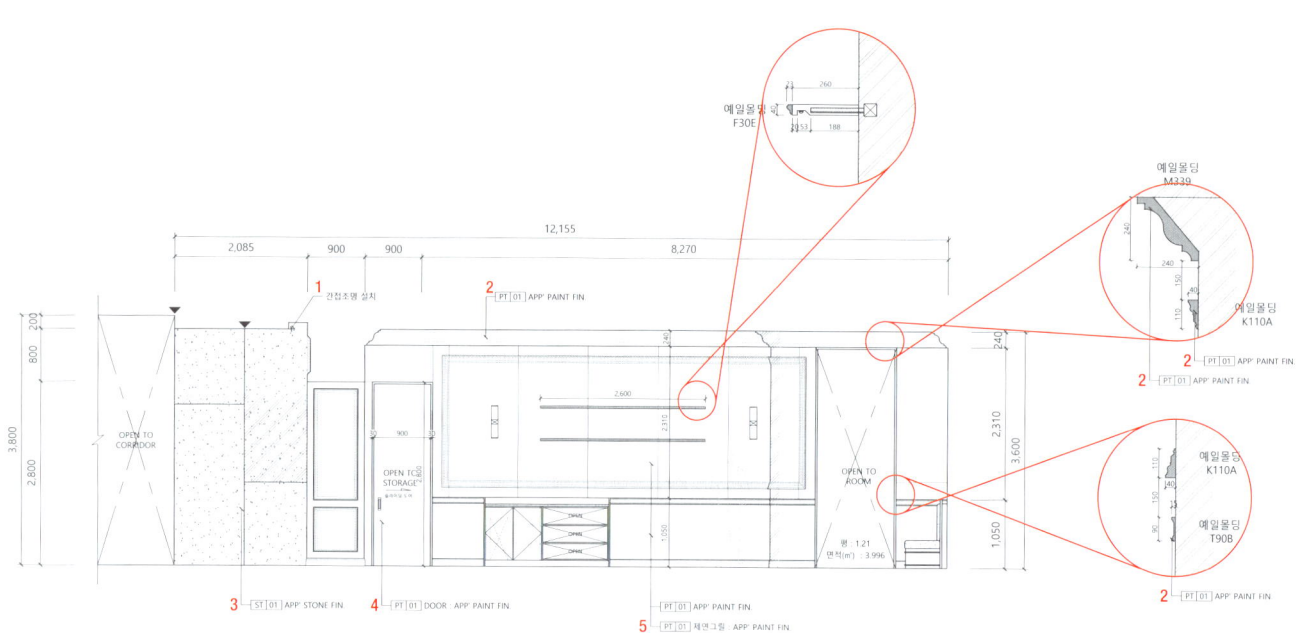

홀 입면 G / hall elevation G

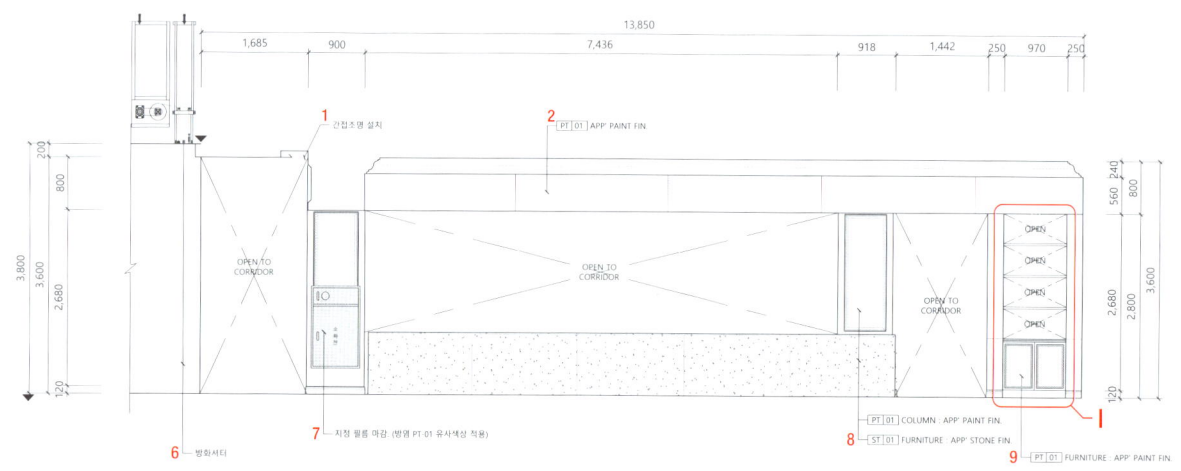

홀 입면 H / hall elevation H

1 간접조명 설치 2 지정 도장 3 지정 석재 4 문 : 지정 도장 5 지정 도장 / 제연 그릴 : 지정 도장 6 방화 셔터 7 지정 필름 8 기둥 : 지정 도장 / 가구 : 지정 석재 9 가구 : 지정 도장 10 선반 간접조명 설치

1 Indirect lighting installation 2 App. paint 3 App. stone 4 Door : App. paint 5 App. paint / Vent grille : App. paint 6 Fire shutter 7 App. film 8 Column : App. paint / Furniture : App. stone 9 Furniture : App. paint
10 Indirect lighting installation for shelves

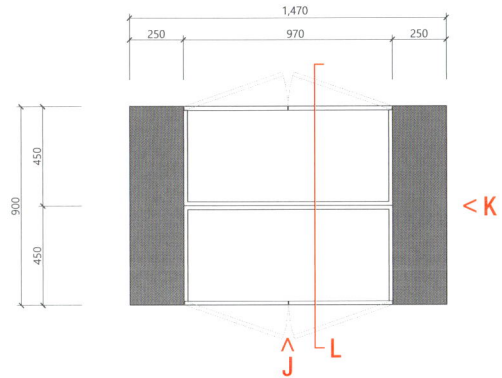

진열장 평면 I / display stand top view I

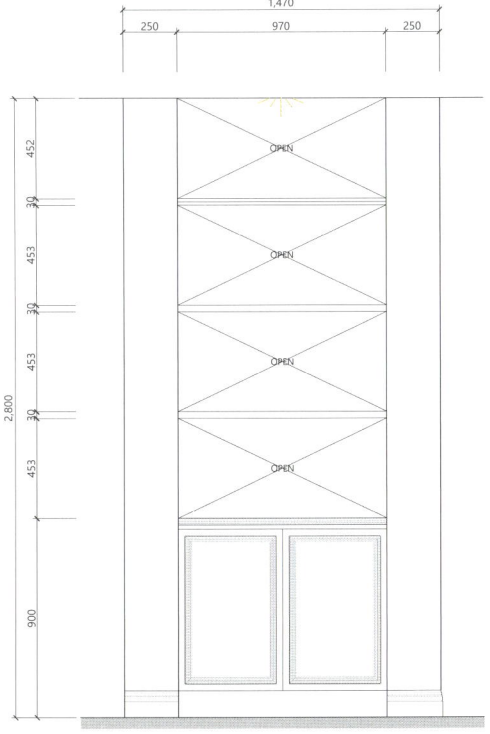

진열장 정면 J / display stand front view J

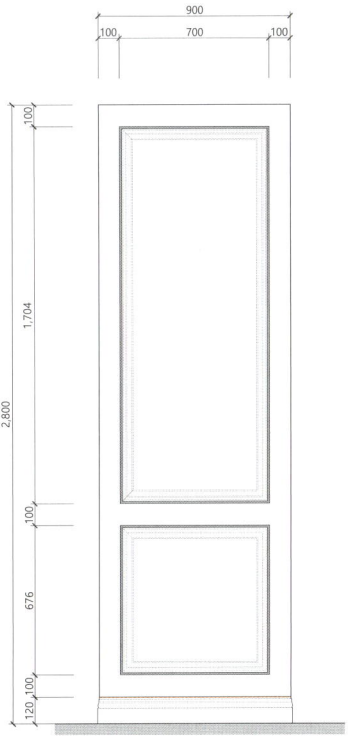

진열장 측면 K / display stand side view K

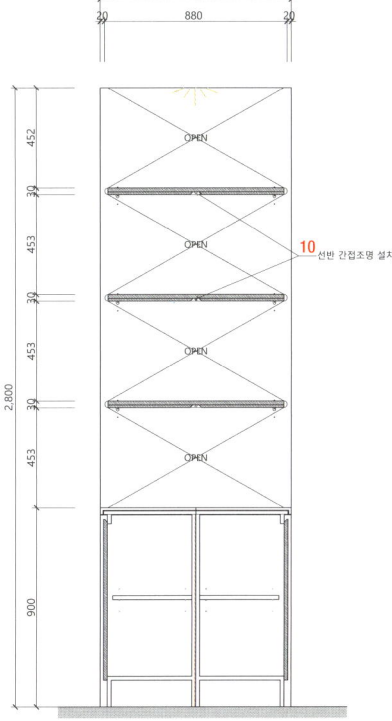

진열장 단면 L / display section L

SECOND ONE LAKE ASAN

Danaham Associate | Seohyeon Yun

A lake filled with pleasure

세컨드원 레이크 아산점은 넓고 평화로운 신정호 호수를 배경으로 탄생한 문화 커뮤니티 공간이다. 세컨드원은 아름다운 경관을 제공하는 것을 넘어 방문객들에게 깊은 감동과 오래 기억될 수 있는 행복을 선사하는 것을 미션으로 한다. 기존의 비전과 코어밸류, 사이트 특징, 그리고 코어 타깃층에 대한 깊은 이해를 바탕으로 다나함은 아산점만의 특별한 브랜드 에센스 'pleasure'를 구축했다. 이번 프로젝트는 세컨드원의 모든 공간에서 브랜드 에센스를 체감할 수 있게 디자인되었다. 신정호의 물결을 내부로까지 잇기 위해 디자이너는 호수의 요소를 디자인에 적용했다. 벽체의 그라데이션과 천장 및 구조물의 곡선은 호수의 물결을 형상화하며 방문객들로 하여금 단순한 관람을 넘어 호수와 하나가 되는 몰입 경험을 할 수 있다. 이곳은 창가부터 내부 좌석까지, 어느 자리에서나 신정호의 경관을 감상할 수 있도록 구조적인 배려가 반영되었다. 모두가 공간의 아름다움을 균등하게 경험할 수 있도록 디자인에 깊은 고민을 담았다. 계절에 따라 변화할 호수, 지역 커뮤니티에게 두 번째 집처럼 편안하고 즐거운 공간이 될 예정이다.

A lake filled with pleasure

Second One Lake Asan is a cultural community space nestled alongside the vast and tranquil Sinjeong Lake. Second One's mission is not only to offer beautiful views but also to give lasting happiness and deep impressions to visitors. Based on the understanding of the brand's vision and core values, site characteristics, and main visitor demographics, the designer developed a unique brand narrative for the café around the keyword 'pleasure'. Every space is designed to reflect Second One's brand narrative. The lake's characteristic features are also incorporated into the design to make the lake flow into the café. Gradation patterned walls and curved ceiling structures symbolize the waves of lake, making visitors feel as if they become part of the lake as they enjoy the view. Second One's layout ensures that visitors can appreciate the lake's scenery from anywhere, whether they sit by the window or in the inner area. The designer paid careful attention to the design so everyone could equally experience the beauty of the space. Seasonal changes in the views of the lake will make the café into a comfortable and enjoyable space that feels like a second home for the local community.

디자인 윤서현 / 다나함 어소시에이트
위치 충청남도 아산시 신정로 520
용도 카페
면적 1,237㎡
마감 바닥 – 에폭시 / 벽 – 스페셜 페인트, 3D 메탈 / 천장 – 페인트
완공 2024. 1
디자인팀 김인희, 박지수
공사팀 문병길, 전동호
사진 스튜디오 톰

Location 520, Sinjeong-ro, Asan-si, Chungcheongnam-do
Use Cafe
Area 1,237㎡
Finishing Floor - Epoxy / Wall - Special paint, 3D metal / Ceiling - Paint
Completion 2024. 1
Photographer Studio Tom

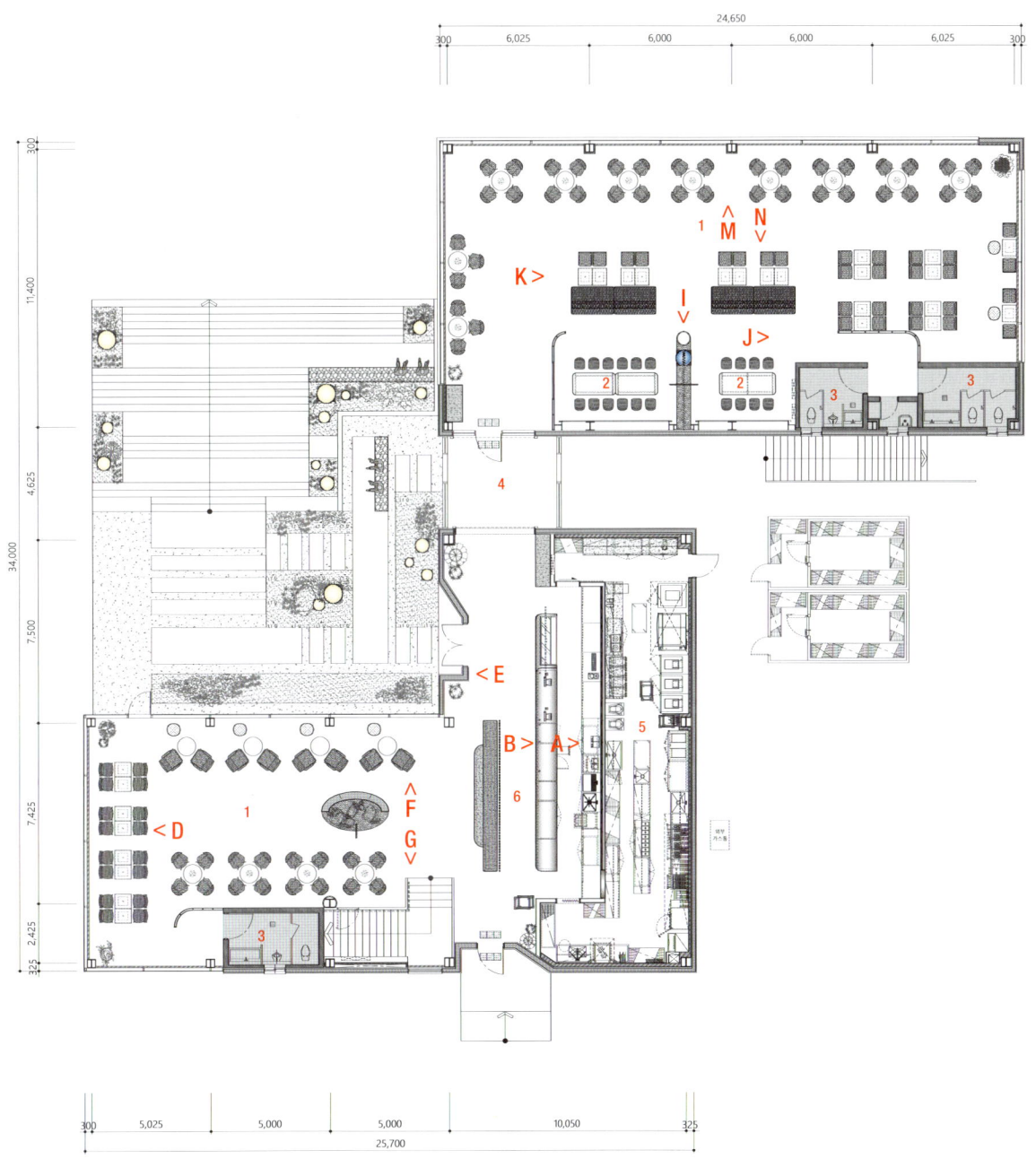

1층 평면도 / 1st floor plan

1 홀 2 프라이빗 존 3 화장실 4 연결 통로 5 주방 6 베이커리 존

1 Hall 2 Private zone 3 Restroom 4 Vestibule 5 Kitchen 6 Bakery zone

1 지정 시트 2 지정 비드블라스트 3 LED 스트립 4 프레임 : 지정 페인트 5 지정 페인트 6 지정 인조석재 / 지정 LPM 패널 7 몰딩 : 지정 페인트 / 걸레받이 : 지정 비드블라스트 8 사인 : 지정 페인트(비조명) 9 □30X30 철제 파이프 / T9.5 석고보드 2겹 / 지정 페인트 10 지정 금속 11 □30X30 철제 파이프 / T5 합판 2겹 / T9 MDF 위 지정 시트 12 L.G.S / 지정 SMC 위 지정 시트 13 □30X30 철제 파이프 / T1.2 철판 보강 / T1.5 스테인리스 스틸 14 C-스터드 65X45X0.8(단열) / T9.5 방수 석고보드 2겹 / 방수재+본드 / 지정 타일

1 App. sheet 2 App. beadblast 3 LED strip 4 Frame : App. paint 5 App. paint 6 App. artificial stone / App. LPM panel 7 Moulding : App. paint / Baseboard : App. beadblast 8 Sign : App. paint (non-illuminated) 9 □30X30 steel pipe / T9.5 gypsum board 2ply / App. paint 10 App. metal 11 □30X30 steel pipe / T5 plywood 2ply / App. sheet on T9 MDF 12 L.G.S / App. sheet on app. SMC 13 □30X30 steel pipe / T1.2 steel reinforcement / T1.5 stainless steel 14 C-stud 65X45X0.8 (insulation) / T9.5 waterproofing gypsum board 2ply / Waterproofing + Bond / App. tile

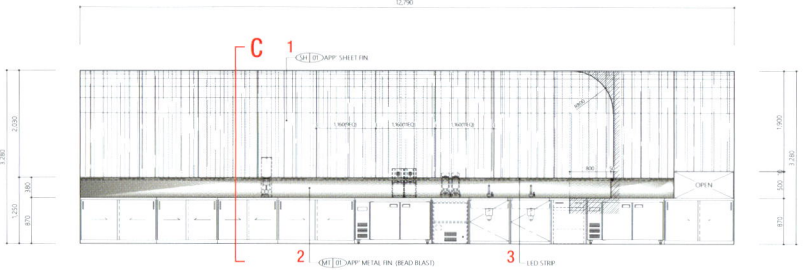

1층 홀 입면 A / 1F hall elevation A

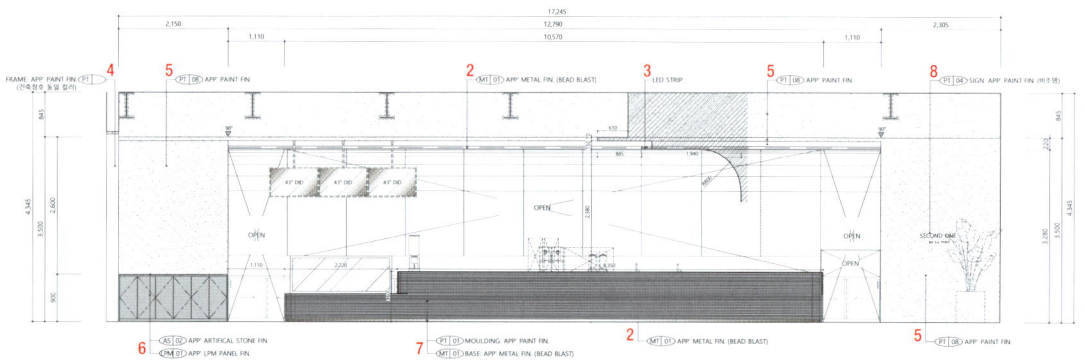

1층 홀 입면 B / 1F hall elevation B

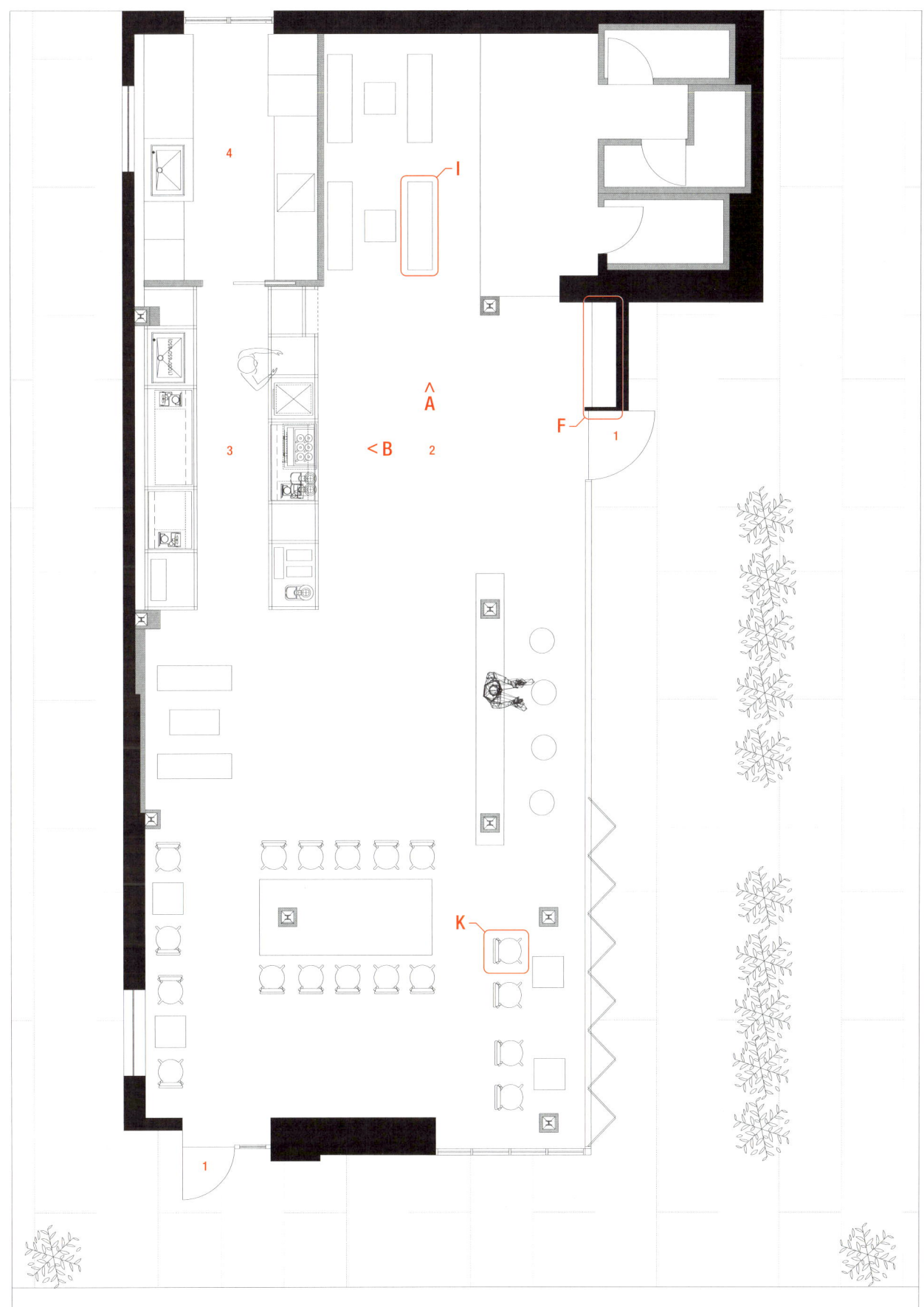

평면도 / floor plan

1 입구 2 홀 3 주방 4 창고

1 Entrance 2 Hall 3 Kitchen 4 Storage

7100
2,600
1 Travertin Amazon
2 Natural wood
3 Aluminum pipe

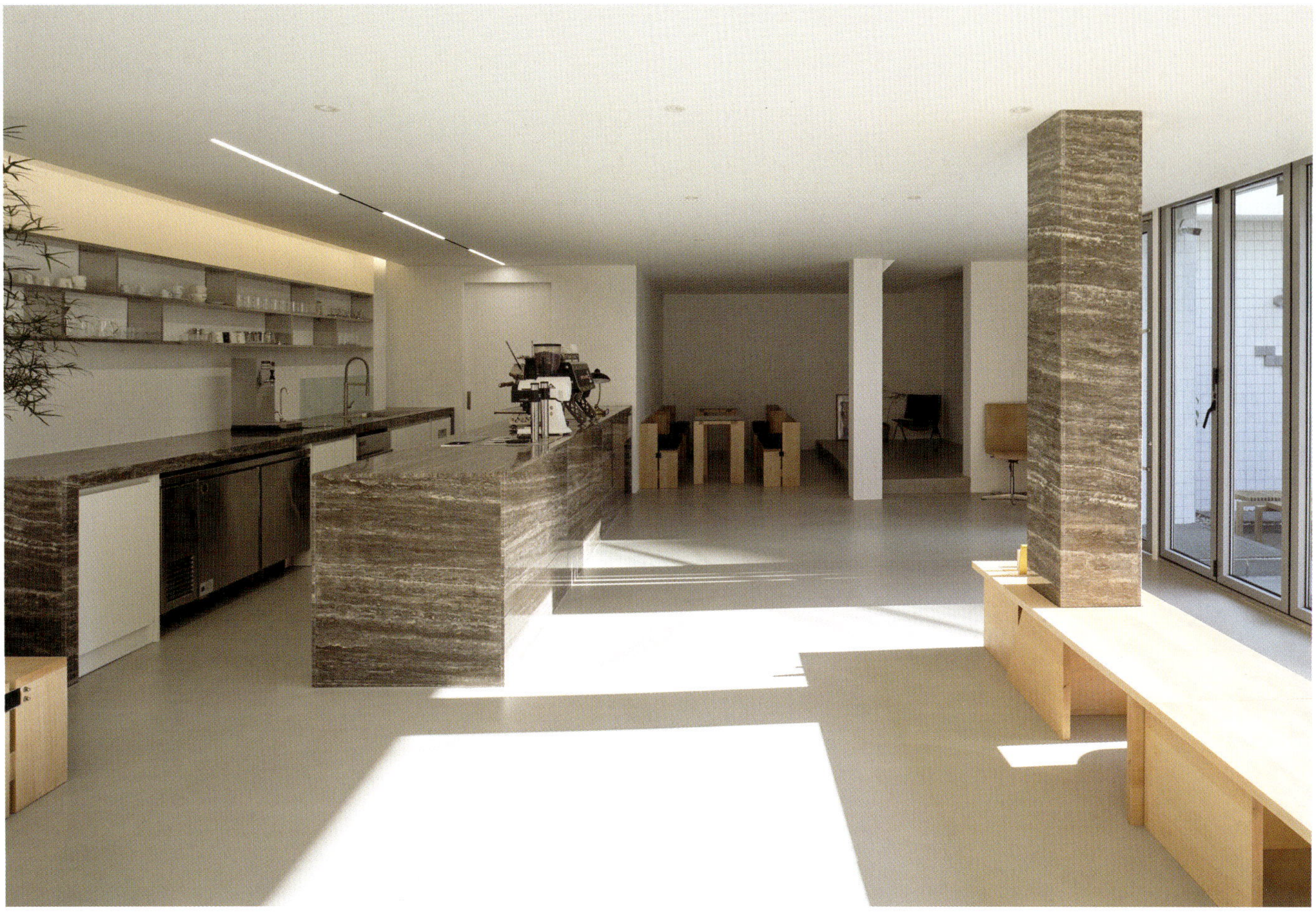

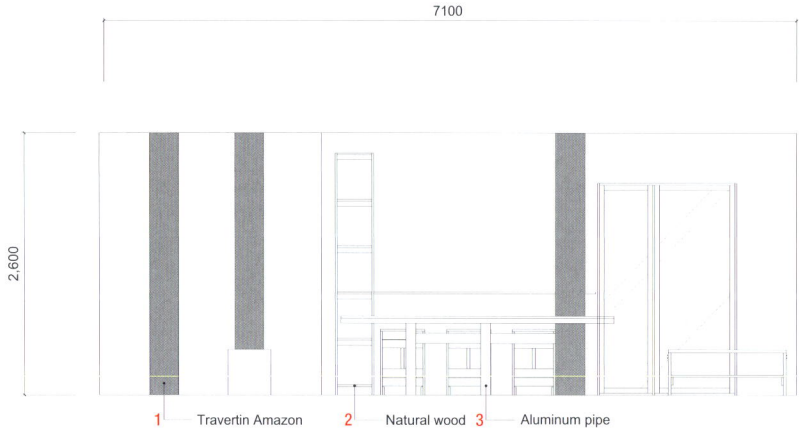

홀 입면 A / hall elevation A

13,170
2,600
2 Natural wood
C
4 SUS vibrations
1 Travertin Amazon

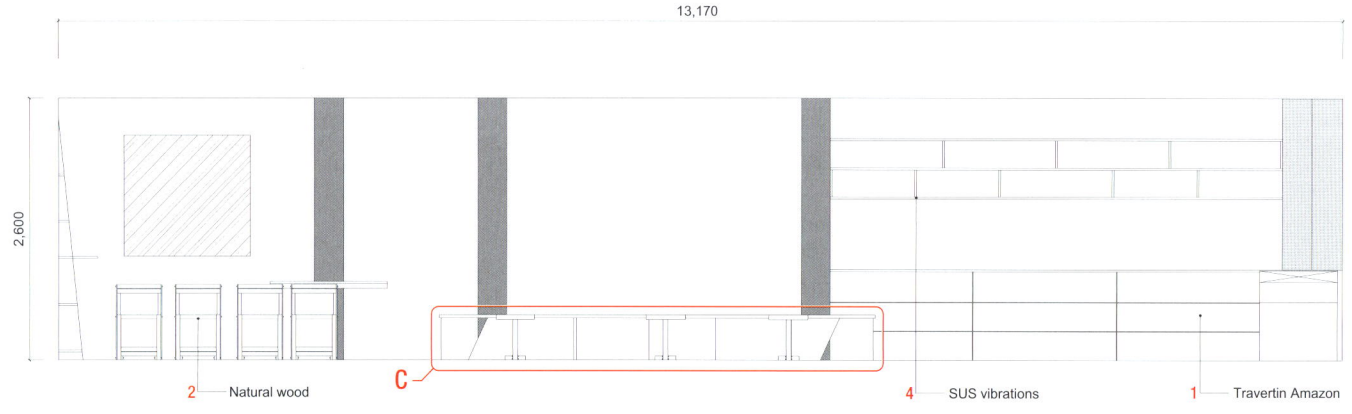

홀 입면 B / hall elevation B

1 대리석(트래버틴 아마존) 2 목재 3 알루미늄 파이프 4 스테인리스 스틸 바이브레이션 5 책 선반 : T2 아연철판 박스 제작

1 Marble (Travertin Amazon) 2 Natural wood 3 Aluminium pipe 4 Stainless steel vibration 5 Bookshelf : Custom-made box with T2 galvanized steel plate

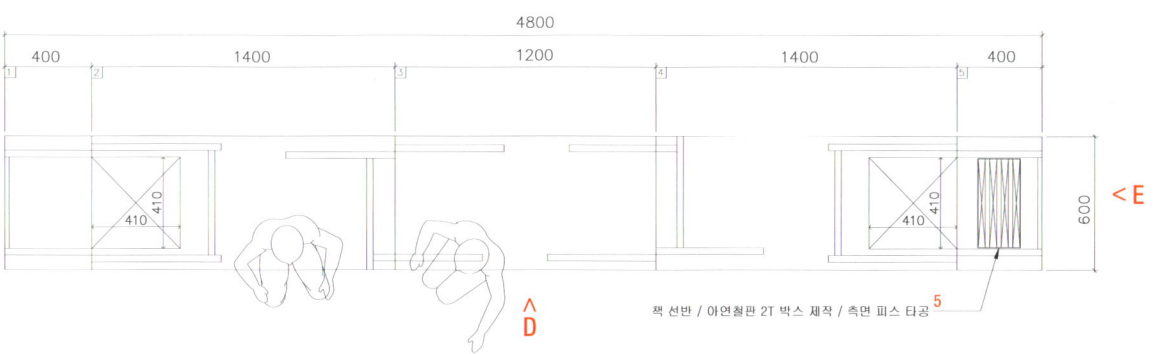

책 선반 / 아연철판 2T 박스 제작 / 측면 피스 타공

벤치 평면 C / bench top view C

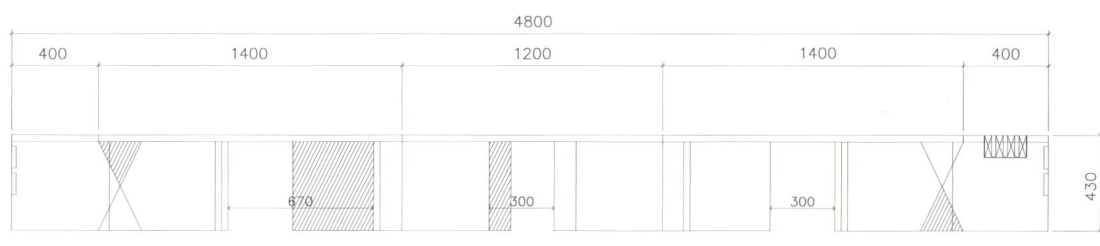
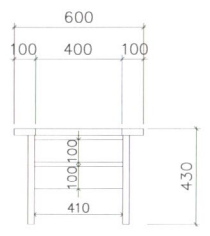

벤치 정면 D / bench front view D

벤치 측면 E / bench side view E

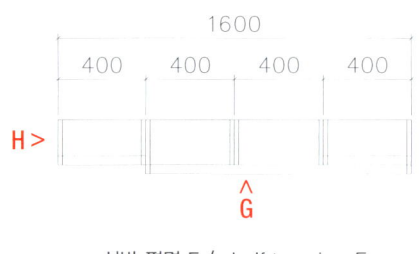

선반 평면 F / shelf top view F

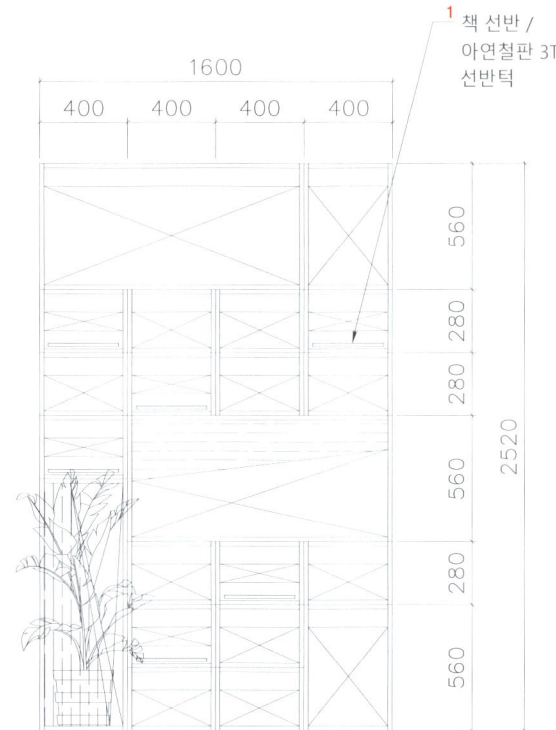

1 책 선반 / 아연철판 3T 선반턱

선반 정면 G / shelf front view G

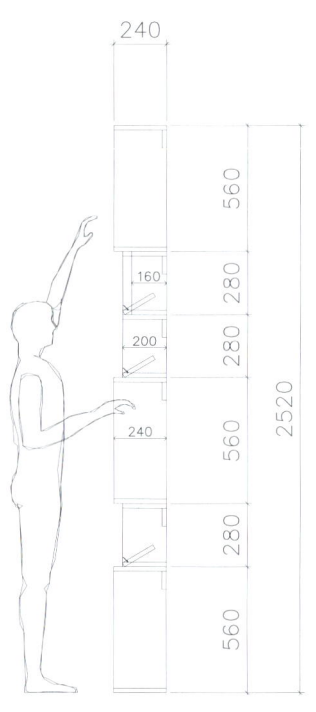

선반 측면 H / shelf side view H

1 책 선반 : T3 아연철판 선반턱

1 Book shelf : T3 galvanized steel plate

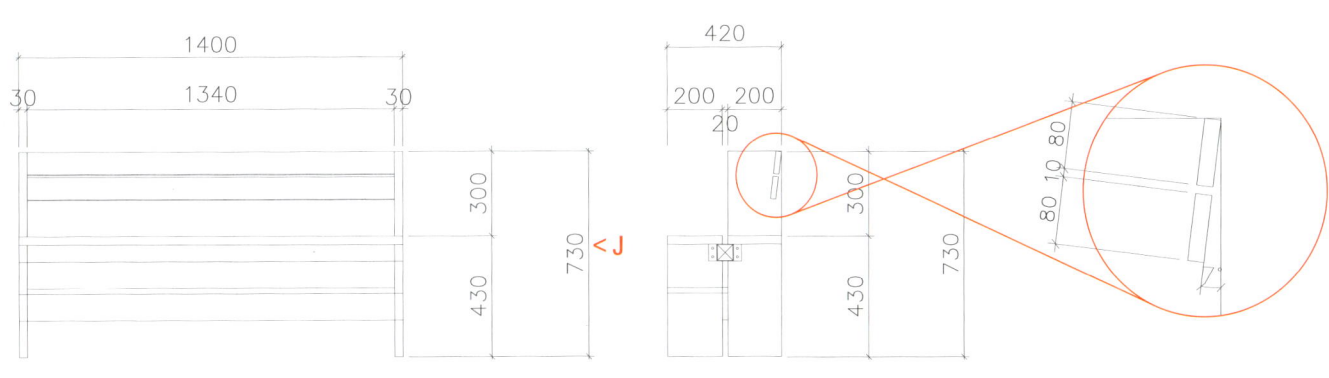

의자 정면 I / chair front view I

의자 측면 J / chair side view J

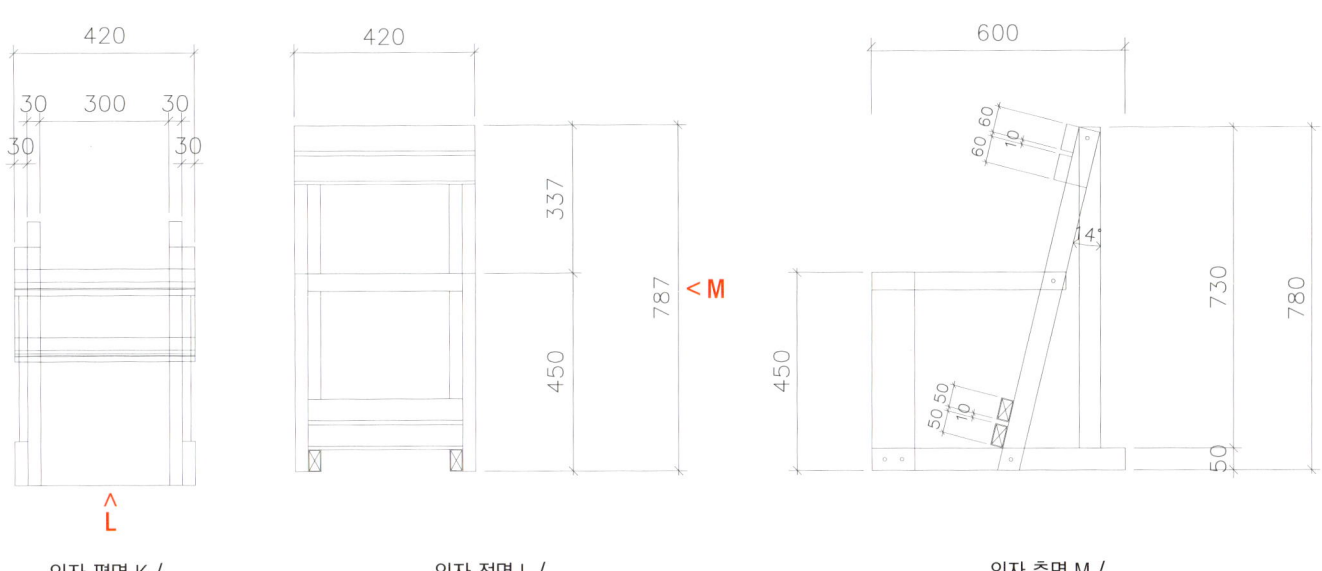

의자 평면 K / chair top view K

의자 정면 L / chair front view L

의자 측면 M / chair side view M

LES MAINS DORÉES

stof. | Seongjae Park

당신의 손에선 오로지 빛만이 가득합니다. – 빅토르 위고

낭만주의 시인 빅토르 위고의 편지에 쓰인 문장에서 시작되어 붙은 '금빛 손들'이라는 의미를 가진 베이커리 카페 레망도레(Les Mains Dorees) 는 서래 마을의 한 귀퉁이에 첫 번째 공간을 선보인다. 점묘화라는 콘셉트로 기획된 이 공간은 각각의 점과 같이, 레망도레 속 공간의 요소들을 각각의 고유한 점으로 존재하게 하여, 파리의 늦은 오후 거리를 물들이는 해와 상점의 불빛들이 금빛으로 빛나는 순간을 기록하고자 하였다. 전체적인 마감재는 따뜻한 컬러의 무늬목과 도장, 파리 거리의 바닥을 모티브로 한 바닥 타일을 바탕으로 화이트 인조대리석, 블랙 에나멜 도장, 모자이크 타일, 가죽, 황동 등 다양한 질감들이 조화를 이루고 있다. 공간의 중심이 되는 블랙 기둥을 둘러 싼 다양한 높낮이의 바의 형태는 파리의 방사형 도시 구조를 차용하여 각각의 매스로 이루어져 있으며, 문을 열고 들어와 빵을 고르고 구매하며 건네 받는 일련의 과정을 각 매스를 통해 순차적으로 경험하게 하여 경험의 과정 또한 깊은 인상으로 심어 주고자 하였다. 마들렌과 휘낭시에의 형태에 맞게 제작된 인조대리석 트레이는 베이커리를 돋보이게 해주며, 마치 전시된 귀한 보석들과 같이 놓여지도록 의도하였다. 창가를 따라 자유롭게 놓여진 좌석들과 테라스를 통한 야외 자석이 마련되어 있으며, 공간의 곳곳에는 황동과 스티치가 놓여진 가죽 마감의 디스플레이 요소를 통해 브랜드의 가치를 전시하여 보여주도록 설계하였다.

Il ne peut sortir de vos mains que de la limiere. – Victor Hugo

The bakery café Les Mains Dorées, meaning 'golden hands', takes its name from a sentence in a letter written by the romantic poet Victor Hugo and unveils its first space in Seorae Village. Planned with the concept of 'pointillism', this space aims to capture moments reminiscent of Parisian late afternoons, where the sunsets and the lights of shops illuminate the streets in golden hues. Warm-toned wood veneer, floor tile inspired by the pavements of Paris, white marble, black enamel, mosaic tiles, leather, brass, and various textures harmonize in the overall finish. The varied heights of bars surrounding the black pillar at the center of the space, inspired by the radial city structure of Paris, create individual masses for visitors to sequentially experience the process of entering, choosing and purchasing bread, and receiving it, aiming to imbue the experience with a profound impression. The artificial marble trays, crafted to match the shapes of madeleines and financiers, are intended to enhance the presentation of bakery items, akin to precious jewels being exhibited. Seats freely arranged along the windows and outdoor seating provided via the terrace, along with display elements featuring brass and stitched leather throughout the space, showcase the brand's values.

디자인 박성재 / 스토프
위치 서울특별시 서초구 사평대로 26길 9-3
용도 베이커리 카페
면적 85m²
마감 무늬목, 타일, 황동, 에나멜 페인트, 가죽
완공 2023. 10
디자인팀 강문희, 전소정
사진 김동규

Location 9-3, Sapyeong-daero 26-gil, Seocho-gu, Seoul
Use Bakery cafe
Area 85m²
Finishing Veneer wood, Tile, Brass, Enamel paint, Leather
Completion 2023. 10
Design team Munhee Kang, Sojeong Jeon
Photographer Donggyu Kim

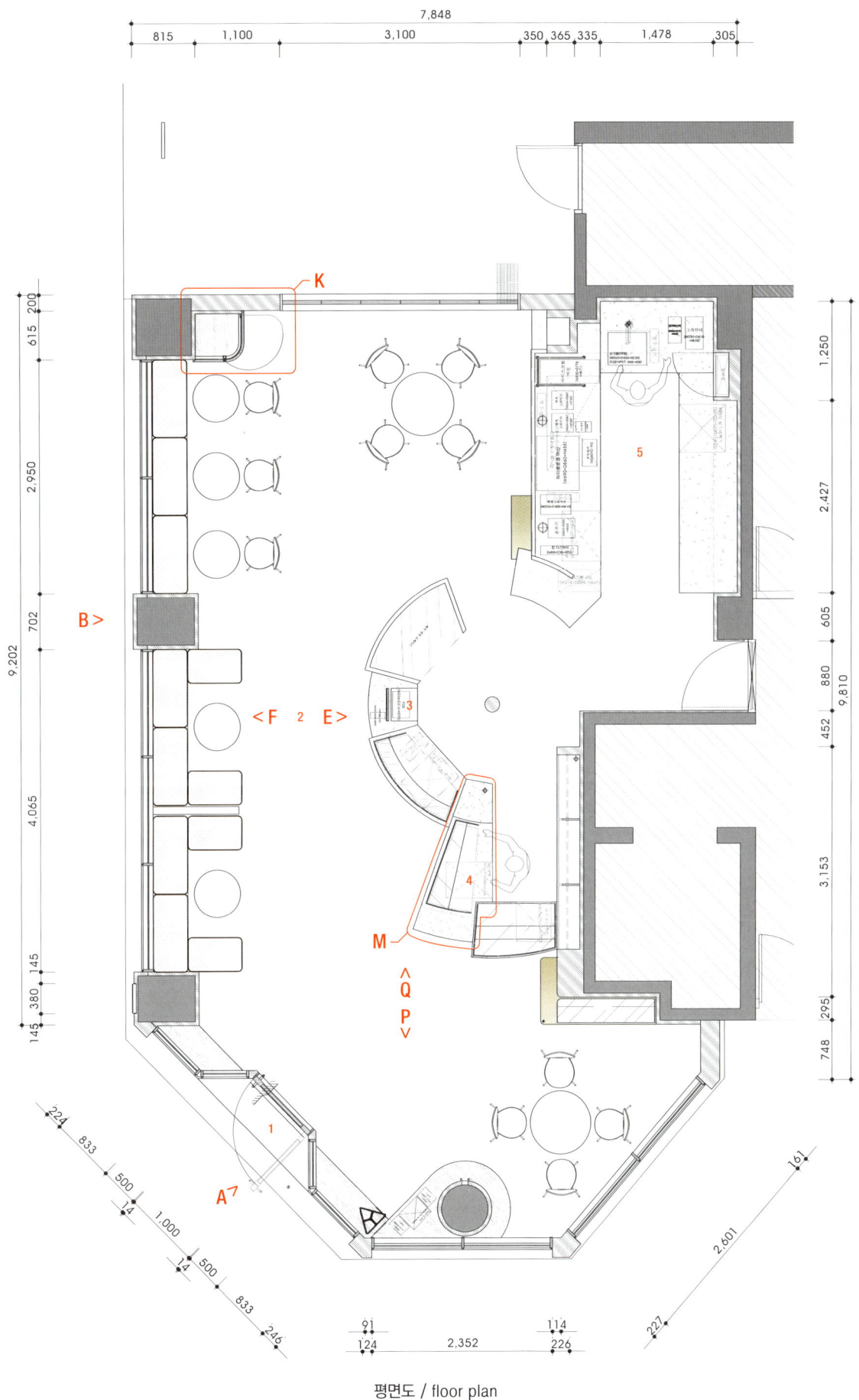

평면도 / floor plan

1 입구 2 홀 3 카운터 4 베이커리 존 5 커피 존

1 Entrance 2 Hall 3 Counter 4 Bakery zone 5 Coffee zone

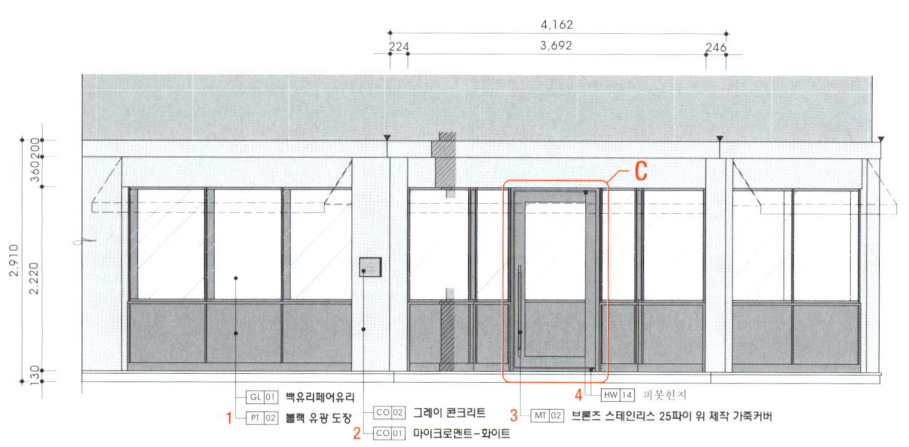

1	GL 01	백유리페어유리
	PT 02	블랙 유광 도장
2	CO 02	그레이 콘크리트
	CO 01	마이크로멘트-화이트
3	MT 02	브론즈 스테인리스 25파이 위 제작 가죽커버
4	HW 14	피봇힌지

파사드 A / facade A

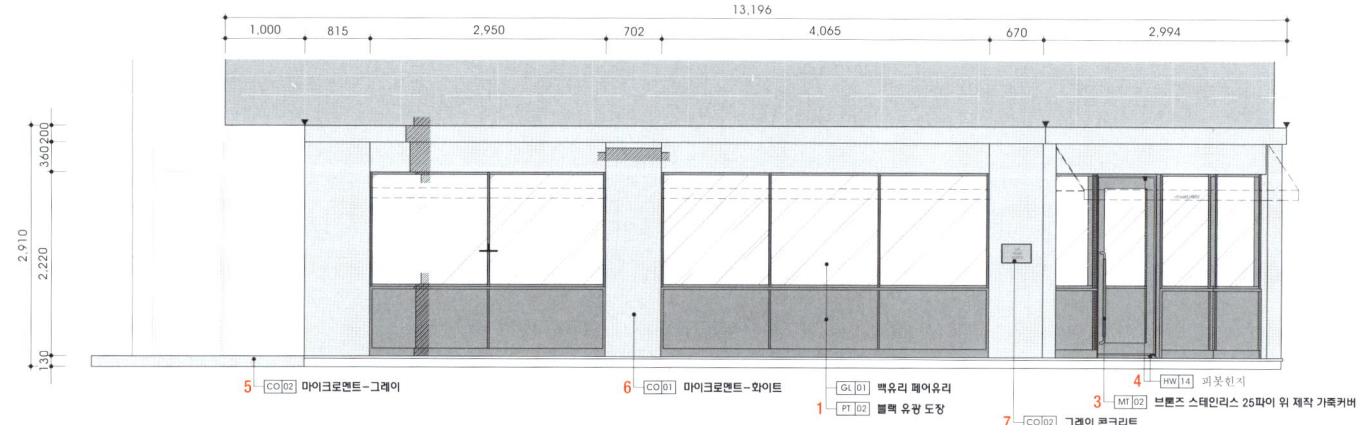

5	CO 02	마이크로멘트-그레이
6	CO 01	마이크로멘트-화이트
1	GL 01	백유리 페어유리
	PT 02	블랙 유광 도장
3	MT 02	브론즈 스테인리스 25파이 위 제작 가죽커버
7	CO 02	그레이 콘크리트
4	HW 14	피봇힌지

파사드 B / facade B

1 저철분 복층유리 / 검은색 유광 도장   2 그레이 콘크리트 / 화이트 마이크로시멘트   3 Ø25 브론즈 스테인리스 스틸 위 제작 가죽 커버   4 피봇 힌지   5 그레이 마이크로시멘트   6 화이트 마이크로시멘트   7 그레이 콘크리트   8 브론즈 스테인리스 스틸 / 브라운 가죽

1 Low-iron paired glass / Black glossy painting   2 Gray concrete / White microcement   3 Order-made leather cover on Ø25 bronze stainless steel   4 Pivot hinge   5 Gray microcement   6 White microcement   7 Gray concrete   8 Bronze stainless steel / Brown leather

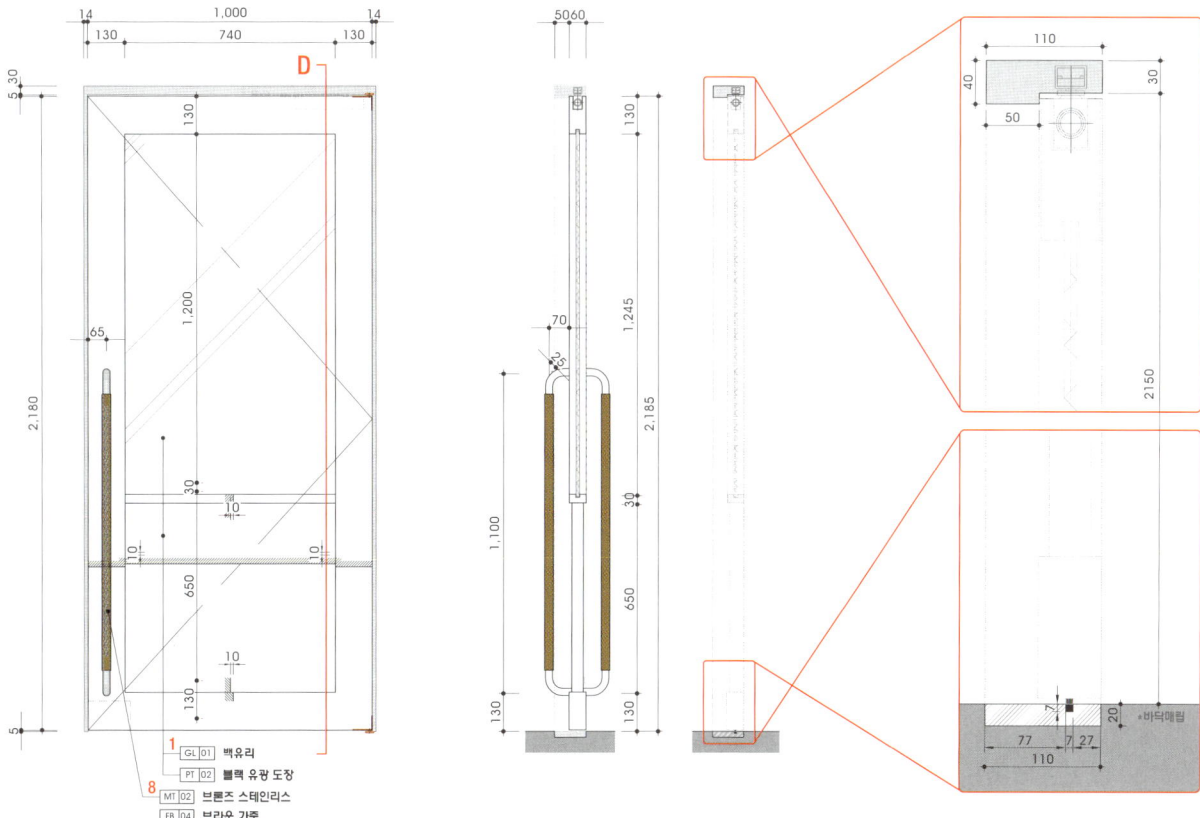

문 정면 C / door front view C

문 단면 D / door section D

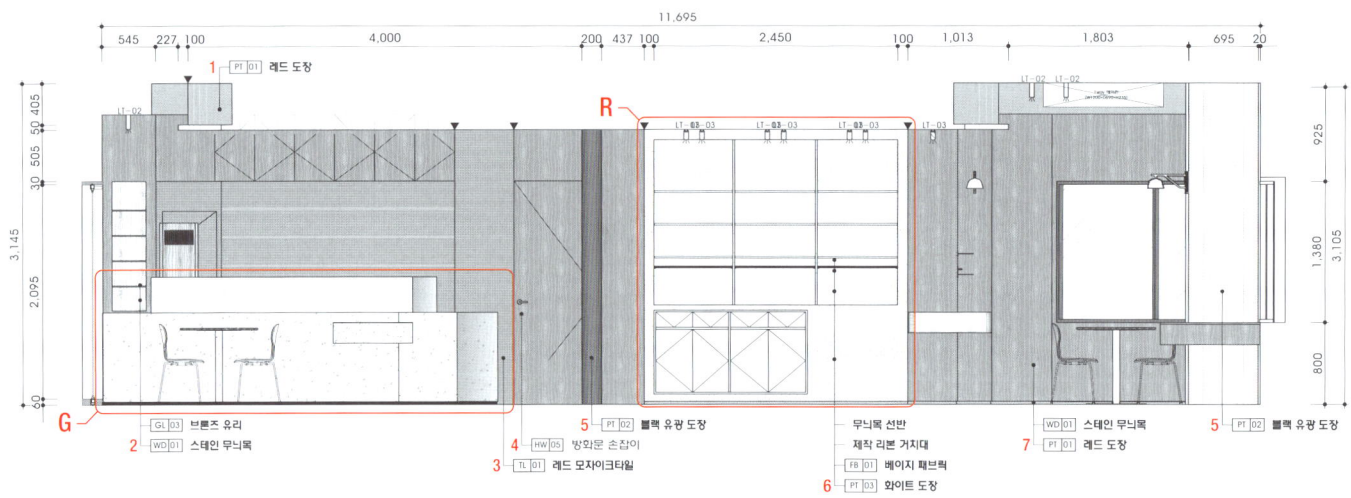

홀 입면 E / hall elevation E

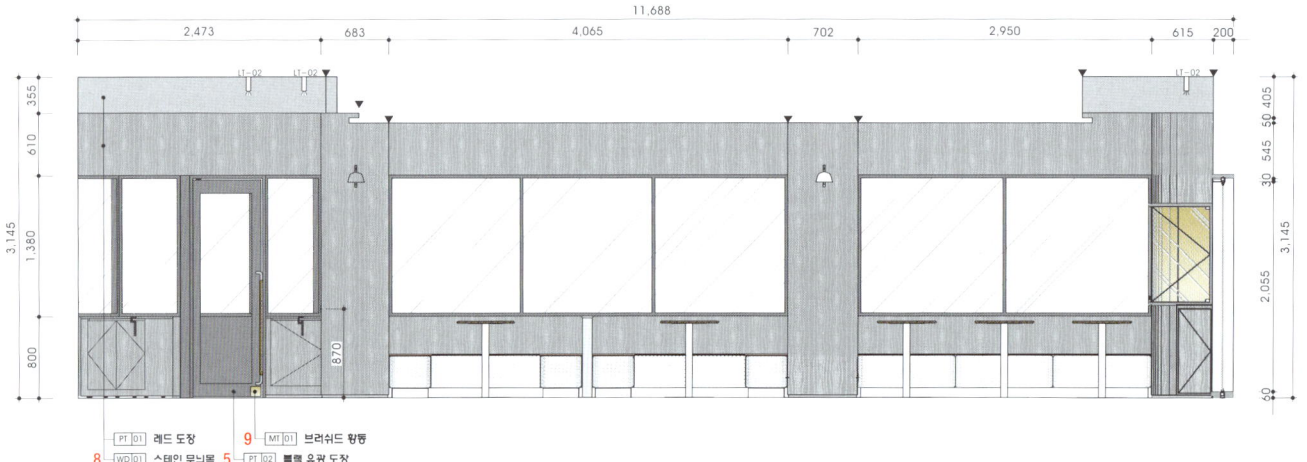

홀 입면 F / hall elevation F

1 빨간색 도장   2 브론즈 유리 / 스테인 무늬목   3 빨간색 모자이크 타일   4 방화문 손잡이   5 검은색 유광 도장   6 무늬목 선반 / 제작 거치대 / 베이지 패브릭 / 흰색 도장   7 스테인 무늬목 / 빨간색 도장   8 빨간색 도장 / 스테인 무늬목   9 브러쉬드 황동   10 흰색 인조대리석   11 매립 콘센트   12 흰색 인조대리석 / 검은색 유광 도장   13 브러쉬드 황동 / 흰색 인조대리석 / 검은색 유광 도장

1 Red painting   2 Bronze glass / Stained wood veneer   3 Red mosaic tile   4 Fire door handle   5 Black glossy painting   6 Wood veneer shelf / Order-made holder   7 Stained wood veneer / Red painting   8 Red painting / Stained wood veneer   9 Brushed brass   10 White artificial stone   11 Embedded outlet   12 White artificial stone / Black glossy painting   13 Brushed brass / White artificial stone / Black glossy painting

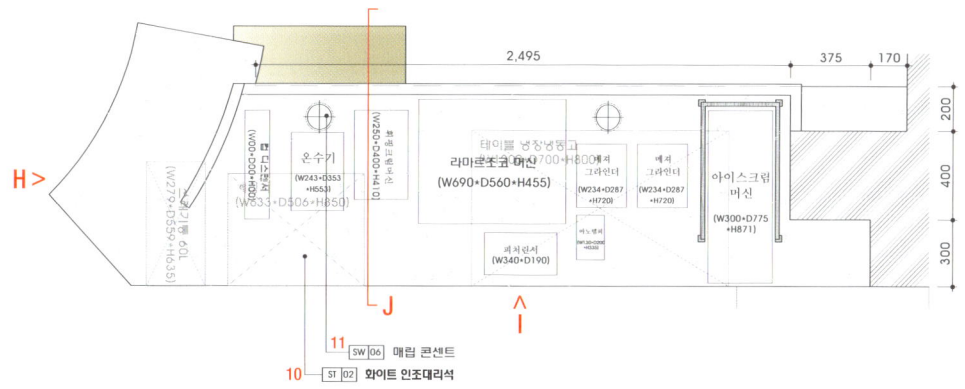
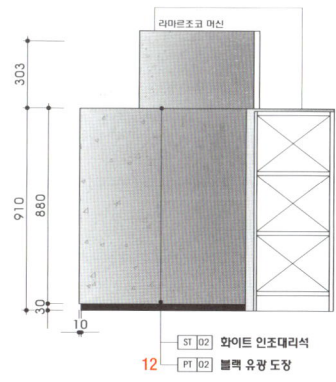

커피 바 평면 G / coffee bar top view G

커피 바 측면 H / coffee bar side view H

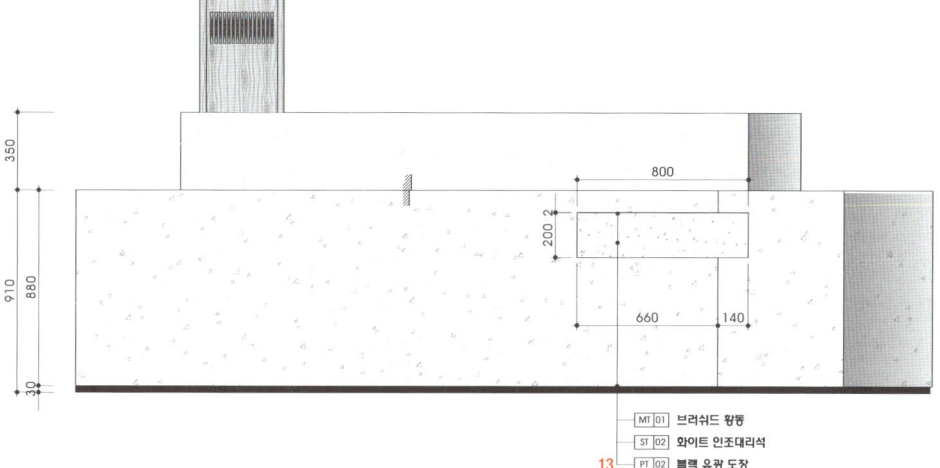
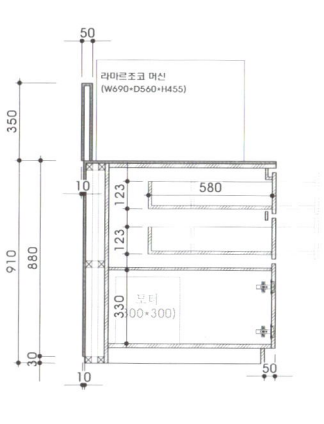

커피 바 정면 I / coffee bar front view I

커피 바 단면 J / coffee bar section J

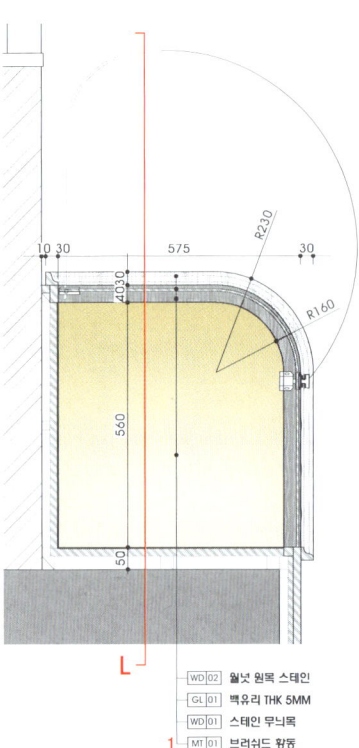

기둥 벽장 평단면 K / coloumn cabinet top section K

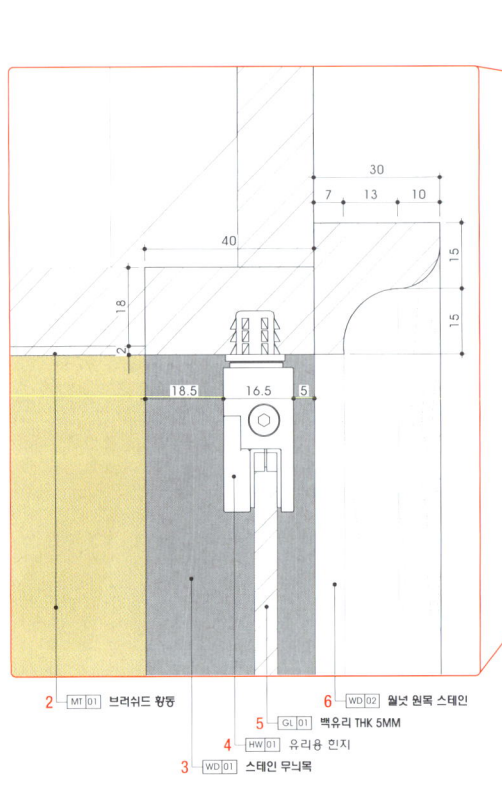

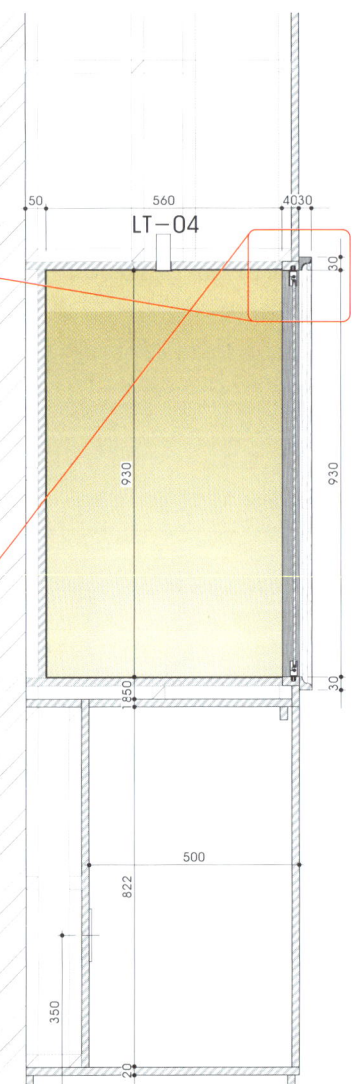

기둥 벽장 단면 L / column cabinet section L

1 스테인 월넛 원목 / T5 저철분유리 / 스테인 무늬목 / 브러쉬드 황동 2 브러쉬드 황동 3 스테인 무늬목 4 유리용 힌지 5 T5 저철분유리 6 스테인 월넛 원목 7 흰색 인조대리석 / 저철분유리 8 저철분유리 9 베이지 인조대리석 / 흰색 인조대리석 / 흰색 PET

1 Stained walnut wood / T5 low-iron glass / Stained wood veneer / Brushed brass 2 Brushed brass 3 Stained wood veneer 4 Hinge for glass 5 T5 low-iron glass 6 Stained walnut wood 7 White artificial stone / Low-iron glass 8 Low-iron glass 9 Beige artificial stone / White artificial stone / White PET

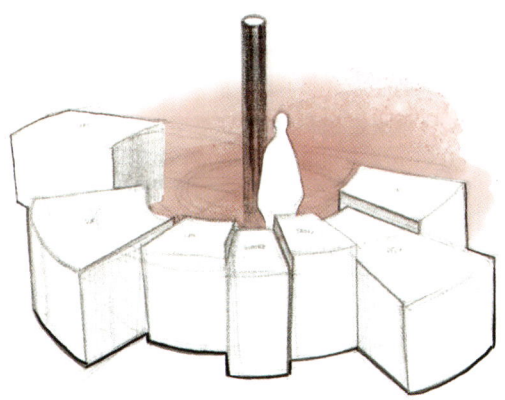

콘셉트 스케치 / concept sketch

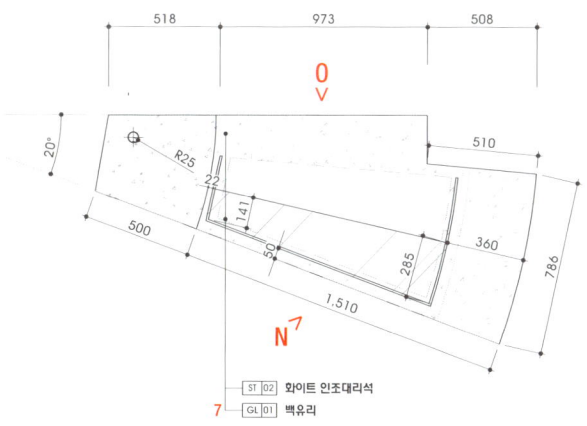

테이블 평면 M / table top view M

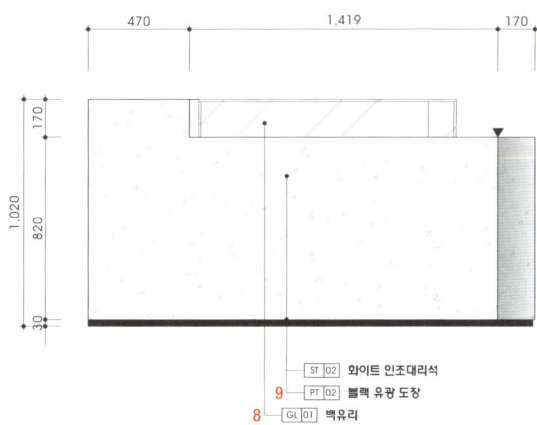

테이블 정면 N / table front view N

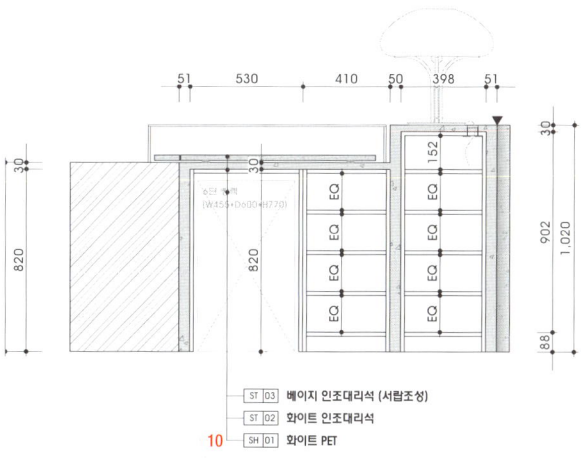

테이블 후면 O / table rear view O

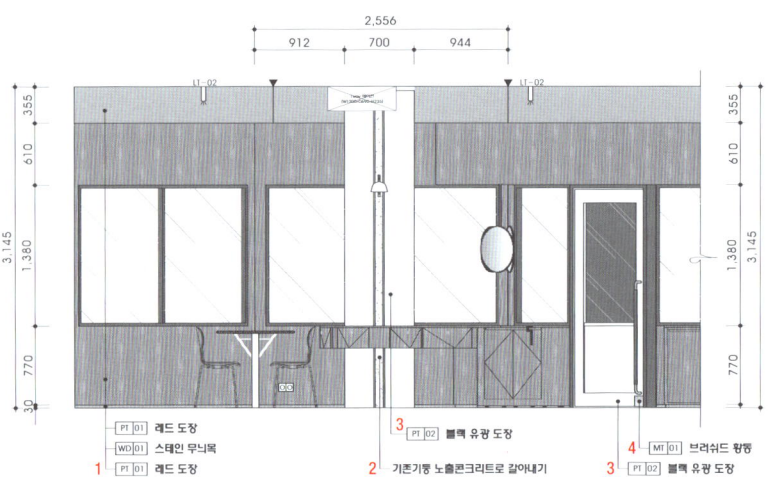

홀 입면 P / hall elevation P

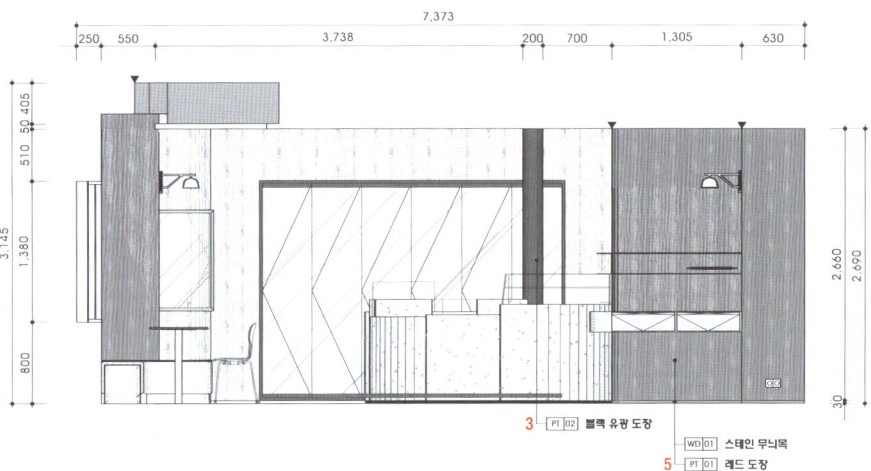

홀 입면 Q / hall elevation Q

1 빨간색 도장 / 스테인 무늬목 / 빨간색 도장 2 노출콘크리트 3 검은색 유광 도장 4 브러쉬드 황동 5 스테인드 무늬목 / 빨간색 도장 6 베이지 패브릭 / 흰색 도장 / 빨간색 도장

1 Red painting / Stained wood veneer / Red painting 2 Exposed concrete 3 Black glossy painting 4 Brushed brass 5 Stained wood veneer / Red painting 6 Beige fabric / White painting / Red painting

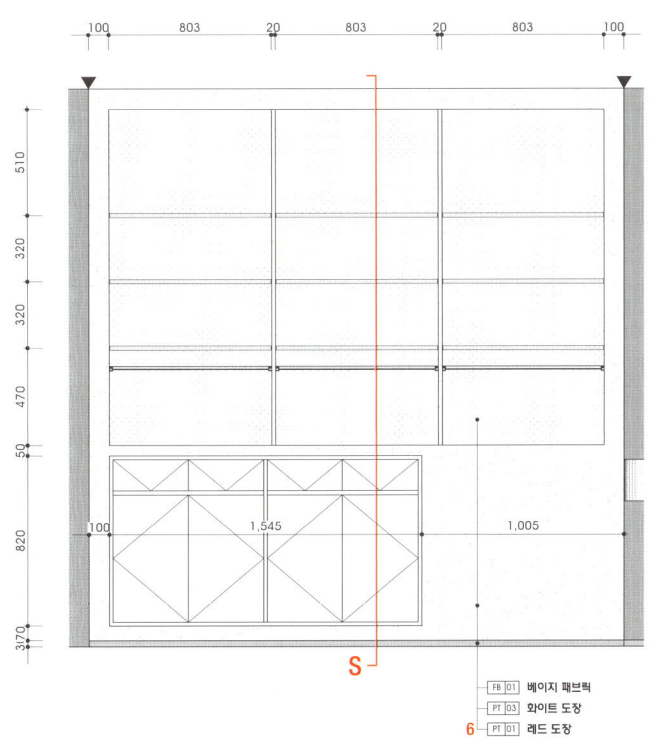

벽장 정면 R / cabinet front view R

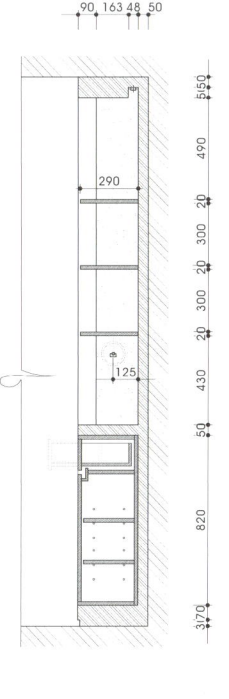

벽장 단면 S / cabinet section S

AISO SOUND

stof. | Seongjae Park

2009년부터 한국 스페셜티 커피 시장의 시작을 함께 해온 아이소 사운드가 기존의 카페와 로스터리 센터를 경기도 용인 남곡리에 확장 이전하여 새롭게 문을 열었다. 디자이너는 오랫동안 유지되어온 브랜드가 새로운 환경과 규모로 시작한다면 그 지속적인 가치관이 공간에 투영되어 경험으로 전달되어야 한다고 생각했고, 관계 맺음을 중요시하는 브랜드의 가치관을 건축 언어로 치환하여 물리적/경험적으로 이어지며 상호작용하는 적극적인 모습을 공간에 담아냈다. 디자이너는 "커피로부터 시작되는 관계"를 공간 내 구조, 환경, 프로그램, 행위 등 서로 영향을 주며 다양하게 형성되는 관계에 대입하고, 형태와 재료, 레이어링과 여백을 통해 표현하였다. 카페는 테라조, 콘크리트, 스테인리스 스틸 등의 무채색 마감재가 주를 이루고 있으며, 온기를 불어 넣는 목재 가구들이 균형을 맞추고 있다. 내부로 진입하면 커피 바 뒷편에 반복적으로 놓여진 거대한 목재 구조가 1, 2층 보이드를 통해 수직적으로 관통하며 각 층을 연결하고 있다. 이 구조는 스피커가 내장되어 소리를 통해 두 층을 연결해주기도 하고, 동시에 수납장 뿐만 아니라 조명의 역할을 하고 있어 커피 바와 전시 테이블들을 비춘다. 홀 공간은 전반적으로 밝고 환대하는 분위기 속 다양한 형태의 좌석들과 엮여 있는 살짝 구부러진 듯한 형태의 격자 선반장이 있는데, 이는 자연스럽게 좌석과 동선을 구분해주어 편안함을 제공하고, 창가 측 천장의 스트레치 실링과 일부 가구는 반사소재로 마감되어 야외의 자연환경을 내부로 끌어들임으로써 생기를 불어넣는 역할을 한다. 2층에는 주로 다른 사람들과 공유할 수 있는 대형 테이블들이 배치되어 있으며, 이는 사람들과 프로그램을 이어주는 매개로서 때에 따라 세미나가 이루어지기도 하고 지역 상품들을 전시하기 위한 디스플레이 테이블이 되기도 한다.

AISO SOUND, a key player in the start of the Korean specialty coffee market since 2009, expanded its existing cafe and roastery center to a new location in Nangok-ri, Yongin, Gyeonggi Province. Designers believed that if a long-standing brand ventures into a new environment and scale, its enduring values should be reflected in the space, conveyed through experiences. They translated the brand's emphasis on relationship-building into architectural language, creating a space where the physical and experiential aspects interact actively. Stof applied the concept of "relationships starting from coffee" to the formation of various connections among elements such as structure, environment, program, and behavior. The cafe features finishes dominated by neutral tones like terrazzo, concrete, and stainless steel. Wooden furniture introduces warmth, balancing the overall ambiance. Upon entering, a massive wooden structure repetitively placed behind the coffee bar vertically penetrates the voids of the first and second floors, connecting them. This structure serves to link the floors through sound with embedded speakers, acting as storage and also providing illumination for the coffee bar and exhibition tables. The hall space, overall bright and welcoming, incorporates various forms of seating and slightly curved grid shelves. These shelves naturally separate seats and circulation paths, offering comfort. The stretch ceiling near the window and some furniture finished with reflective materials draw the outdoor natural environment inward, creating a lively atmosphere. On the second floor, large tables designed for shared use are placed, serving as a medium for connecting people and programs. At times, these tables host seminars, and they also double as display tables for showcasing local products.

디자인 박성재 / 스토프
위치 경기도 용인시 처인구 양지면 은이로 72
용도 베이커리 카페, 로스터리
면적 955㎡
마감 바닥 - 타일, 테라코타 타일 / 벽 - 콘크리트 / 천장 - 스트레치 실링 시스템
완공 2023. 12
디자인팀 김혜진, 백성현
사진 김동규

Location 72, Euni-ro, Yangji-myeon, Cheoin-gu, Yongin-si, Gyeonggi-do
Use Bakery cafe, Roastery
Area 955㎡
Finishing Floor - Tile, Terracotta tile / Wall - Concrete / Ceiling - Stretch ceiling system
Completion 2023. 12
Design team Hyejin Kim, Seonghyun Baek
Photographer Donggyu Kim

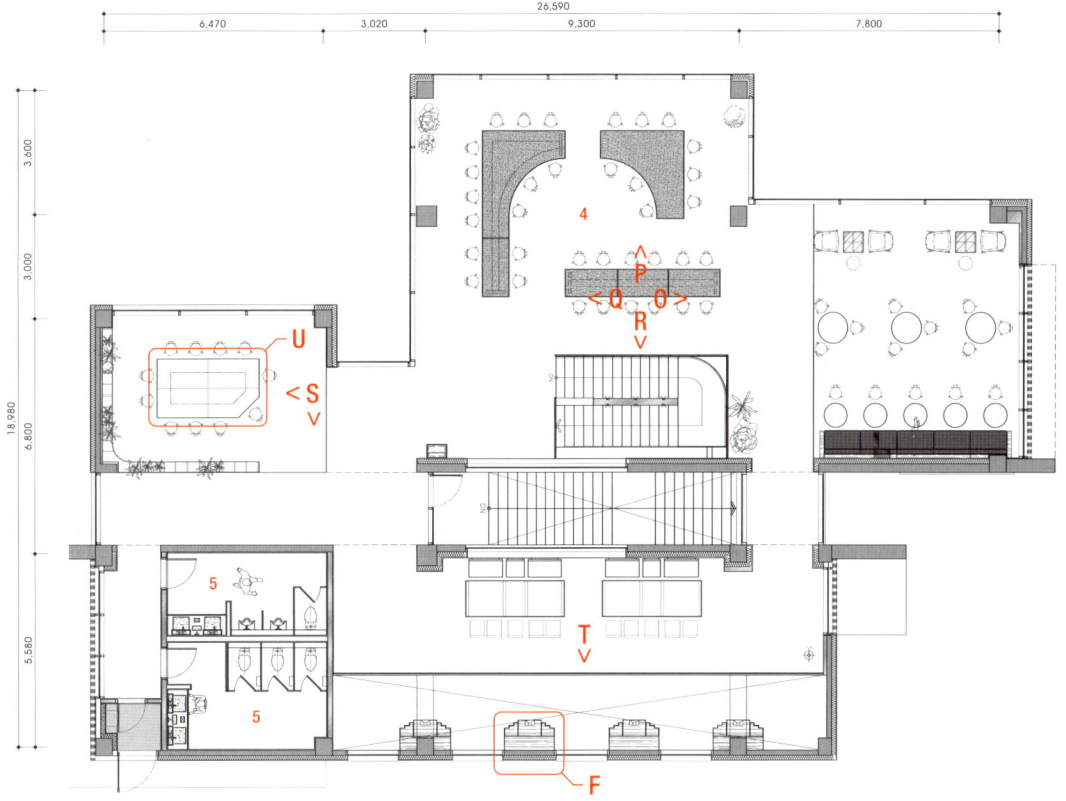

2층 평면도 / 2nd floor plan

1층 평면도 / 1st floor plan

1 입구 2 커피 바 3 베이커리 존 4 홀 5 화장실

1 Entrance 2 Coffee bar 3 Bakery zone 4 Hall 5 Restroom

1 LED 간접조명 2 자동문 3 블랙 리놀륨 4 건축 콘크리트 위 투명 코팅 / T5 SUS 미러 / 에쉬 원목 위 지정색 오픈포어 도장 5 스피커 / 스테인리스 비드 블라스트 6 스피커 / 스테인리스 비드 블라스트 / 전면 바 7 합판 절단면 / 콩자갈 노출포장 / 합판 절단면 8 자작나무 합판 위 지정색 오픈포어 도장 / 45X45 에쉬 원목 위 지정색 오픈포어 도장 / 블랙 리놀륨 9 블랙 리놀륨 / 납작머리 볼트 / 60X40 에쉬 원목 위 지정색 오픈포어 도장 10 자작나무 합판 위 지정색 오픈포어 도장 / 45X45 에쉬 원목 위 지정색 오픈포어 도장 11 자작나무 합판 위 지정색 오픈포어 도장 / 콩자갈 노출포장 / 자작나무 합판 위 지정색 오픈포어 도장 / 각재 보강

1 LED indirect lighting 2 Automatic door 3 Black linoleum 4 Clear coating on concrete / T5 SUS mirror / App. color openpore painting on ash wood 5 Speaker / Stainless bead blast 6 Speaker / Stainless bead blast / Bar 7 Plywood cutting surface / Pebble wash / Plywood cutting surface 8 App. color openpore painting on birch plywood / App. color openpore painting on 45X45 ash wood / Black linoleum 9 Black linoleum / Low head cap screws / App. color openpore painting on 60X40 ash wood 10 App. color openpore painting on birch plywood / App. color openpore painting on 45X45 ash wood 11 App. color openpore painting on birch plywood / Pebble wash / App. color openpore painting on birch plywood / Square timber reinforcement

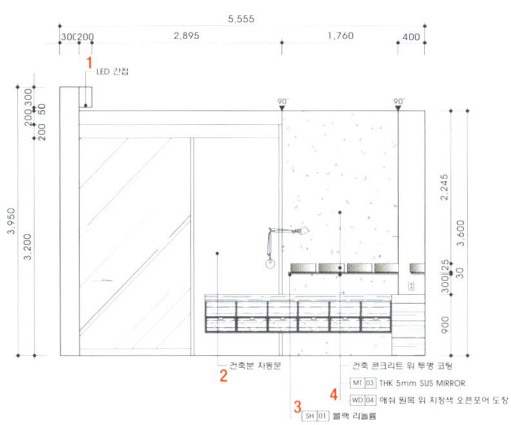

1층 홀 입면 A / 1F hall elevation A

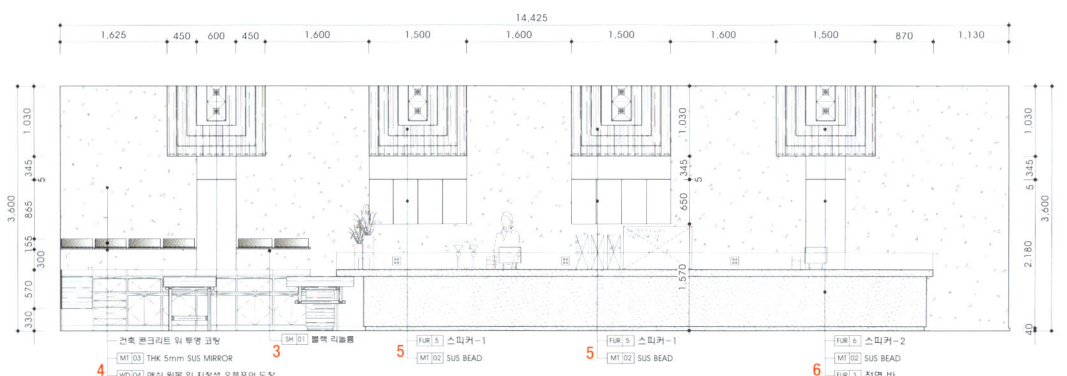

1층 홀 입면 B / 1F hall elevation B

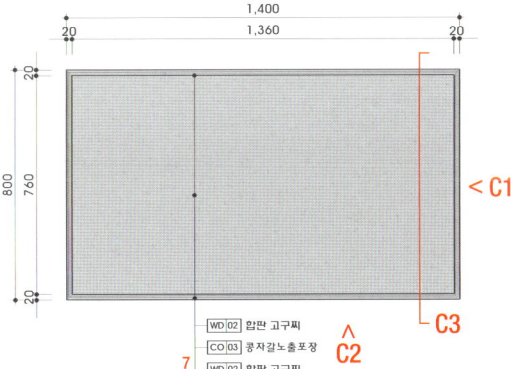

베이커리 테이블 평면 C / bakery table top view C

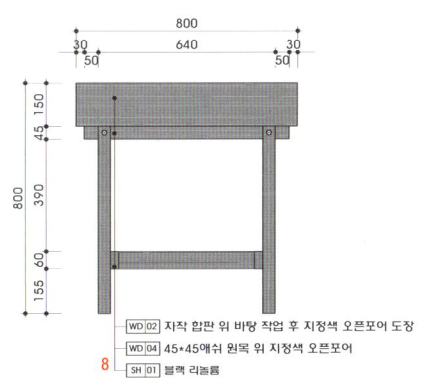

베이커리 테이블 측면 C1 / bakery table side view C1

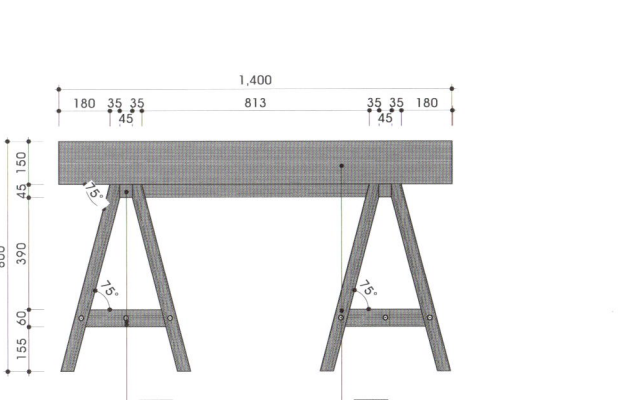

베이커리 테이블 정면 C2 / bakery table front view C2

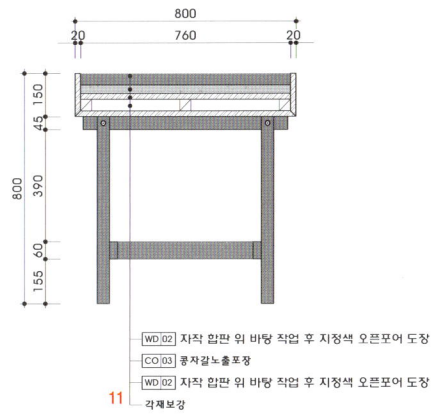

베이커리 테이블 단면 C3 / bakery table section C3

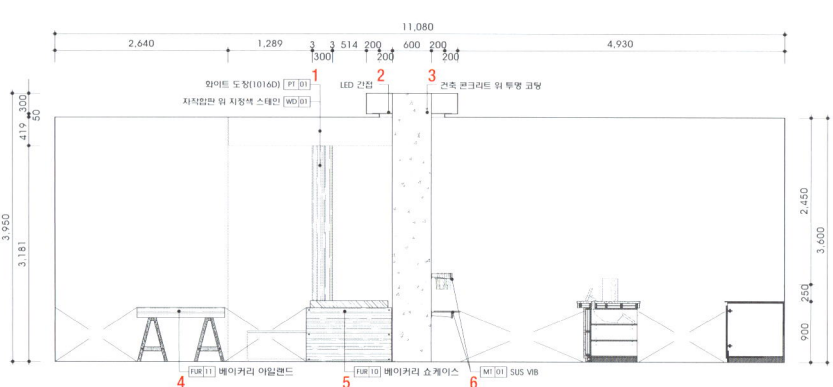

1층 홀 입면 D / 1F hall elevation D

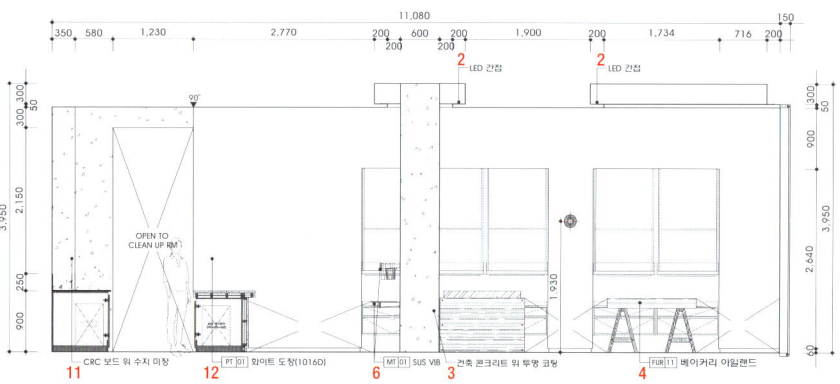

1층 홀 입면 E / 1F hall elevation E

1 흰색 도장 / 자작나무 합판 위 지정색 스테인 2 LED 간접조명 3 건축콘크리트 위 투명 코팅 4 베이커리 아일랜드 테이블 5 베이커리 쇼케이스 6 SUS 바이브레이션 7 스테인리스 비드 블라스트 8 자작나무 합판 위 지정색 스테인 9 스피커 매입 10 자작나무 합판 위 지정색 스테인 / 광확산 아크릴 11 CRC 보드 위 수지 미장 12 흰색 도장

1 White painting / App. color stain on birch plywood 2 LED indirect lighting 3 Clear coating on concrete 4 Bakery island table 5 Bakery showcase 6 SUS vibration 7 Stainless bead blast 8 App. color stain on birch plywood 9 Embedding speaker 10 App. color stain on birch plywood / Light diffuser acrylic 11 Resin plaster finish on CRC board 12 White painting

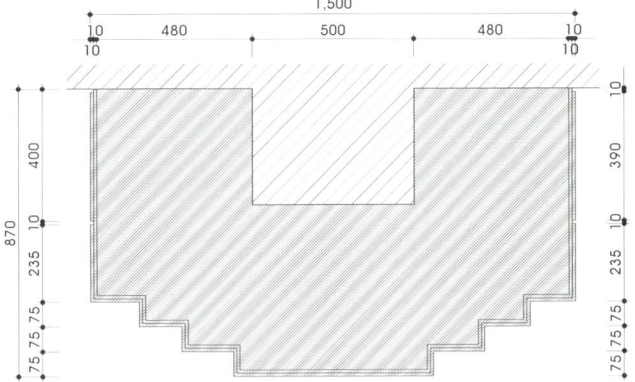

스피커 단면 F1 / speaker section F1

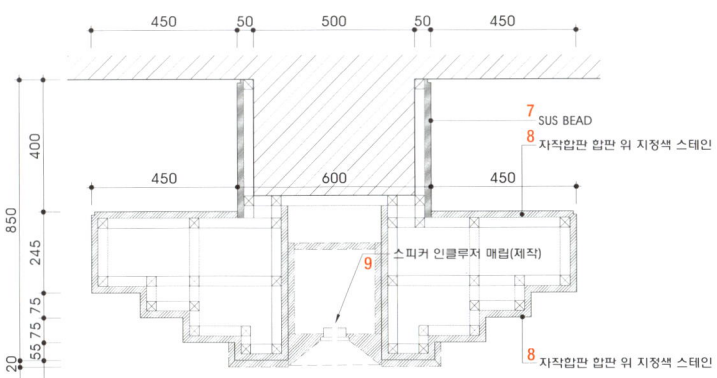

스피커 단면 F2 / speaker section F2

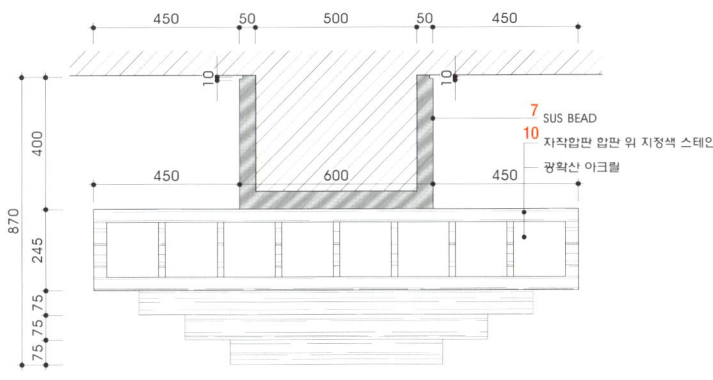

스피커 단면 F3 / speaker section F3

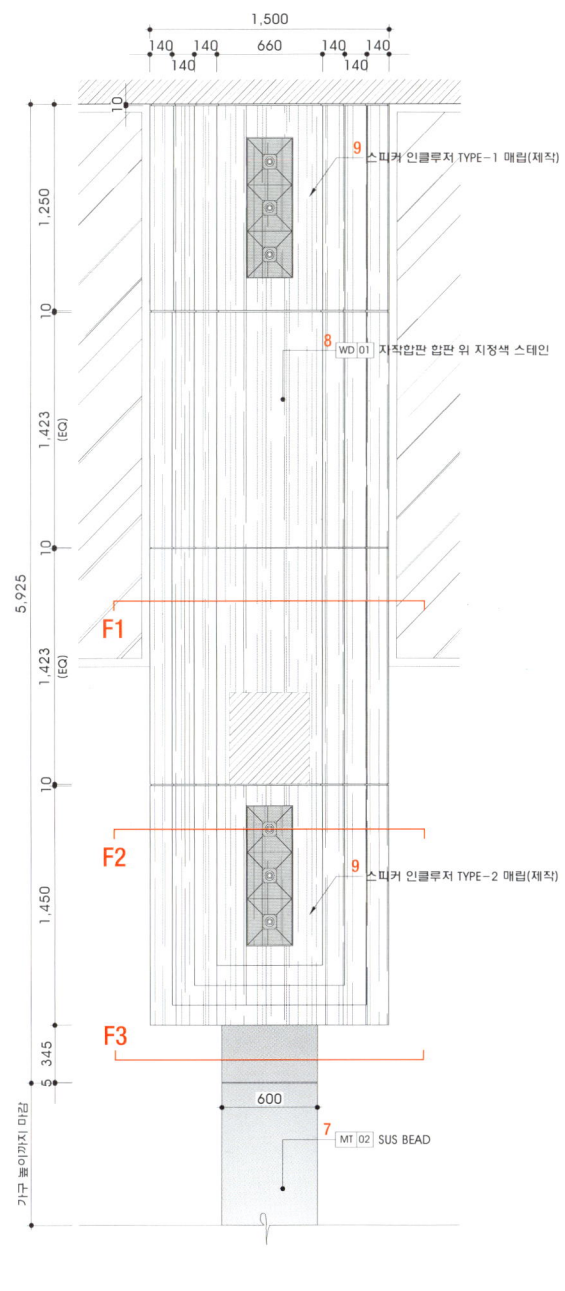

스피커 정면 F / speaker front view F

1 CRC 보드 위 수지 미장 2 흰색 도장 3 LED 간접조명 / 유광 타일 4 LED 간접조명 5 브래킷 6 T5 SUS 바이브레이션 / T5 SUS 바이브레이션 / 건축 콘크리트 위 투명 코팅 7 자작나무 합판 위 지정색 스테인 8 자동문 9 자작나무 합판 위 지정색 스테인 / 합판 절단면 / SUS 바이브레이션 10 합판 절단면 / Ø20 SUS 파이프 11 자작나무 합판 위 지정색 스테인 / LED 12 30X30 SUS 파이프 보강 / T1.2 SUS 바이브레이션 13 T5 SUS 바이브레이션 / 30X30 SUS 파이프 보강 14 30X30 SUS 파이프 보강 / 자작나무 합판 위 지정색 스테인 / 합판 절단면 15 진회색 도장 16 Ø20 SUS 파이프 / 자작나무 합판 위 지정색 스테인 / SUS 바이브레이션 17 합판 절단면 / SUS 바이브레이션 / 30X30 SUS 파이프 보강 / 자작나무 합판 위 지정색 스테인 18 서랍 레일

1 Resin plaster finish on CRC board 2 White painting 3 LED indirect lighting / Glossy tile 4 LED indirect lighting 5 Bracket 6 T5 SUS vibration / T5 SUS vibration / Clear coating on concrete 7 App. color stain on birch plywood 8 Automatic door 9 App. color stain on birch plywood / Plywood cutting surface / SUS vibration 10 Plywood cutting surface / Ø20 SUS pipe 11 App. color stain on birch plywood / LED 12 30X30 SUS pipe reinforcement / T1.2 SUS vibration 13 T5 SUS vibration / 30X30 SUS pipe reinforcement 14 30X30 SUS pipe reinforcement / App. color stain on birch plywood / Plywood cutting surface 15 Dark gray painting 16 Ø20 SUS pipe / App. color stain on birch plywood / SUS vibration 17 Plywood cutting surface / SUS vibration / 30X30 SUS pipe reinforcement / App. color stain on birch plywood 18 Drawer rail

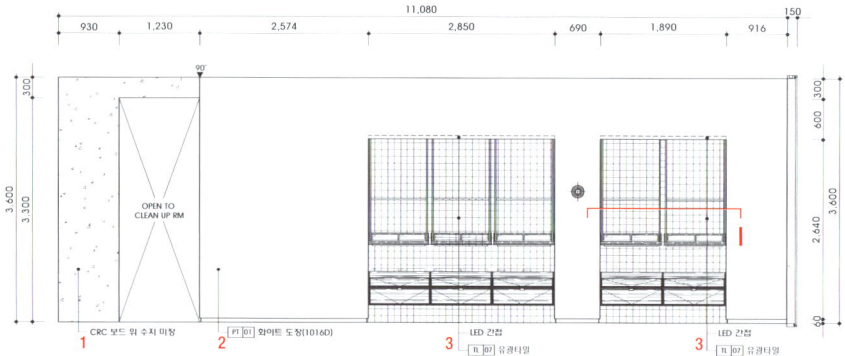

1층 홀 입면 G / 1F hall elevation G

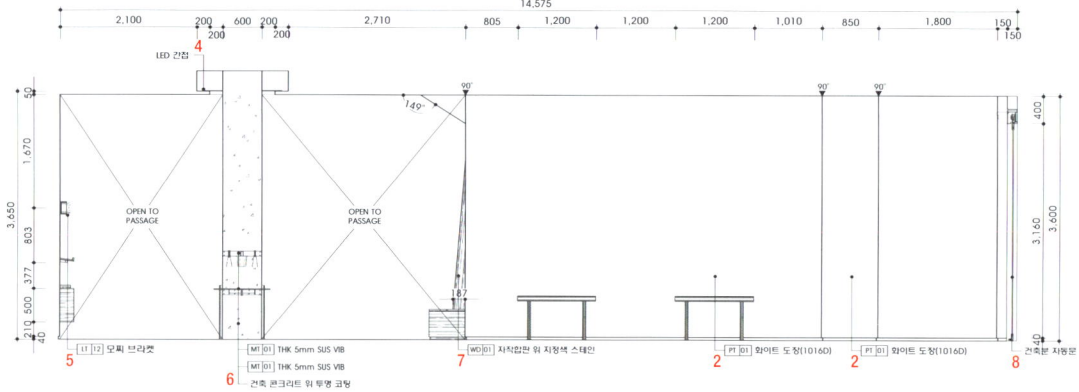

1층 홀 입면 H / 1F hall elevation H

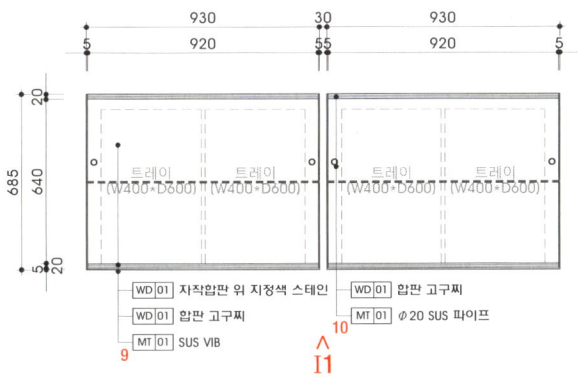

베이커리 선반 단면 I1 / bakery shelves section I1

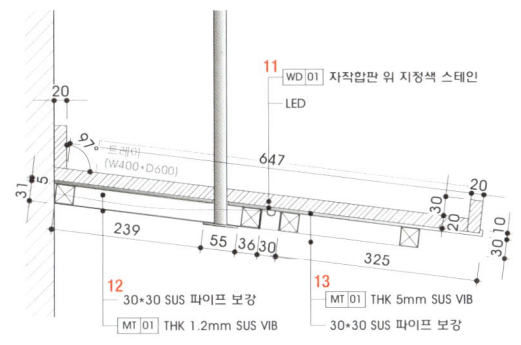

베이커리 선반 상세 I4 / bakery shelves detail I4

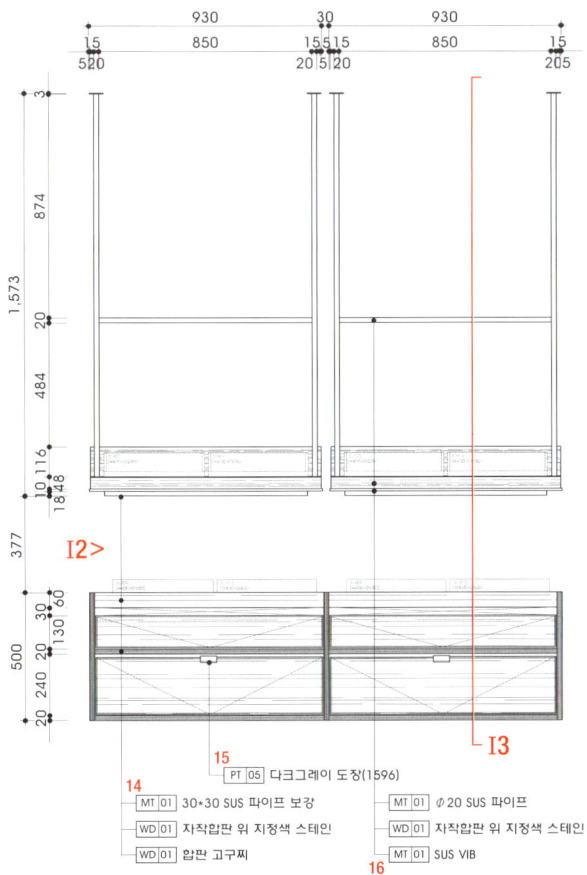

베이커리 선반 정면 I1 / bakery shelves front view I1

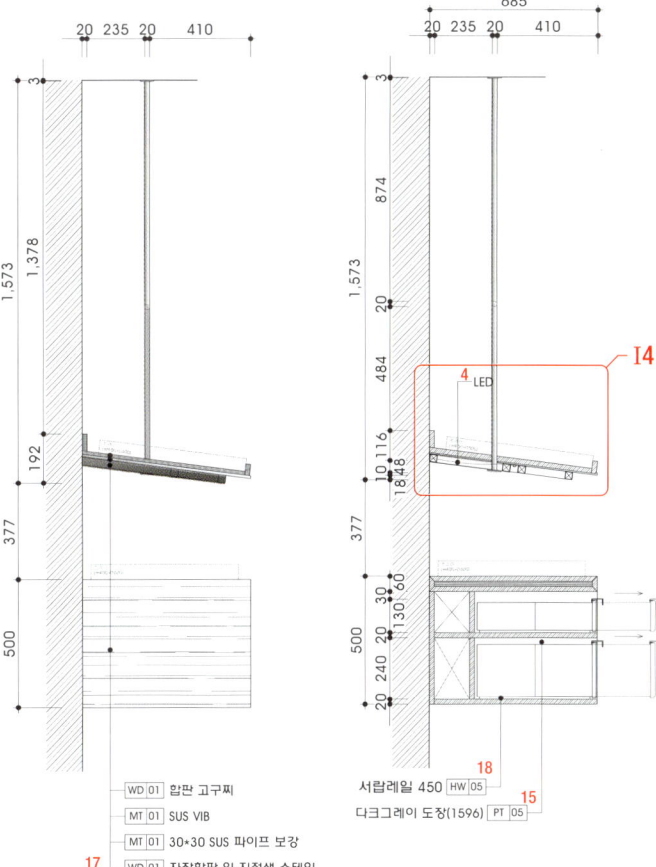

베이커리 선반 측면 I2 / bakery shelves side view I2

베이커리 선반 단면 I3 / bakery shelves section I3

1 커튼박스 2 회색 질감 도장 3 T1.6 갈바륨 위 연회색 도장 4 10T 평철 위 연회색 도장 / T1.6 갈바륨 위 연회색 도장 5 LED 간접조명 6 건축 콘크리트 위 투명 코팅 / 제작 소파 / 2구 콘센트 7 타일 / 타일 8 SUS 미러 9 흰색 도장 10 회색 도장 / 흰색 인조대리석 11 합판 절단면 / 자작나무 합판 위 지정색 스테인 12 2mm 줄눈 13 자작나무 합판 위 지정색 스테인 14 회색 도장 / 합판 절단면 / 자작나무 합판 위 지정색 스테인 / 자작나무 합판 위 지정색 스테인 15 회색 도장 16 1구 콘센트 17 테이블 램프 / 회색 도장

1 Curtain box 2 Gray textured painting 3 Light gray painting on T1.6 galvalume 4 Light gray painting on 10T flat steel / Light gray painting on T1.6 galvalulme 5 LED indirect lighting 6 Clear coating on concrete / Custom-made sofa / Dual outlet 7 Tile / Tile 8 SUS mirror 9 White painting 10 Gray painting / White engineered marble 11 Plywood cutting surface / App. color stain on birch plywood 12 2mm joint reveal 13 App. color stain on birch plywood 14 Gray painting / Plywood cutting surface / App. color stain on birch plywood / App. color stain on birch plywood 15 Gray painting 16 Single outlet 17 Table lamp / Gray painting

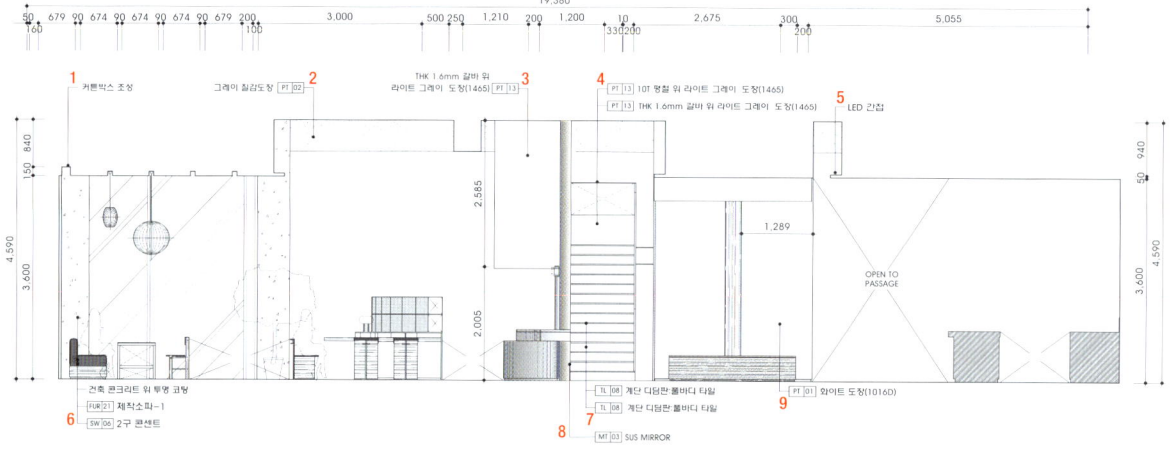

1층 홀 입면 J / 1F hall elevation J

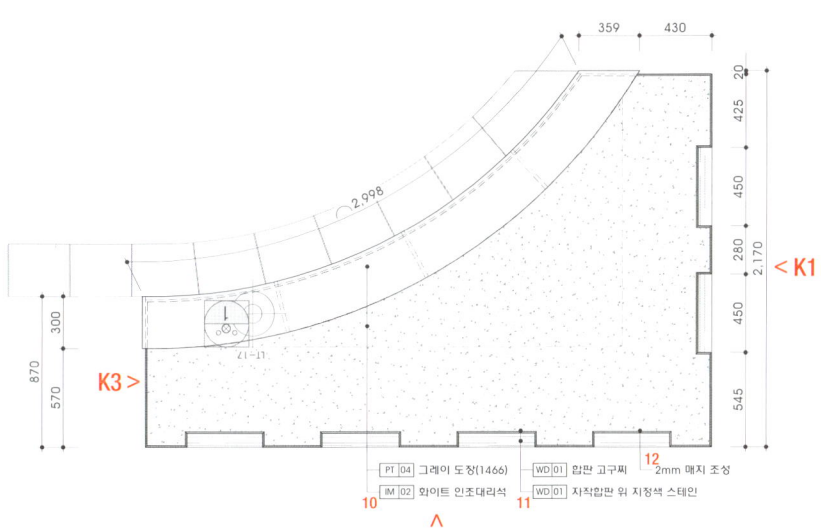

테이블 평면 K / table top view K

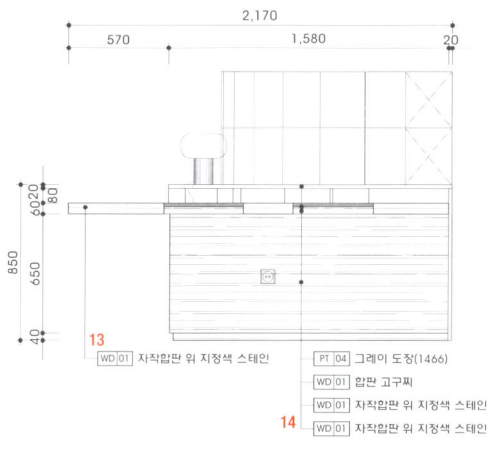

테이블 측면 K1 / table side view K1

테이블 정면 K2 / table front view K2

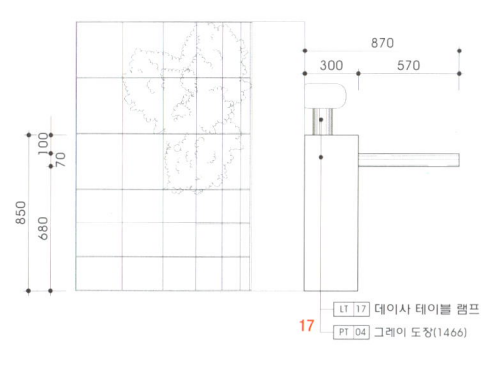

테이블 측면 K3 / table side view K3

1 아이보리 브릭 타일 / 자동문 2 10T 평철 위 연회색 도장 / T1.6 갈바륨 위 연회색 도장 / SUS 미러 3 콘크리트 타일 4 회색 질감 도장 / 흰색 도장 5 LED 간접조명 6 커튼박스 7 콘크리트 위 투명 코팅 8 건축 창호 9 T1.6 SUS 미러 / 1구 콘센트 10 SUS 미러 11 피봇 힌지 / SUS 미러 12 T1.6 SUS 미러

1 Ivory brick tile / Automatic door 2 Light gray painting on 10T flat steel / Light gray painting on T1.6 galvalume / SUS mirror 3 Concrete tile 4 Gray textured painting / White painting 5 LED indirect lighting 6 Curtain box 7 Clear coating on concrete 8 Architecture windows 9 T1.6 SUS mirror / Single outlet 10 SUS mirror 11 Pivot hinge / SUS mirror 12 T1.6 SUS mirror

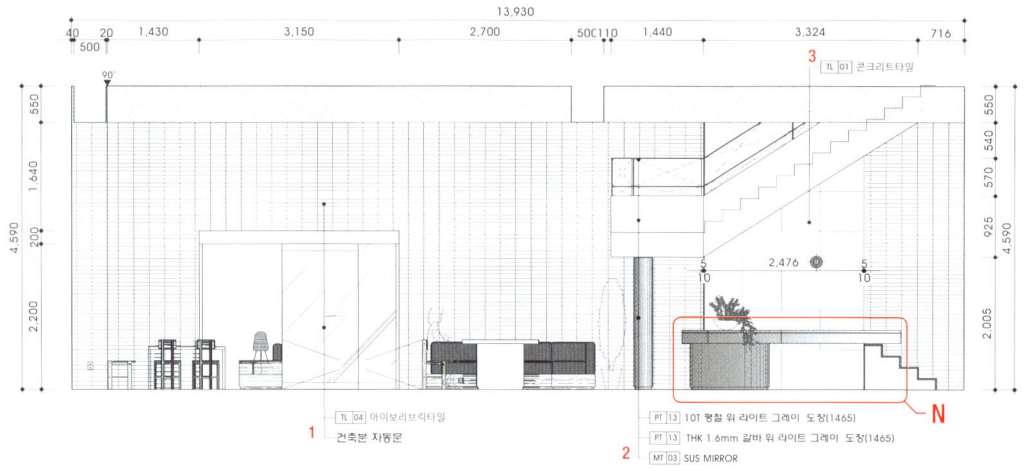

1층 홀 입면 L / 1F hall elevation L

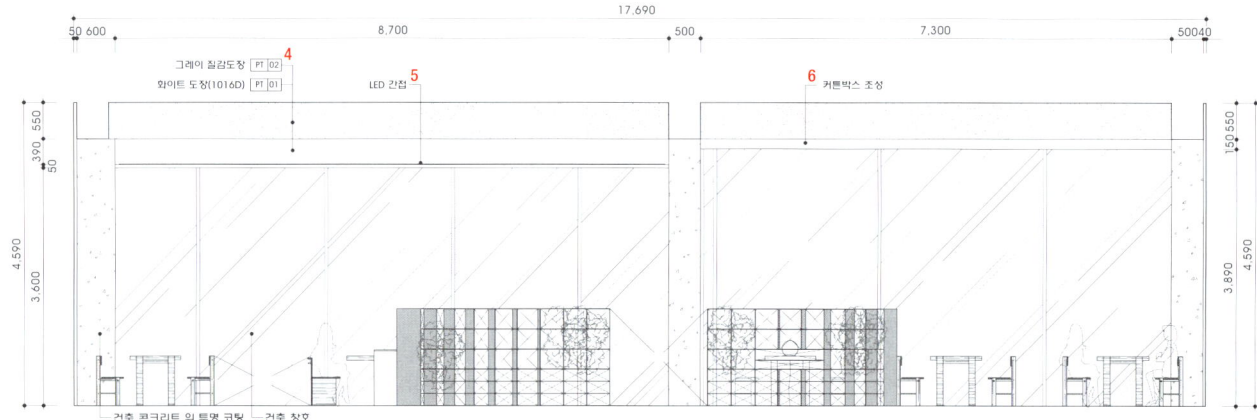

1층 홀 입면 M / 1F hall elevation M

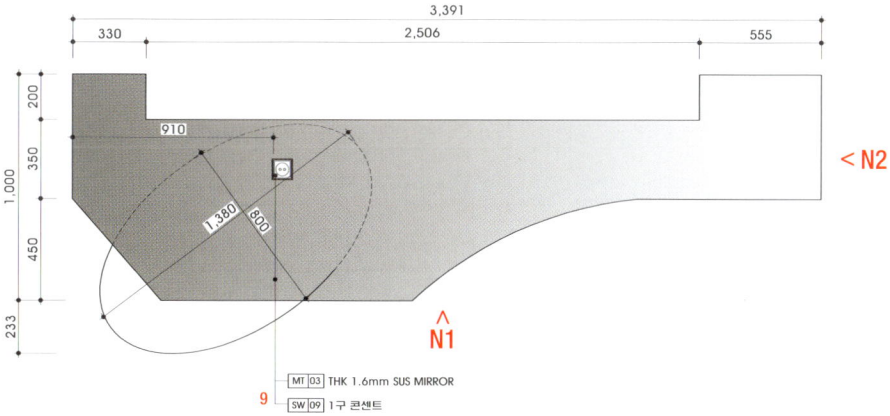

테이블 평면 N / table top view N

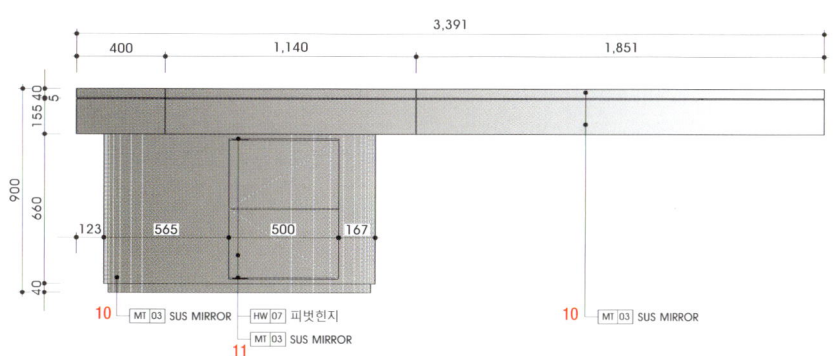

테이블 정면 N1 / table front view N1

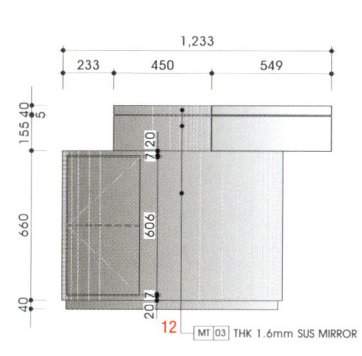

테이블 측면 N2 / table side view N2

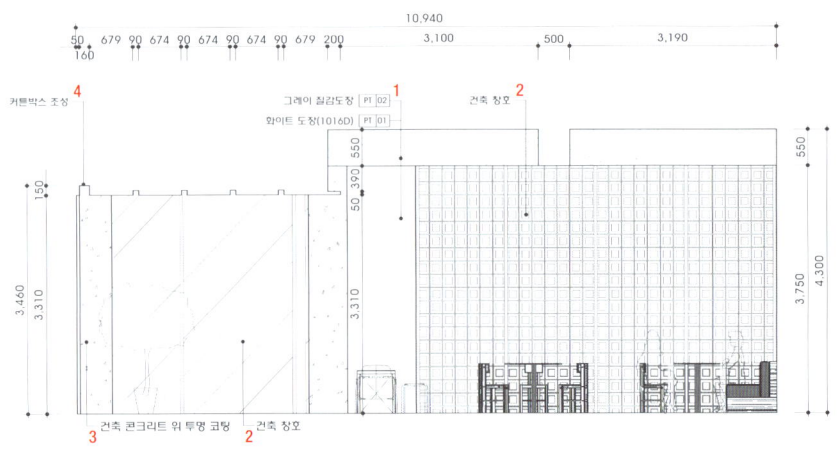

2층 홀 입면 O / 2F hall elevation O

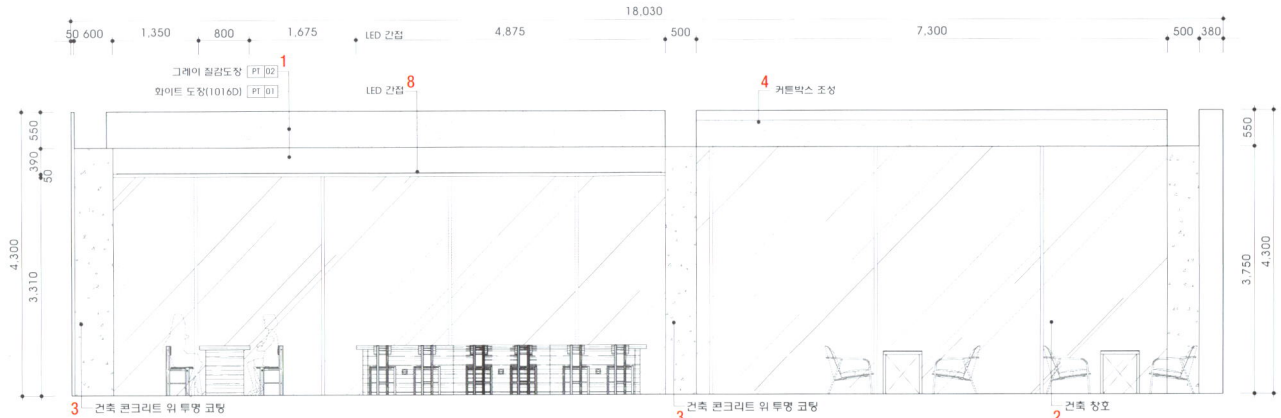

2층 홀 입면 P / 2F hall elevation P

1 회색 질감 도장 / 흰색 도장   2 건축 창호   3 콘크리트 위 투명 코팅   4 커튼박스   5 10T 평철 위 라이트 블루 무광 도장   6 아이보리 브릭 타일   7 아이보리 브릭 타일 / 10T 평철 위 연회색 도장   8 LED 간접조명

1 Gray textured painting / White painting   2 Architecture windows   3 Clear coating on concrete   4 Curtain box   5 Light blue matt painting on 10T flat steel   6 Ivory brick tile   7 Ivory brick tile / Light gray painting 10T flat steel   8 LED indirect lighting

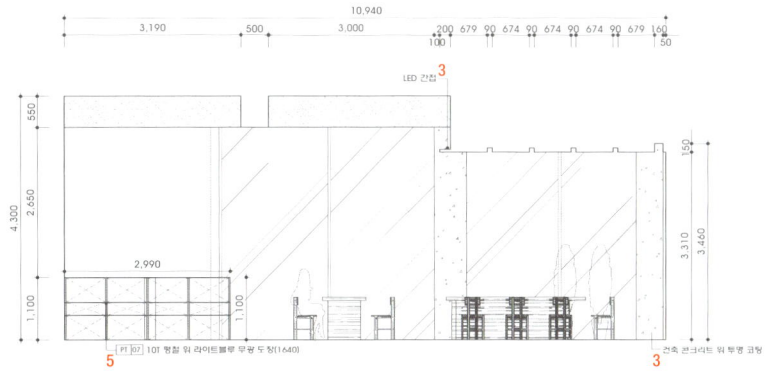

2층 홀 입면 Q / 2F hall elevation Q

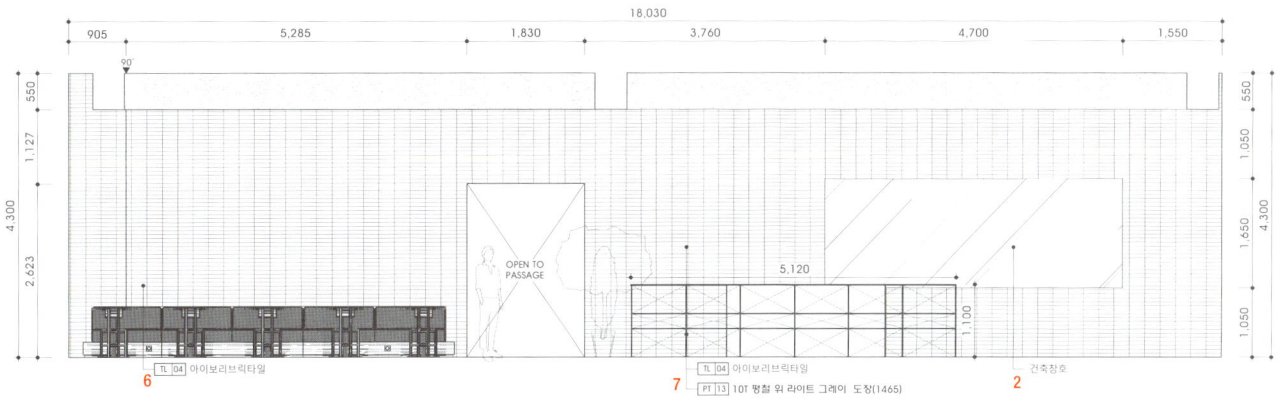

2층 홀 입면 R / 2F hall elevation R

1 회색 질감 도장 / 콘크리트 위 투명 코팅 2 회색 질감 도장 / 흰색 도장 3 회색 질감 도장 4 2구 콘센트 5 콘크리트 위 투명 코팅 / 핸드레일 6 블랙 리놀륨 7 T5 열연강판 8 합판 위 블랙 리놀륨 9 합판 위 블랙 리놀륨 / 열연강판 10 열연강판

1 Gray textured painting / Clear coating on concrete 2 Gray textured painting / White painting 3 Gray textured painting 4 Dual outlet 5 Clear coating on concrete / Handrail 6 Black linoleum 7 T5 hot rolled steel sheet 8 Black linoleum on plywood 9 Black linoleum on plywood / Hot rolled steel plate 10 Hot rolled steel plate

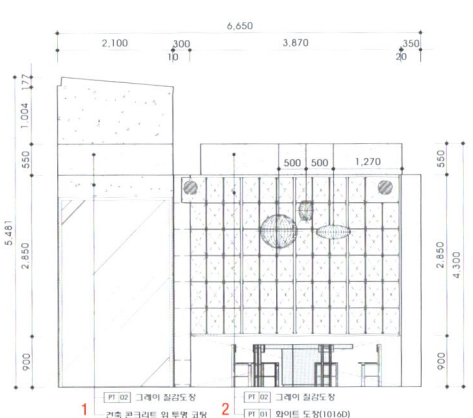

2층 단체석 입면 S1 / 2F group room elevation S1

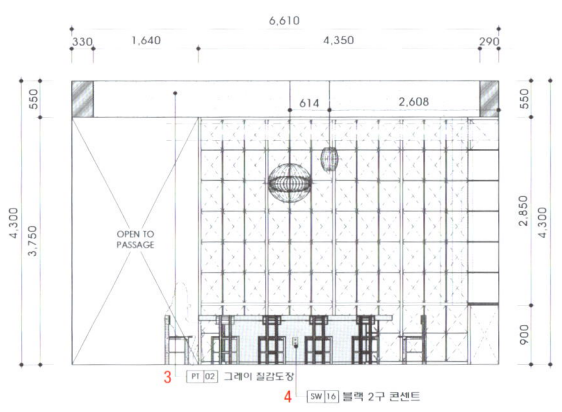

2층 단체석 입면 S2 / 2F group room elevation S2

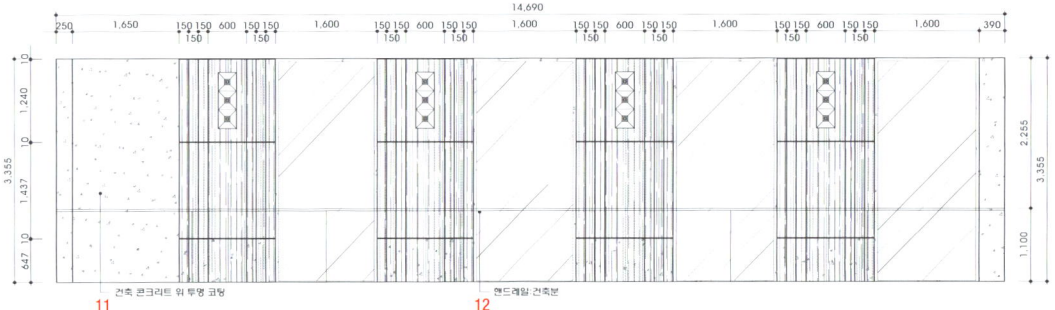

2층 홀 입면 T / 2F hall elevation T

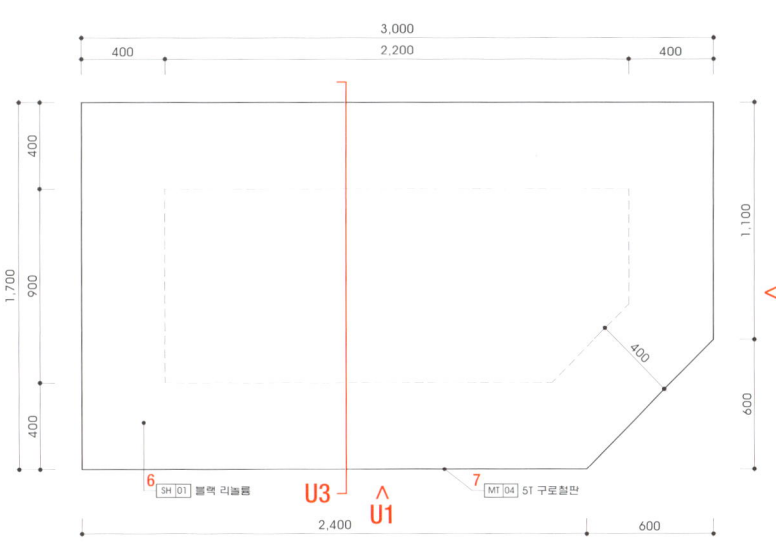

테이블 평면 U / table top view U

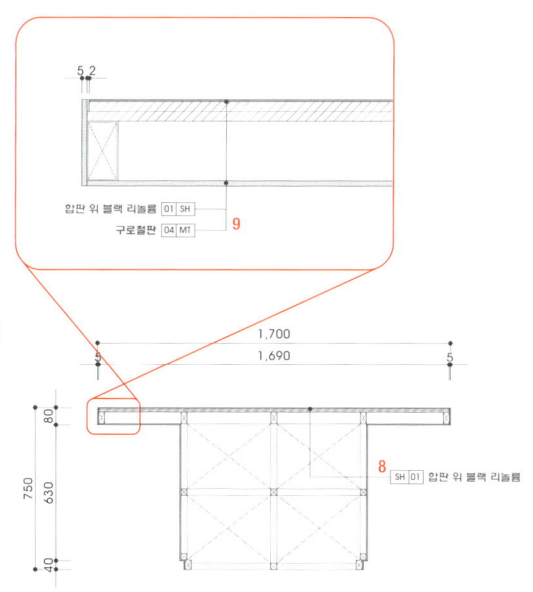

테이블 단면 U3 / table section U3

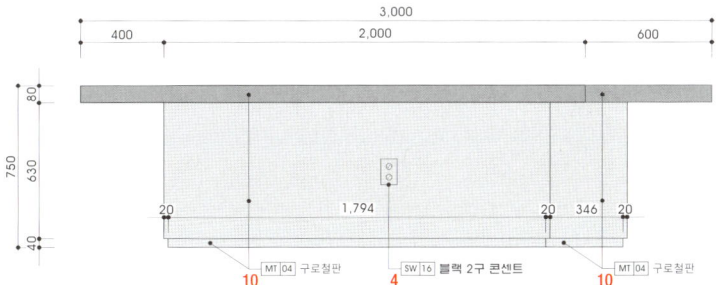

테이블 정면 U1 / table front view U1

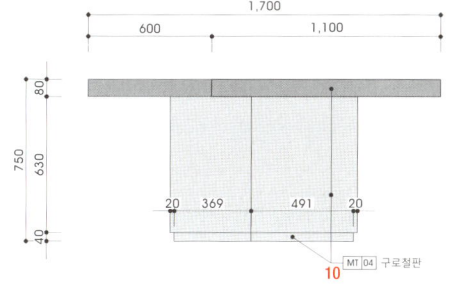

테이블 측면 U2 / table side view U2

HORONG

Design studio maoom | Minkyu Choi

호롱 베이커리는 촛불처럼 은은하게 빛나는 호롱에서 시작된 온화한 이야기를 품고 있다. 호롱 베이커리는 바쁜 도시 속에서 잠시 쉴 수 있는 따뜻한 공간을 선물하고자 한다. 고덕 신도시 개발 구역의 한편, 삭막한 도시 풍경 사이에 자리한 이 베이커리는 단조로운 일상 속에서 마음을 녹일 수 있는 쉼의 공간을 꿈꾼다. 뉴트럴한 베이지 톤의 색감과 넓은 창을 통해 들어오는 자연광이 조화를 이루며 더욱 아늑하고 평온한 분위기를 자아낸다. 이러한 설계는 방문객들이 공간의 온화함 속에서 편안함을 느끼고, 마치 집과 같은 안정감을 경험할 수 있도록 의도되었다. 특히 커다란 기둥들이 웅장하게 자리하고 있지만, 그 자체가 주는 안정감 덕분에 사람들은 한층 더 편안하게 머물 수 있는 효과를 준다. 기둥과 천장에는 큰 원형 디테일이 반복적으로 사용되어 공간에 웅장함과 동시에 부드러운 분위기를 더해준다. 빛을 이용해 원형 구조물들에 입체감을 더함으로써 리듬감 있게 표현된 이 디테일은 시각적인 즐거움을 제공하며, 공간에 활력을 불어넣는다. 바닥을 따라 배열된 테이블과 의자들은 따스한 나무 소재로 이루어져 있으며, 각 자리들은 독립적인 섬처럼 배치되어 있어 개인적인 휴식을 가능하게 한다. 창가에 자리한 의자들은 햇살을 온전히 즐길 수 있도록 배치되어 있으며, 자연광과 어우러진 아늑한 분위기가 공간에 온기를 더한다. 호롱 베이커리에서 밝혀진 작은 불빛들이 하나둘 모여 사람들의 일상 속에 더 큰 온기와 아늑함으로 스며들기를 소망한다.

Horong Bakery embodies a heartwarming story reminiscent of a lantern glowing softly like a candle, offering a warm sanctuary within the bustling city. Nestled in a corner of the Godeok New Town development area, surrounded by stark urban landscapes, the bakery becomes a refuge for people seeking a warm respite from their monotonous life. The neutral, beige-toned color scheme harmonizes with natural light streaming through wide windows, creating a cozy and serene atmosphere. This cozy atmosphere allows visitors to feel at home. Massive pillars are seemingly imposing, but they provide a sense of stability that helps visitors feel more relaxed. Large circular details repeatedly appearing on the pillars and ceiling imbue the space with a sense of grandeur and elegance. By using light to create depth in a circular structure, these details introduce a sense of rhythm that gives pleasure to the eye while adding energy to the space. Tables and chairs arranged in line with the flooring are made of warm wood materials. The tables are positioned like an independent island for visitors who want to enjoy their rest privately. Window seats are strategically placed to fully embrace sunlight. The intimate feeling created by natural light adds warmth. Small lanterns kindled at Horong Bakery will gradually increase in number and eventually fill people's daily lives with warmth and comfort.

디자인 최민규 / 디자인스튜디오 마움
위치 경기도 평택시 함박산2길 47
용도 베이커리 카페
면적 236m²
마감 바닥 - 타일 / 벽 - 스페셜 페인트, 대리석, 페인트 / 천장 - 페인트
완공 2023. 12
디자인팀 김연종, 이정환, 이상윤, 하종기
사진 진성기 / SOULGRAPH

Location 47, Hambaksan-2gil, Pyeongtaek-si, Gyeonggi-do
Use Bakery cafe
Area 236m²
Finishing Floor - Tile / Wall - Special paint, Marble, Paint / Ceiling - Paint
Completion 2023. 12
Photographer Sungkee Jin / SOULGRAPH

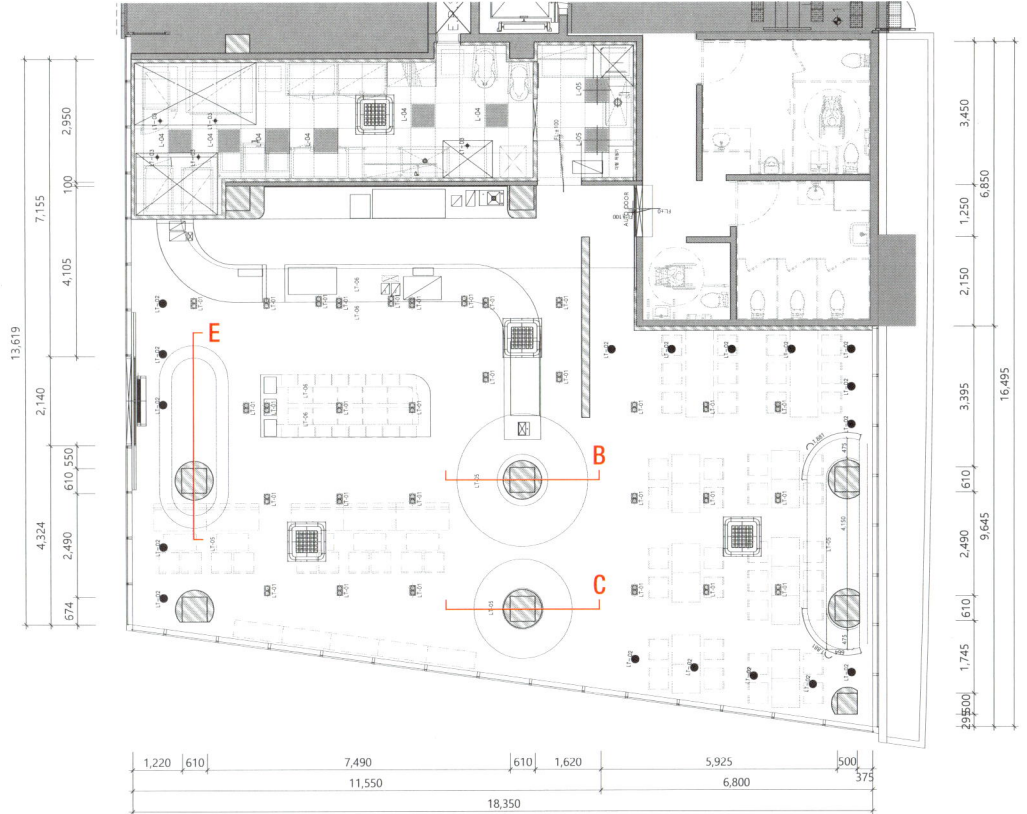

천장도 / ceiling plan

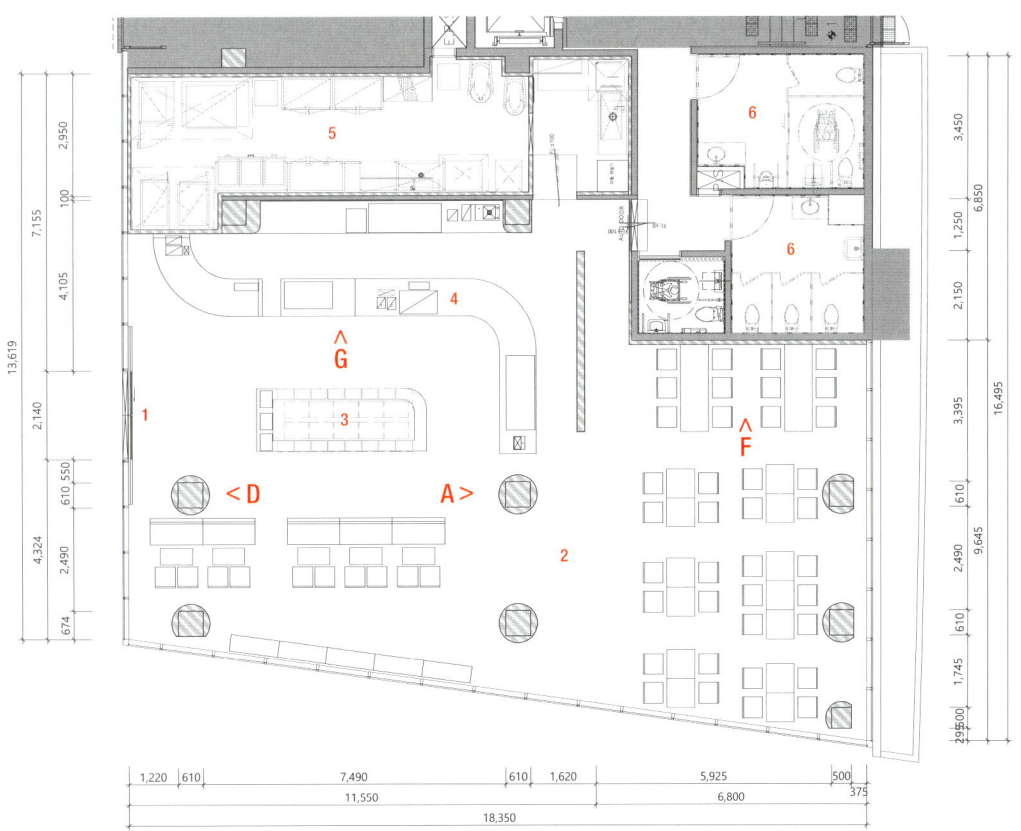

평면도 / floor plan

1 입구 2 홀 3 베이커리 존 4 카운터 5 주방 6 화장실

1 Entrance 2 Hall 3 Bakery zone 4 Counter 5 Kitchen 6 Restroom

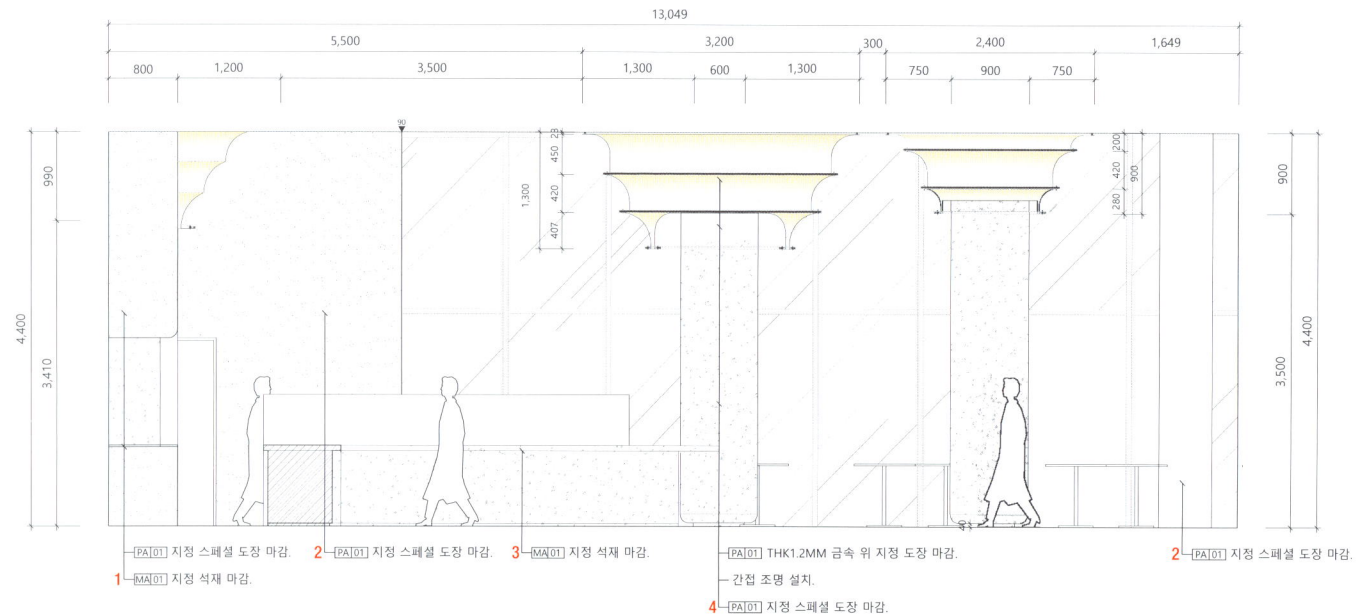

1 지정 스페셜 페인트 / 지정 석재 2 지정 스페셜 페인트 3 지정 석재 4 T1.2 금속 위 지정 도장 / 간접조명 / 지정 스페셜 페인트 5 T1.2 갈바륨 위 지정 스페셜 페인트 / 간접조명 : LED 바 줄조명 6 T1.2 갈바륨 위 지정 스페셜 페인트 / □50X50X1.6T 철제 파이프

1 App. special paint / App. stone 2 App. special paint 3 App. stone 4 App. painting on T1.2 metal / App. special paint 5 App. special paint on T1.2 galvalume / Indirect lighting LED bar linear lighting 6 App. special paint on T.12 galvalume / □50X50X1.6T steel pipe

홀 입면 A / hall elevation A

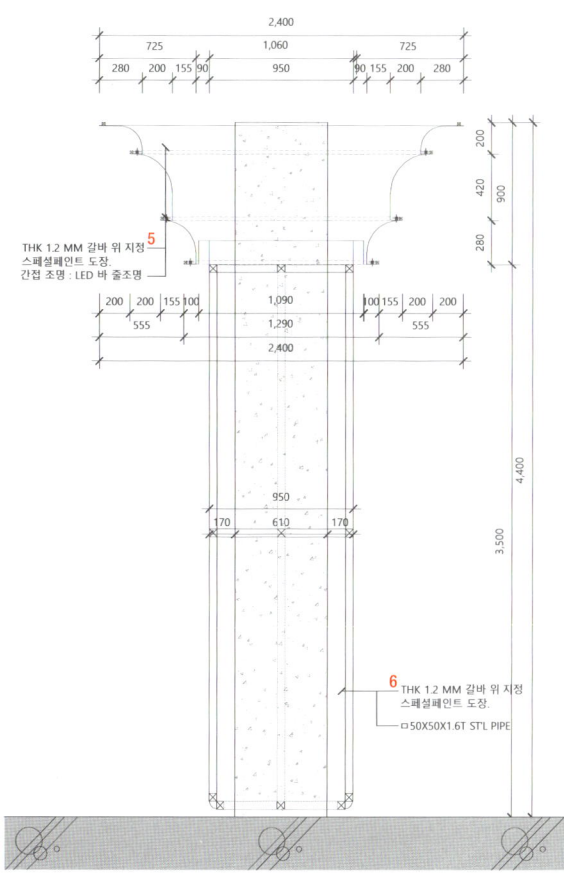

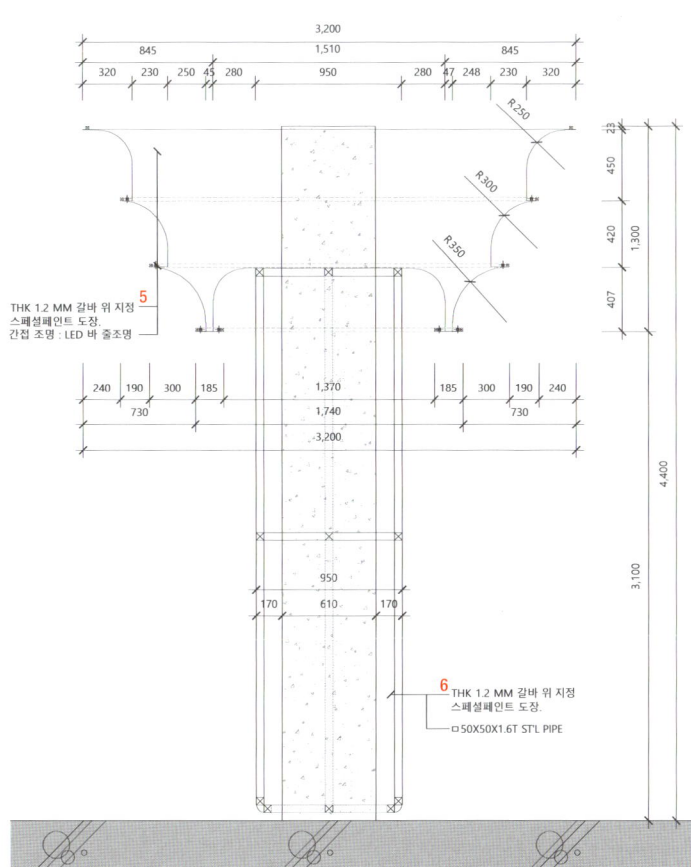

기둥 단면 B / column section B

기둥 단면 C / column section C

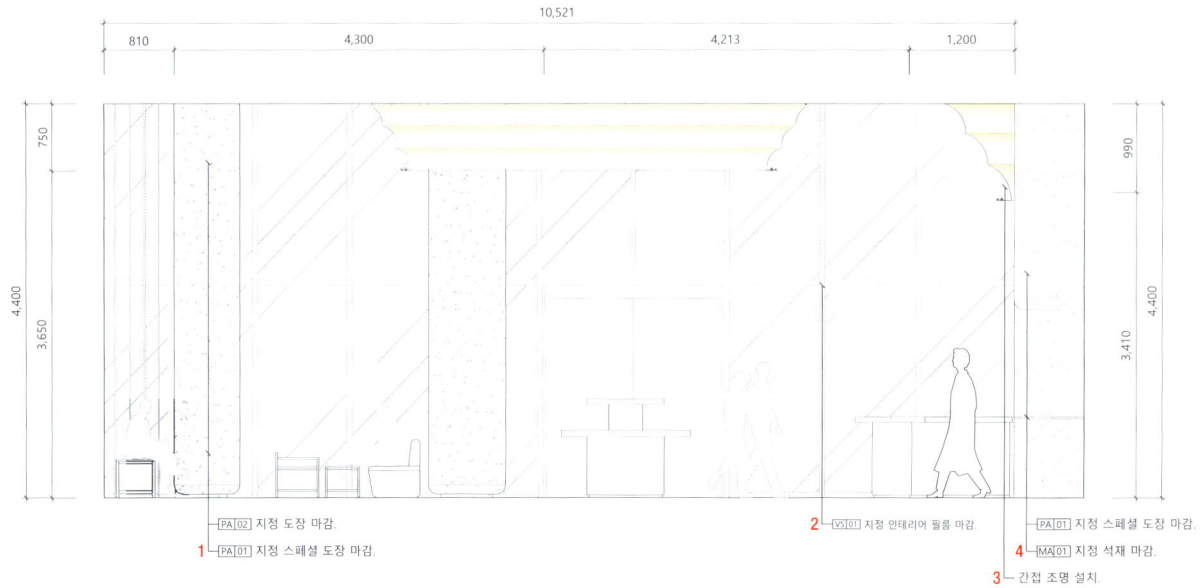

1 지정 도장 / 지정 스페셜 페인트 2 지정 인테리어 필름 3 간접조명 4 지정 스페셜 페인트 / 지정 석재 5 T1.2 갈바륨 위 지정 스페셜 페인트 / 간접조명 : LED 바 줄조명 6 T1.2 갈바륨 위 지정 스페셜 페인트 / □50X50X1.6T 철제 파이프 7 □50X50X1.6T 철제 파이프 / T1.2 갈바륨 위 지정 스페셜 페인트 8 기존 기둥 위 지정 스페셜 페인트

1 App. painting / App. special paint 2 App. interior film 3 Indirect lighting 4 App. special paint / App. stone 5 App. special paint on T1.2 galvalume / Indirect lighting : LED bar linear lighting 6 App. special paint on T.12 galvalume / □50X50X1.6T steel pipe 7 □50X50X1.6T steel pipe / App. special paint on T.12 galvalume 8 App. special paint on existing column

홀 입면 D / hall elevation D

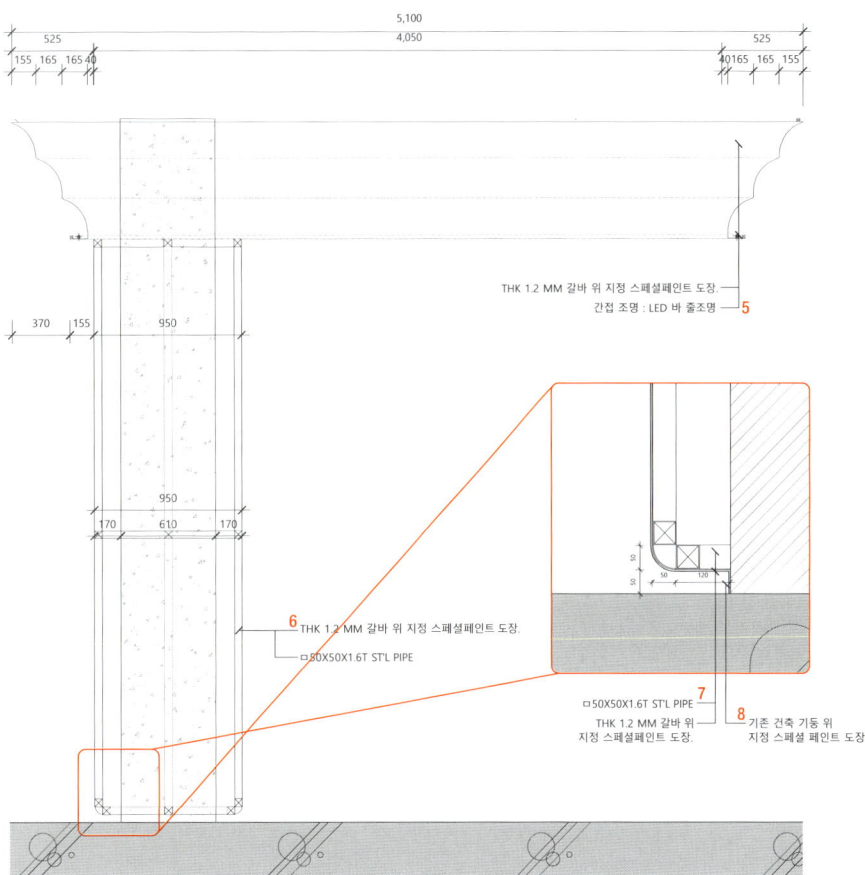

기둥 단면 E / column section E

1 지정 도장 2 문 : 지정 도장 3 간접조명 4 지정 도장 / 지정 스페셜 페인트 / 지정 석재 5 지정 파티클 보드 6 □30X30X1.6T 철제 파이프 / 지정 석재

1 App. painting 2 Door : App. painting 3 Indirect lighting 4 App. painting / App. special paint / App. stone 5 App. particle board 6 □30X30X1.6T steel pipe / App. stone

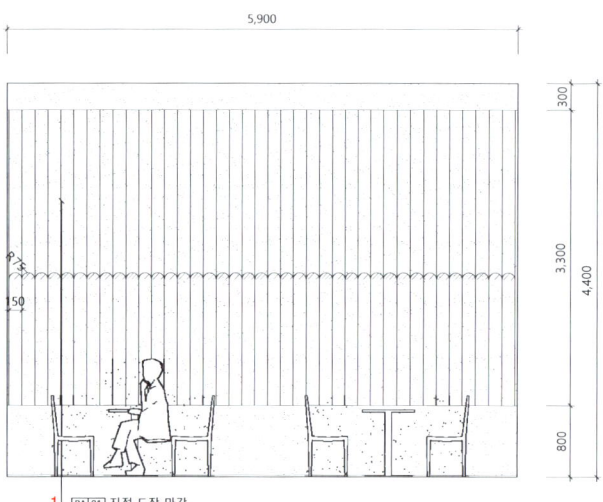

홀 입면 F / hall elevation F

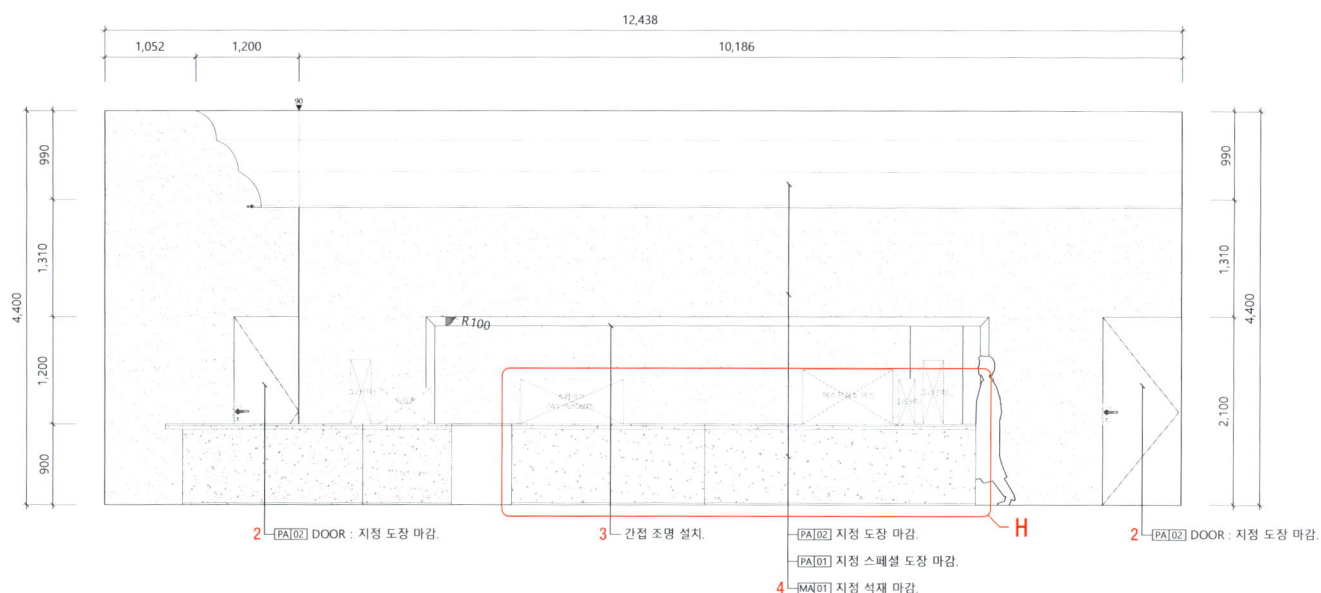

홀 입면 G / hall elevation G

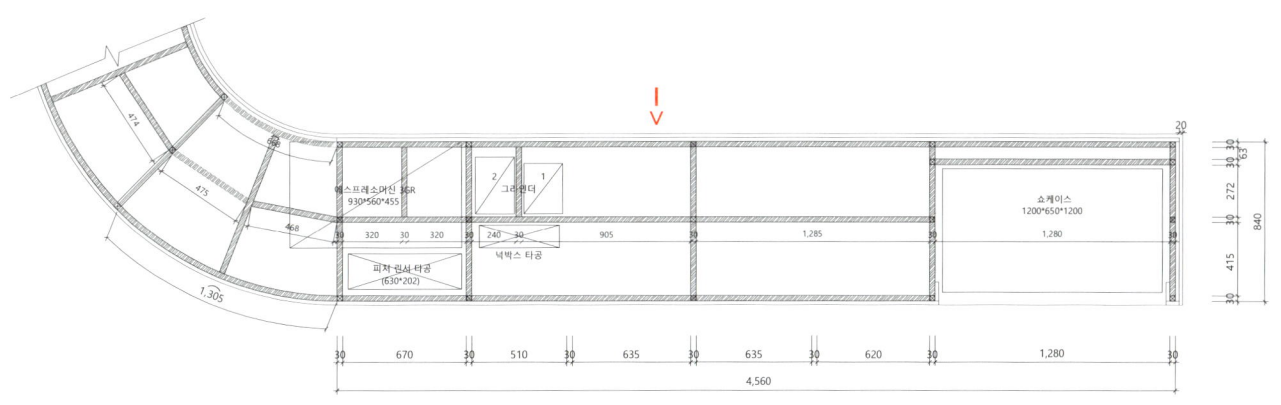

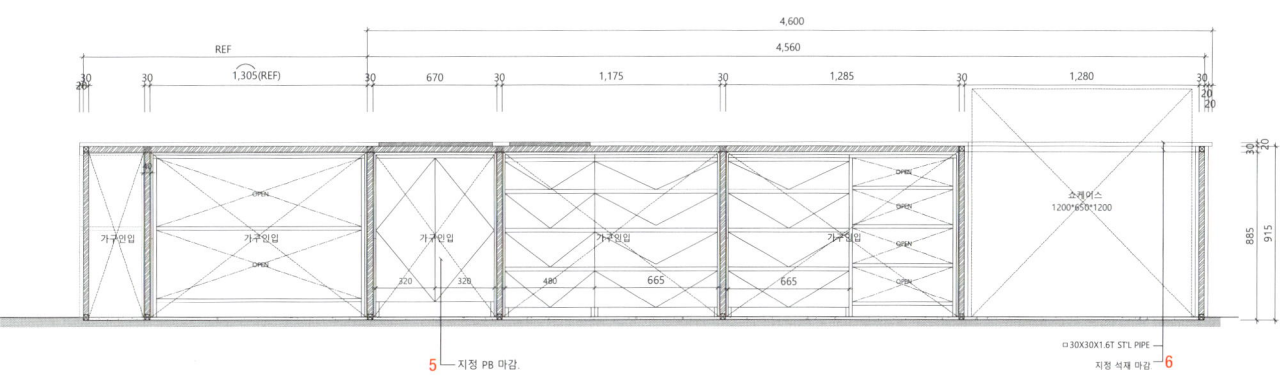

커피 바 평면 H / coffee bar top view H

커피 바 후면 I / coffee bar rear view I

FILLMATE

LABOTORY | Jinho Jung

필메이트는 고객의 일상에 특별한 시작을 제공하는 브랜드로, 친구와 같은 존재를 지향하며 강남에 첫 플래그십 스토어를 선보였다. 필메이트는 'Dive into demitasse'라는 콘셉트 아래, 에스프레소의 깊은 맛을 탐험하며 다양한 향기를 즐길 수 있는 고유한 경험을 제공한다. 이 플래그십 스토어는 단순한 공간 프로그래밍을 넘어서 에스프레소의 세 가지 구성 요소인 에스프레소, 크레마, 아로마를 공간으로 재해석했다. 첫 번째 층은 에스프레소 바로, 빠른 서비스와 강한 인상을 주는 곳이다. 두 번째 층인 크레마 층은 베이커리와 함께 따뜻하고 친근한 대화가 가능한 공간으로 꾸며졌으며, 마지막 층은 아로마 층으로, 방문객이 편안하게 시간을 보낼 수 있는 환대의 공간으로 디자인되었다. 디자이너는 프랜차이즈 커피 시장에서의 지속 가능성을 심도있게 고려하여 이 프로젝트를 진행했다. 각 층마다 독특한 캐릭터를 부여함으로써 유연한 전개가 가능하게 하였고, 각 지역의 특성과 타겟 고객의 필요에 맞추어 각 매장이 개별적인 캐릭터의 비중을 조절할 수 있도록 설계했다. 이를 통해 각 매장은 다른 모습을 보여주면서도 필메이트만의 일관된 세계관을 유지할 수 있다. 이 새로운 플래그십 스토어를 통해 커피 애호가들에게 단순한 음료를 넘어서는 깊이 있는 문화 경험을 선사하여 커피 한 잔이 주는 작은 기쁨이 일상의 소중한 순간으로 변모되길 기대한다.

Fillmate is a brand that offers a special start to customers' daily lives, aiming to be like a friend. They have opened their first flagship store in Gangnam. Under the concept of 'Dive into demitasse', Fillmate offers a unique experience where you can explore the deep flavors of espresso and enjoy the different aromas. The flagship store goes beyond simple spatial programming, reinterpreting the three components of espresso - espresso, crema, and aroma - into the space. The first floor is the espresso bar, swhere fast service and strong impressions are made. The second level, the crema level, is designed as a space for warm, friendly conversation with a bakery, and the final level, the aroma level, is designed as a hospitality space where visitors can relax and spend time. By giving each floor a unique character, designers have allowed for flexibility and designed each store to be able to adjust the proportion of its individual character to suit the specifics of each location and the needs of its target audience. This allows each store to have a different look, while still maintaining a consistent worldview of Fillmate. With these new flagship stores, we're bringing coffee lovers a deeper cultural experience that goes beyond just a beverage. We look forward to transforming the small joys of a cup of coffee into the cherished moments of everyday life.

디자인 정진호 / 라보토리
위치 서울특별시 강남구 테헤란로1길 28
용도 카페
면적 356㎡
마감 바닥 - 마이크로시멘트, 우드플로링 / 벽 - 스페셜 페인트, 화이트 오크 무늬목, 대리석 / 천장 - 페인트
완공 2023. 12
디자인팀 유슬기, 이서정, 김유화, 권명주
사진 최용준

Location 28, Teheran-ro 1-gil, Gangnam-gu, Seoul
Use Cafe
Area 356㎡
Finishing Floor - Microcement, Wood flooring / Wall - Special paint, White oak wood veneer, Marble / Ceiling - Paint
Completion 2023. 12
Design team Seulgi Yoo, Seojeong Lee, Youhwa Kim, Myungjoo Kwon
Photographer Yongjoon Choi

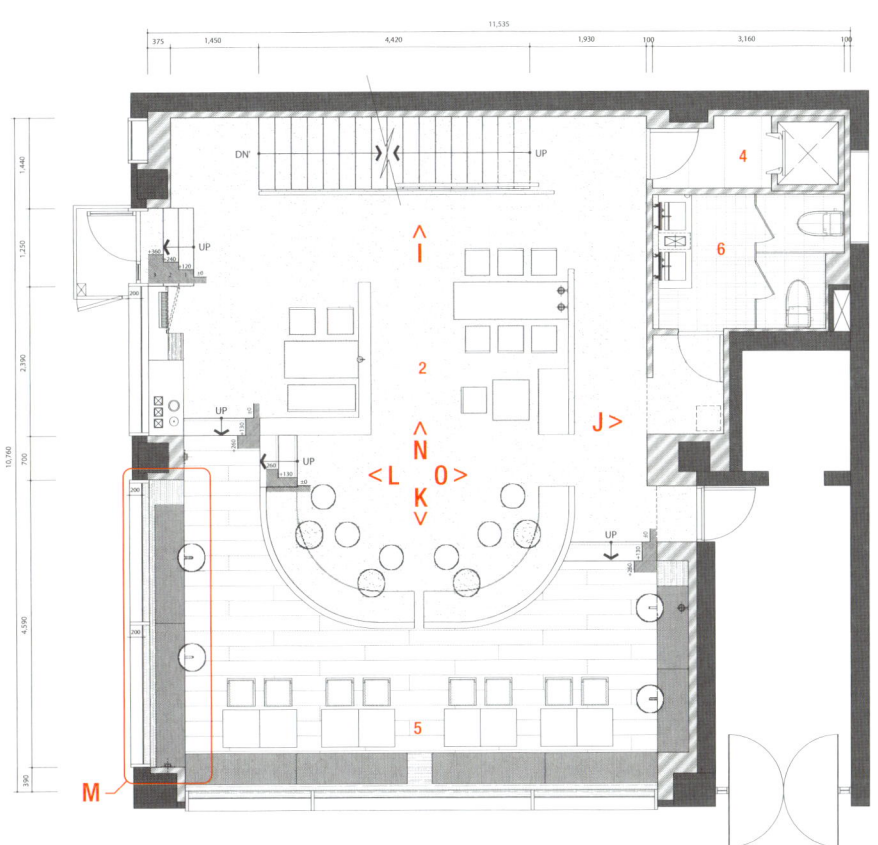

2층 평면도 / 2nd floor plan

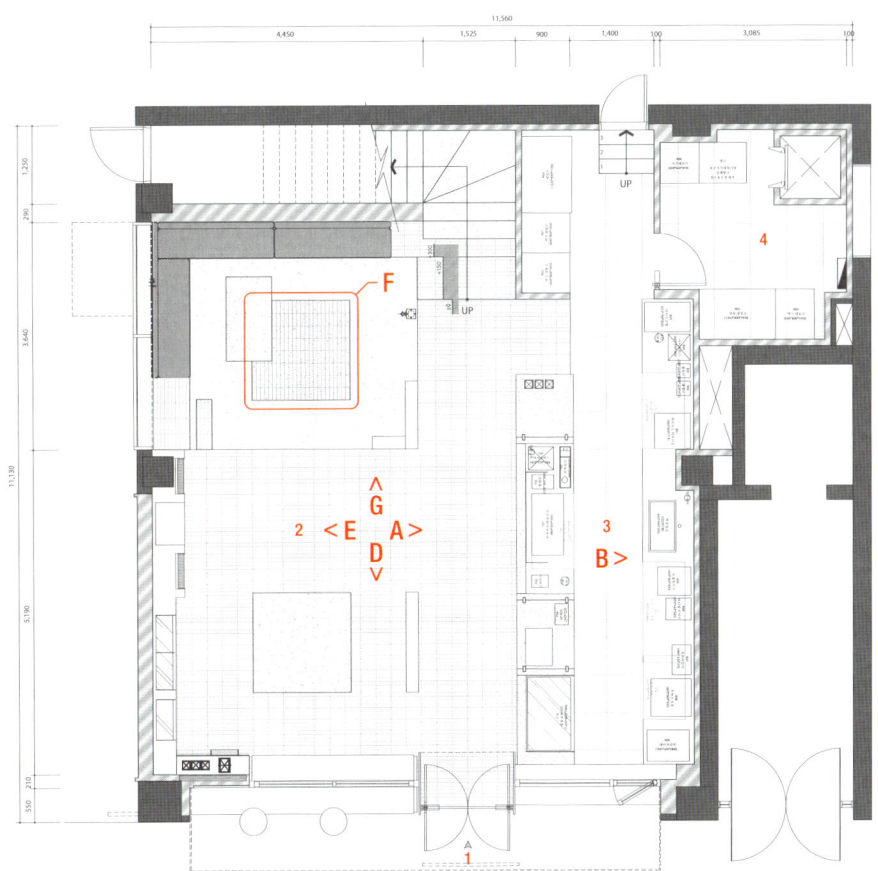

1층 평면도 / 1st floor plan

1 입구 2 홀 3 에스프레소 바 4 백룸 5 데스크 스테이션 6 화장실

1 Entrance 2 Hall 3 Espresso bar 4 Back room 5 Desk station 6 Restroom

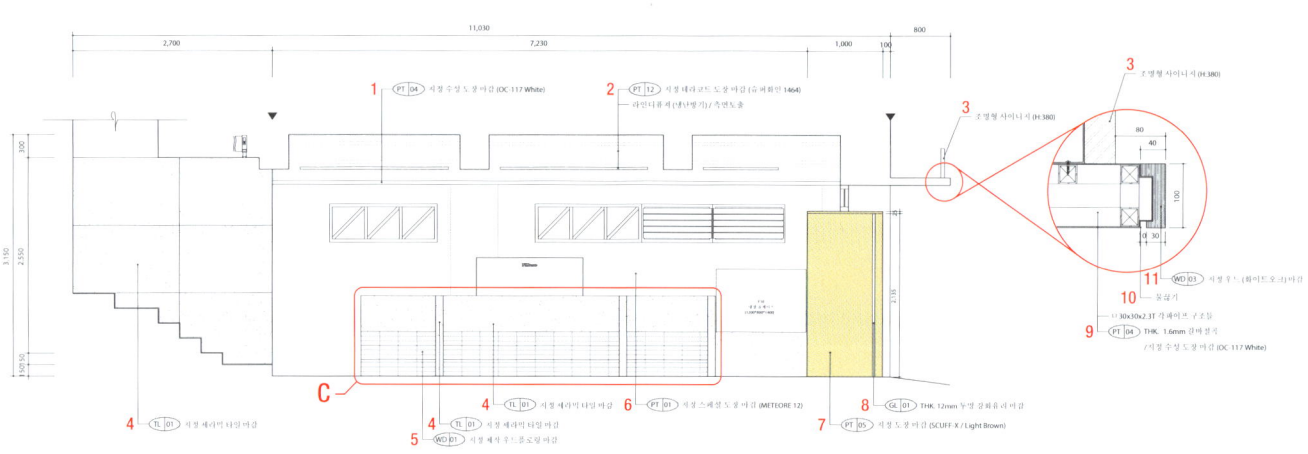

1층 홀 입면 A / 1F hall elevation A

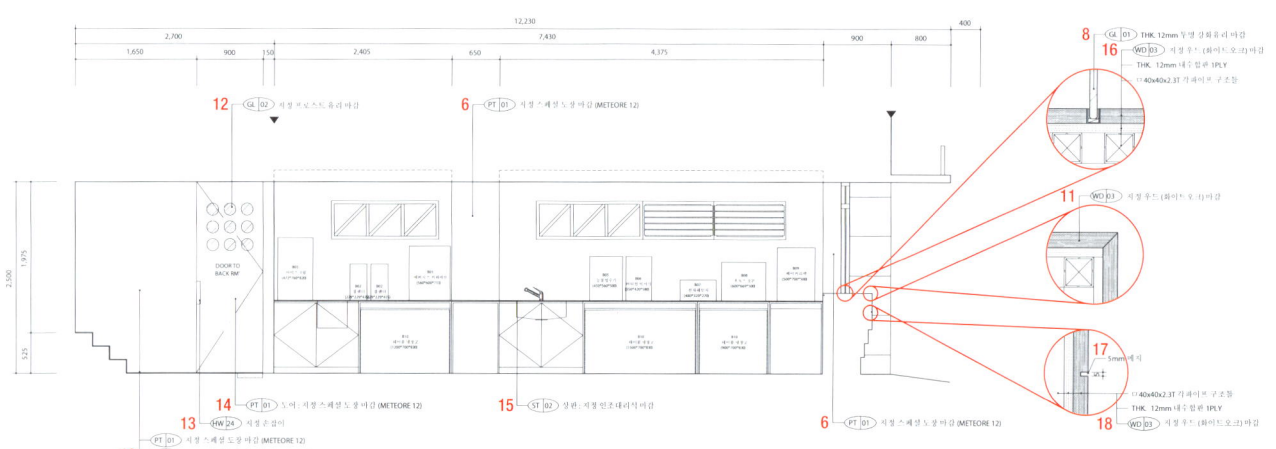

1층 홀 입면 B / 1F hall elevation B

1 지정 수성도장 2 지정 테라코트 도장 / 라인 디퓨저 3 조명형 사이니지 4 지정 세라믹 타일 5 지정 제작 우드플로링 6 지정 스페셜 페인트 7 지정 도장 8 T12 투명 강화유리 9 □30X30X2.3T 각파이프 구조틀 / T1.6 갈바륨 절곡 위 지정 수성도장 10 물끊기 11 지정 화이트 오크 12 지정 스페셜 페인트 / 10mm 걸레받이 : 지정 수성도장 13 지정 손잡이 14 문 : 지정 스페셜 페인트 15 상판 : 지정 인조대리석 16 지정 화이트 오크 / T12 내수합판 □40X40X2.3T 각파이프 구조틀 17 5mm 줄눈 18 □40X40X2.3T 각파이프 구조틀 / T12 내수합판 / 지정 화이트 오크 19 지정 세라믹 타일 / 냅킨통 : T1.6 SUS 바이브레이션 20 지정 인조대리석 / T5 SUS 바이브레이션 21 지정 원목 22 3mm 줄눈 : 지정 도장 23 T5 SUS 바이브레이션 24 비조명형 사이니지 25 지정 세라믹 타일 / 지정 제작 우드플로링

1 App. water paint 2 App. terracoat painting / Line diffuser 3 Illuminated signage 4 App. ceramic tile 5 Custom-made wood flooring 6 App. special paint 7 App. painting 8 T12 clear tempered glass 9 □30X30X2.3T square pipe structure / App. water paint on T1.6 galvalume bending 10 Drip edge 11 App. white oak 12 App. special paint / 10mm baseboard : App. water paint 13 App. handle 14 Door : App. special paint 15 Top : App. special paint 16 App. white oak / T12 concrete panel plywood □40X40X2.3T square pipe structure 17 5mm joint reveal 18 □40X40X2.3T square pipe structure / T12 concrete panel plywood / App. white oak 19 App. ceramic tile / Napkin box : T1.6 SUS vibration 20 App. engineered marble / T5 SUS vibration 21 App. solid wood 22 3mm joint reveal : App. paint 23 T5 SUS vibration 24 Non-illuminated sign 25 App. ceramic tile / Custom-made wood flooring

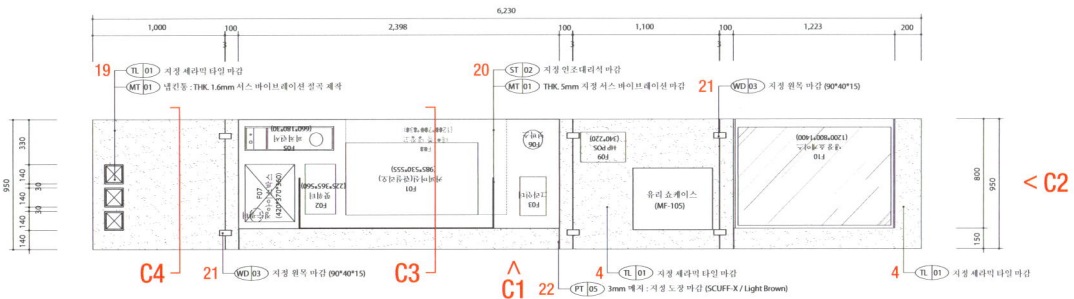

카운터 평면 C / counter top view C

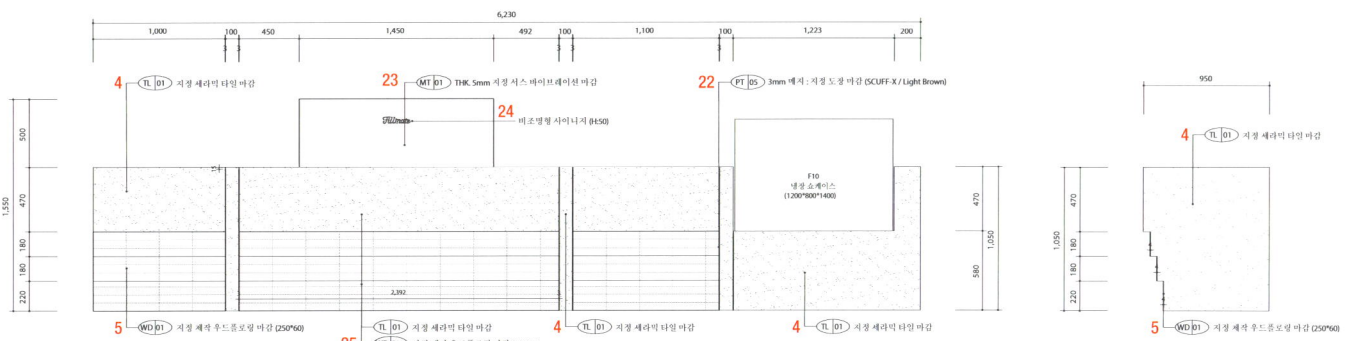

카운터 정면 C1 / counter front view C1

카운터 측면 C2 / counter side view C2

1 T5 SUS 바이브레이션   2 지정 세라믹 타일 / T12 합판 / □30X30X2.3T 각파이프 구조틀   3 지정 인조대리석 / 지정 우레탄 도장 / 상판 보강   4 □30X30X2.3T 각파이프 구조틀 / T12 합판 / 지정 세라믹 타일   5 □50X50X2.3T 각파이프 구조틀 / T12 합판 / 제작 우드플로링   6 지정 세라믹 타일 / □30X30X2.3T 각파이프 구조틀 / T12 합판   7 지정 도장   8 지정 세라믹 타일   9 □30X30X2.3T 각파이프 구조틀 / T12 합판 / 제작 우드플로링   10 LED 바 간접조명   11 라인디뮤저   12 지정 화이트 오크 / 지정 도장   13 지정 화이트 오크 / T12 투명 강화유리   14 지정 테라코트 도장 / 지정 무늬목   15 지정 하드웨어   16 T12 투명 강화유리 / 지정 화이트오크 / 지정 스페셜 페인트   17 제작 손잡이 / T12 투명 강화유리   18 지정 화이트 오크 / 지정 화이트 오크   19 지정 무늬목 / 지정 무늬목   20 지정 리프트업 창호   21 지정 테라코트 도장 / 지정 무늬목   22 제작 스피커   23 지정 우드 인테리어 필름   24 선반 : 지정 투명 백유리 / 지정 무늬목 / 지정 무늬목 / 지정 무늬목   25 비조명형 사이니지   26 지정 프로스트 백유리 / 지정 거친 스페셜 페인트 / 제작 우드플로링   27 지정 패브릭

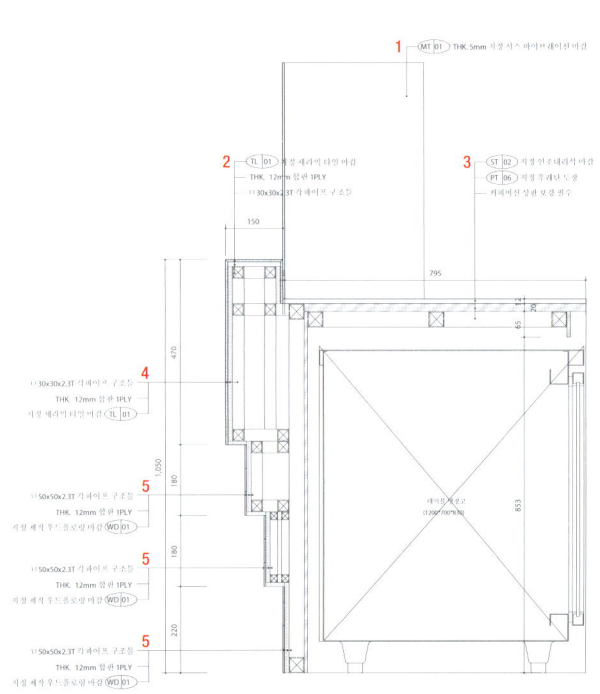

카운터 단면 C3 / counter section C3

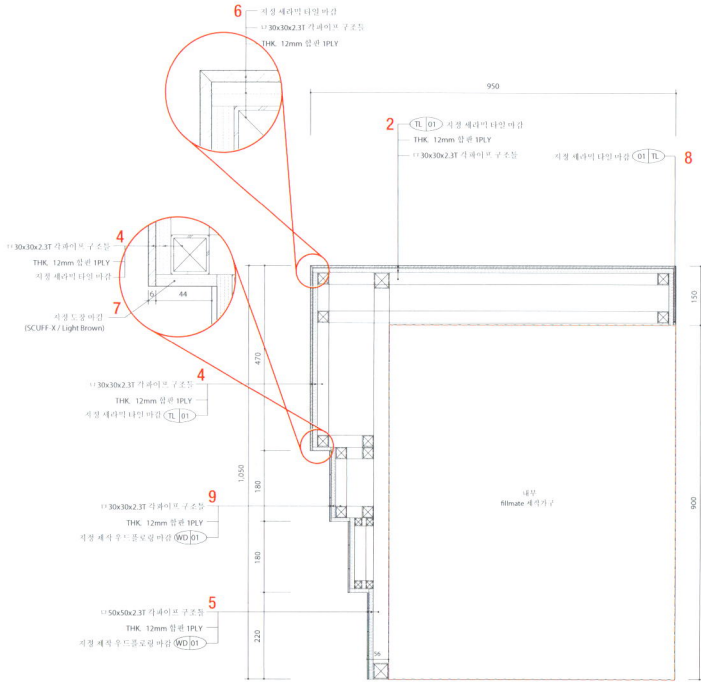

카운터 단면 C4 / counter section C4

1 T5 SUS vibration 2 App. ceramic tile / T12 plywood / □30X30X2.3T square pipe structure 3 App. engineered marble / App. urethane painting / Table top reinforcement 4 □30X30X2.3T square pipe structure / T12 plywood / App. ceramic tile 5 □50X50X2.3T square pipe structure / T12 plywood / Custom-made wood flooring 6 App. ceramic tile / □30X30X2.3T square pipe structure / T12 plywood 7 App. painting 8 App. ceramic tile 9 □30X30X2.3T square pipe structure / T12 plywood / Custom-made wood flooring 10 LED bar indirect lighting 11 Line diffuser 12 App. white oak / App. painting 13 App. white oak / T12 clear tempered glass 14 App. terracoat painting / App. wood veneer 15 App. hardware 16 T12 clear tempered glass / App. white oak / App. special paint 17 Custom-made handle / T12 clear tempered glass 18 App. white oak / App. white oak 19 App. wood veneer / App. wood veneer 20 App. lift-up windows 21 App. terracoat painting / App. wood veneer 22 Custom-made speaker 23 App. wood interior film 24 Shelf : App. clear low-iron glass / App. wood veneer / App. wood veneer / App. wood veneer 25 Non-illuminated signage 26 App. frost glass / App. rough surface special paint / Custom-made wood flooring 27 App. fabric

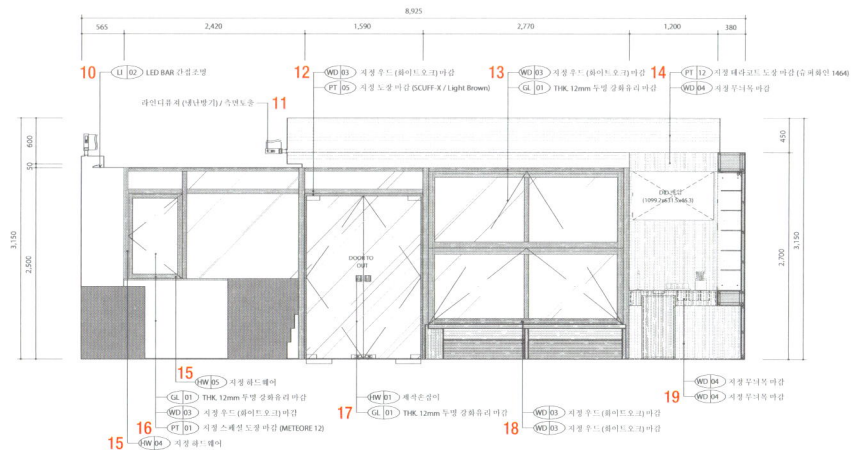

1층 홀 입면 D / 1F hall elevation D

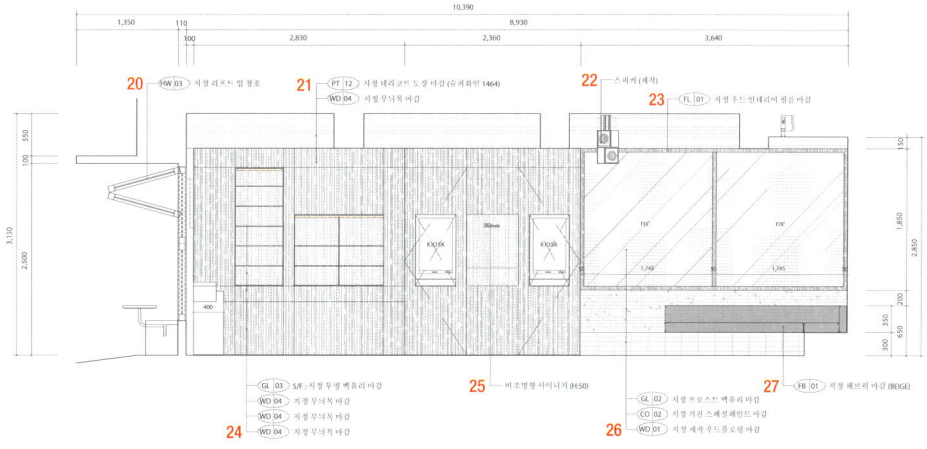

1층 홀 입면 E / 1F hall elevation E

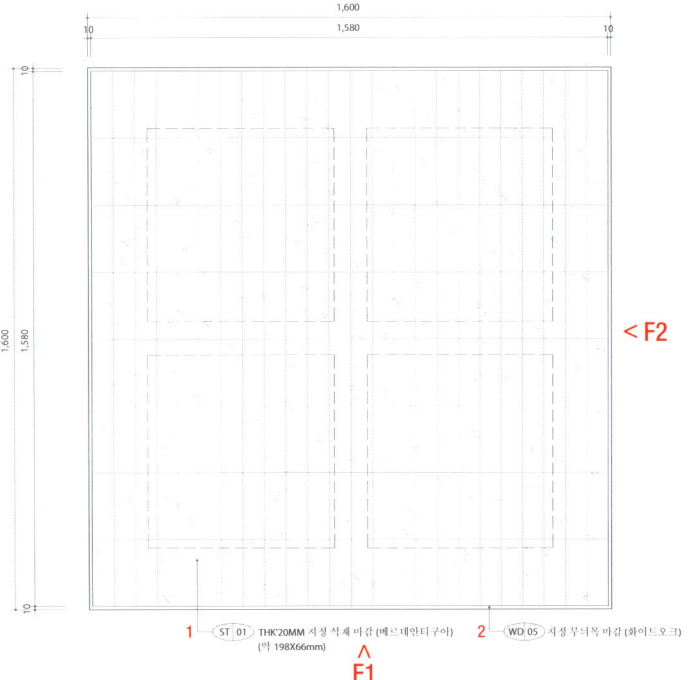

테이블 평면 F / table top view F

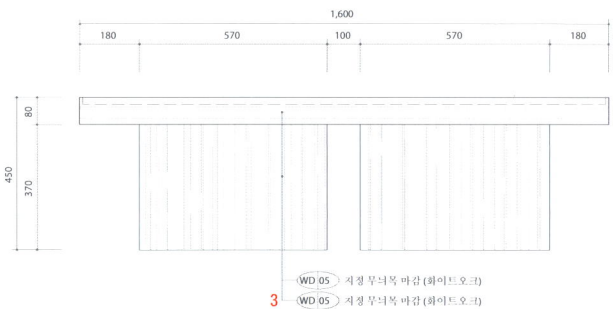

테이블 정면 F1 / table front view F1

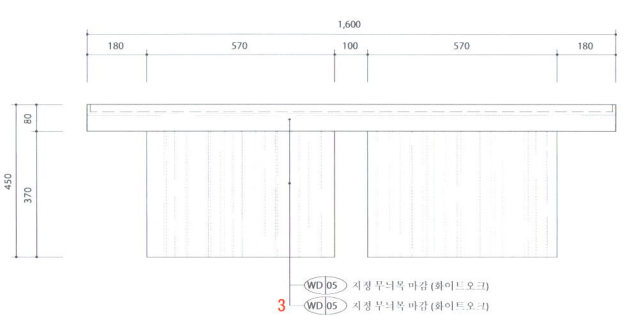

테이블 측면 F2 / table side view F2

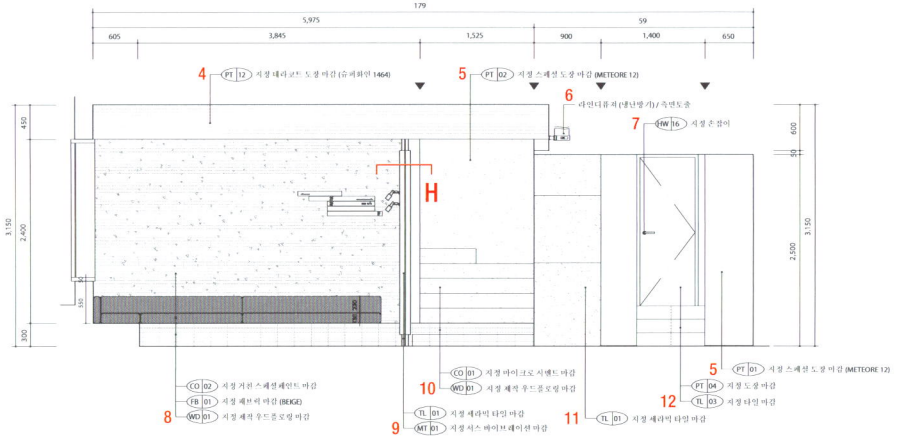

1층 홀 입면 G / 1F hall elevation G

1 T20 지정 석재 2 지정 무늬목(화이트 오크) 3 지정 무늬목(화이트 오크) / 지정 무늬목(화이트 오크) 4 지정 테라코트 도장 5 지정 스페셜 페인트 6 라인디퓨저 7 지정 손잡이 8 지정 거친 스페셜 페인트 / 지정 패브릭 / 제작 우드플로링 9 지정 세라믹 타일 / 지정 SUS 바이브레이션 10 지정 마이크로 시멘트 / 제작 우드플로링 11 지정 세라믹 타일 12 지정 도장 / 지정 타일 13 T2 SUS 바이브레이션 14 ㅁ100X100X2.3T 각파이프 보강 / T1.6 갈바륨 절곡 / 에폭시 본드 / 지정 세라믹 타일 15 ㅁ40X40X2.3T 각파이프 보강 / T1.6 SUS 바이브레이션 16 T1.6 SUS 바이브레이션 17 지정 세라믹 타일 / T2 SUS 바이브레이션 / 지정 세라믹 타일 18 건축구조물 19 T3 평철 앵커 고정 / ㅁ40X40X2.3T 각파이프 보강 / T1.6 SUS 바이브레이션 20 지정 마이크로시멘트

1 T20 app. stone 2 App. wood veneer (white oak) 3 App. wood veneer (white oak) / App. wood veneer (white oak) 4 App. terracoat painting 5 App. special paint 6 Line diffuser 7 Custom-made handle 8 App. rough surface special paint / App. fabric / Custom-made wood flooring 9 App. ceramic tile / App. SUS vibration 10 App. microcement / Custom-made wood flooring 11 App. ceramic tile 12 App. painting / App. tile 13 T2 SUS vibration 14 ㅁ100X100X2.3T square pipe reinforcement / T1.6 galvalume bending / Epoxy bond / App. ceramic tile 15 ㅁ40X40X2.3T square pipe reinforcement / T1.6 SUS vibration 16 T1.6 SUS vibration 17 App. ceramic tile / T2 SUS vibration / App. ceramic tile 18 Architecture structure 19 T3 flat steel anchor joint / ㅁ40X40X2.3T square pipe reinforcement / T1.6 SUS vibration 20 App. microcement

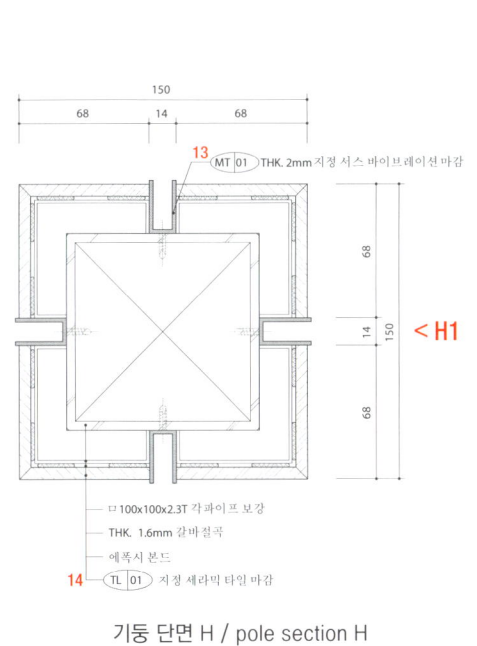

기둥 단면 H / pole section H

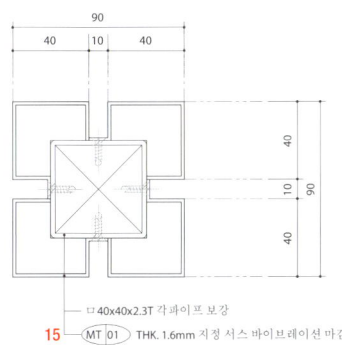

기둥 단면 H2 / pole section H2

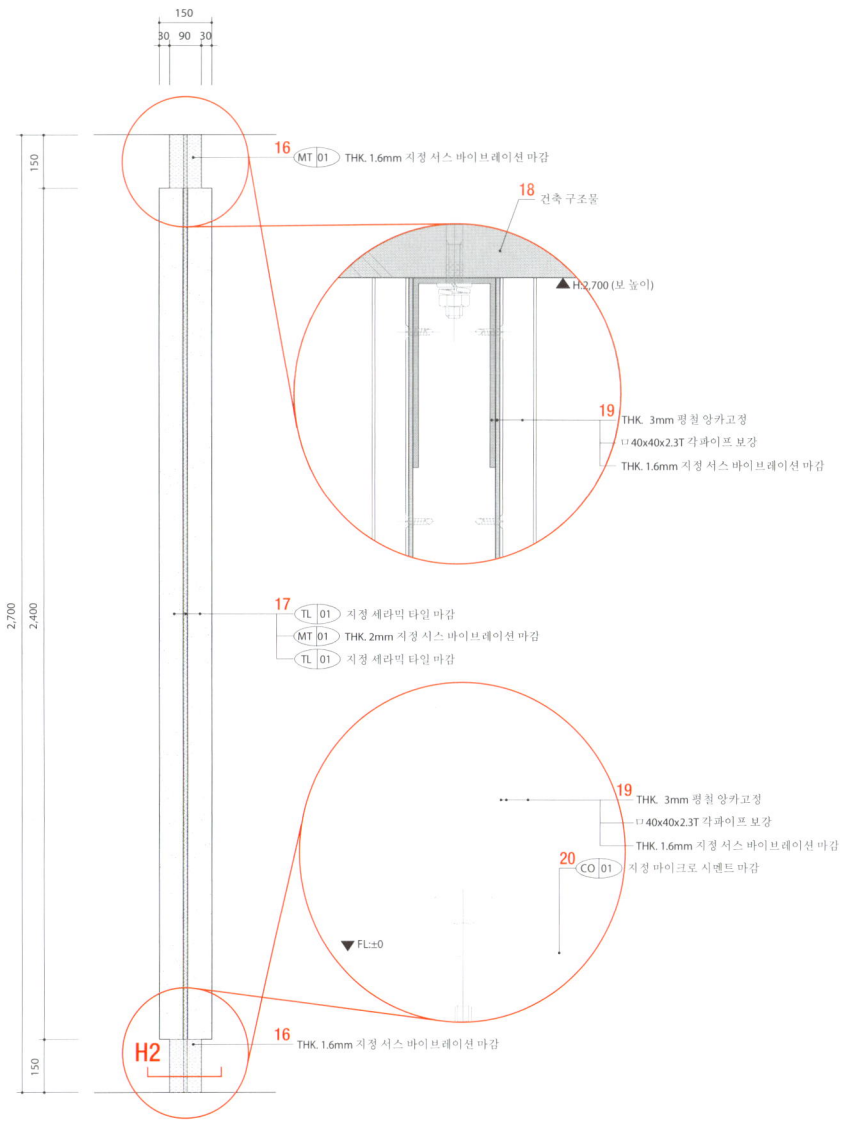

기둥 정면 H1 / pole front view H1

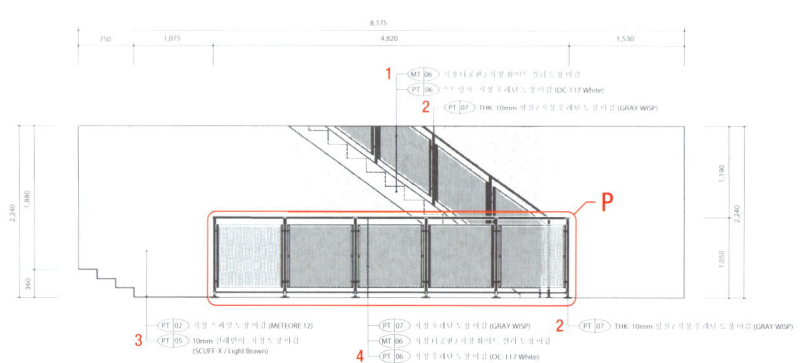

2층 홀 입면 I / 2F hall elevation I

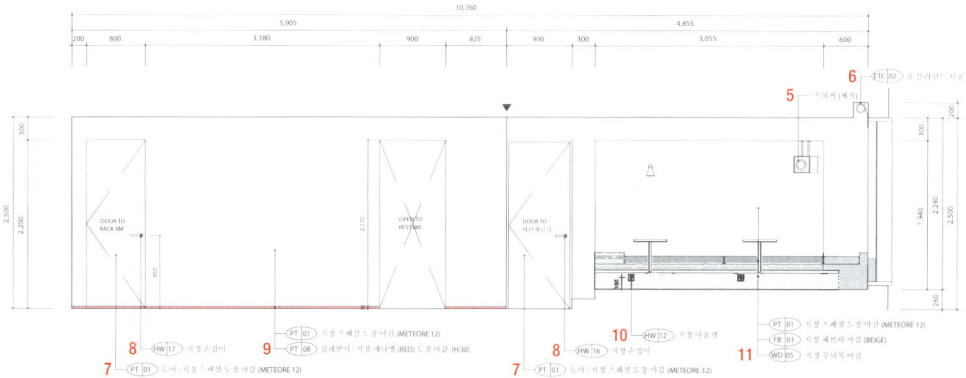

2층 홀 입면 J / 2F hall elevation J

1 지정 타공판 위 지정 흰색 도장 / 스트링거 : 지정 우레탄 도장 2 T10 평철 위 지정 우레탄 도장 3 지정 스페셜 페인트 / 10mm 걸레받이 : 지정 도장 4 지정 우레탄 도장 / 지정 타공판 위 지정 흰색 도장 / 지정 우레탄 도장 5 제작 스피커 6 롤 블라인드 7 문 : 지정 스페셜 페인트 8 지정 손잡이 9 지정 스페셜 페인트 / 걸레받이 : 지정 에나멜 도장 10 지정 콘센트 11 지정 스페셜 페인트 / 지정 패브릭 / 지정 무늬목 12 지정 테라코트 도장 / 지정 수성도장 13 라인디퓨저 14 지정 우드플로링 15 지정 스페셜 페인트 / 지정 무늬목 16 핸드레일 : 지정 우레탄 도장

1 App. white painting on perforated sheet / Stringer : App. urethane painting 2 App. urethane painting on T10 perforated sheet 3 App. special paint / 10mm baseboard : App. painting 4 App. urethane painting / App. white painting on perforated sheet / App. urethane painting 5 Custom-made speaker 6 Roll blind 7 Door : App. special paint 8 Custon-made handle 9 App. special paint / Baseboard : App. enamel painting 10 App. outlet 11 App. special paint / App. fabric / App. wood veneer 12 App. terracoat painting / App. water paint 13 Line diffuser 14 App. wood flooring 15 App. special paint / App. wood veneer 16 Handrail : App. urethane painting

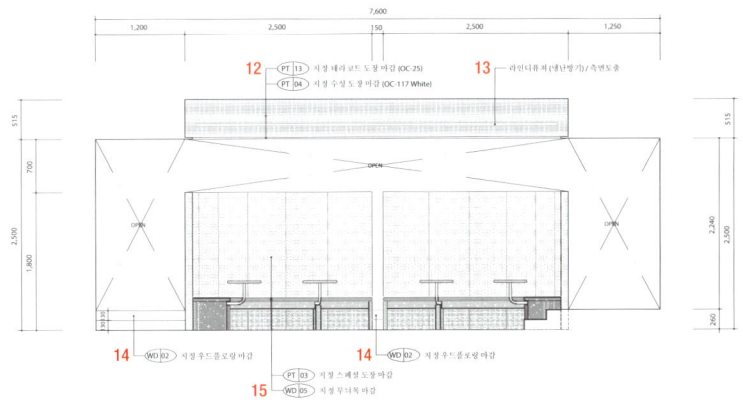

2층 홀 입면 K / 2F hall elevation K

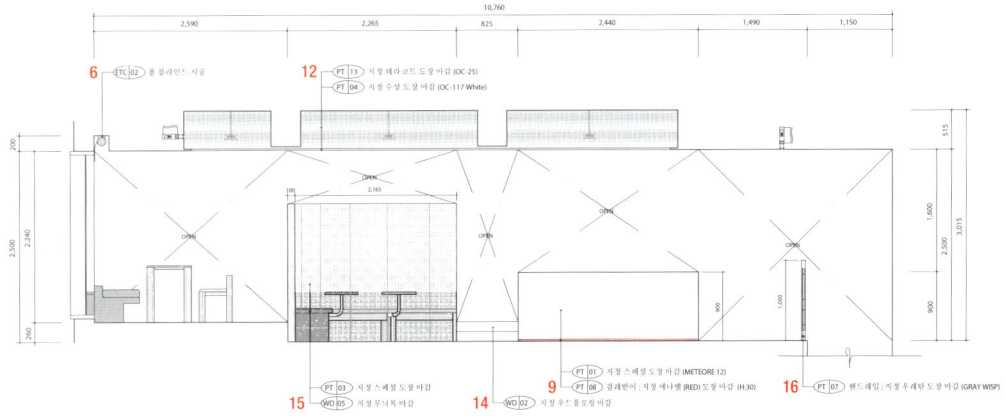

2층 홀 입면 L / 2F hall elevation L

1 T12 투명 강화유리 / T18 MDF 위 지정 무늬목   2 창호 프레임 : T1.6 갈바륨 절곡 위 지정 우드 인테리어 필름   3 지정 무늬목 / 지정 패브릭   4 테이블 상판 : 지정 화이트 오크 원목   5 지정 패브릭 / 지정 무늬목 / 지정 무늬목   6 Ø32 지정 SUS 바이브레이션   7 지정 콘센트   8 T12 투명 강화유리   9 T18 MDF 위 지정 무늬목   10 T9 MDF 위 지정 무늬목 / T12 합판 / □30X30X2.3T 각파이프 구조틀   11 지정 패브릭 / T9 MDF 위 지정 무늬목   12 테이블 상판 : 지정 화이트 오크 원목   13 회전 테이블 힌지   14 T9 MDF 위 지정 무늬목   15 충격 방진 패드   16 Ø32 지정 SUS 바이브레이션 / 지정 우드플로링 / T12 합판 / □50X50X2.3T 각파이프 구조틀   17 지정 테라코트 도장 / 지정 수성도장   18 지정 스페셜 페인트 / 걸레받이 : 지정 에나멜 도장   19 지정 스페셜 페인트   20 라인디뮤저   21 핸드레일 : 지정 우레탄 도장   22 지정 스페셜 페인트 / 지정 무늬목   23 롤 블라인드

1 T12 clear tempered glass / App. wood veneer on T18 MDF   2 Windows frame : App. wood interior film on T1.6 galvalume bending   3 App. wood veneer / App. fabric   4 Table top : App. white oak wood   5 App. fabric / App. wood veneer / App. wood veneer   6 Ø32 app. SUS vibration   7 App. outlet   8 T12 clear tempered glass   9 App. wood veneer on T18 MDF   10 App. wood veneer on T9 MDF / T12 plywood / □30X30X2.3T square pipe structure   11 App. fabric / App. wood veneer on T9 MDF   12 Table top : App. white oak wood   13 Revolving table hinge   14 App. wood veneer on T9 MDF   15 Anti-vibration pad   16 Ø32 app. SUS vibration / App. wood flooring / T12 plywood / □50X50X2.3T square pipe structure   17 App. terracoat painting / App. water paint   18 App. special paint / Baseboard : App. enamel painting   19 App. special paint   20 Line diffuser   21 Handrail : App. urethane painting   22 App. special painting / App. wood veneer   23 Roll blind

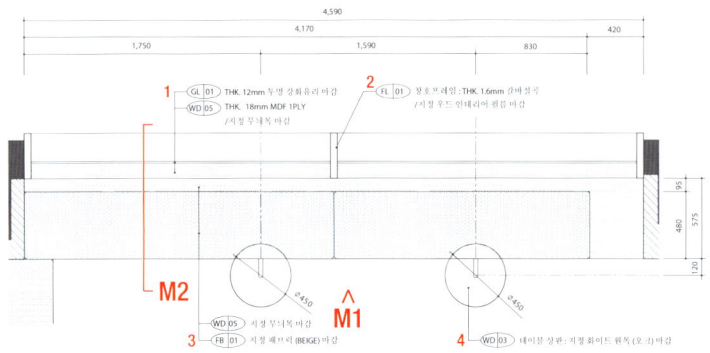

소파 평면 M / sofa top view M

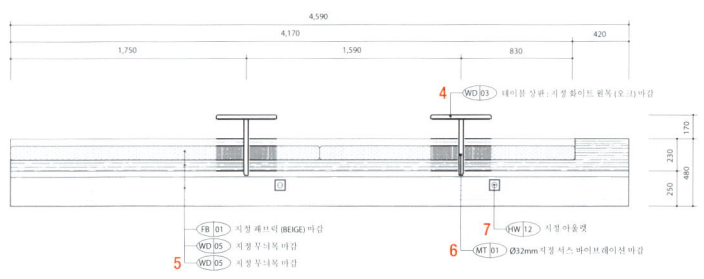

소파 정면 M1 / sofa front view M1

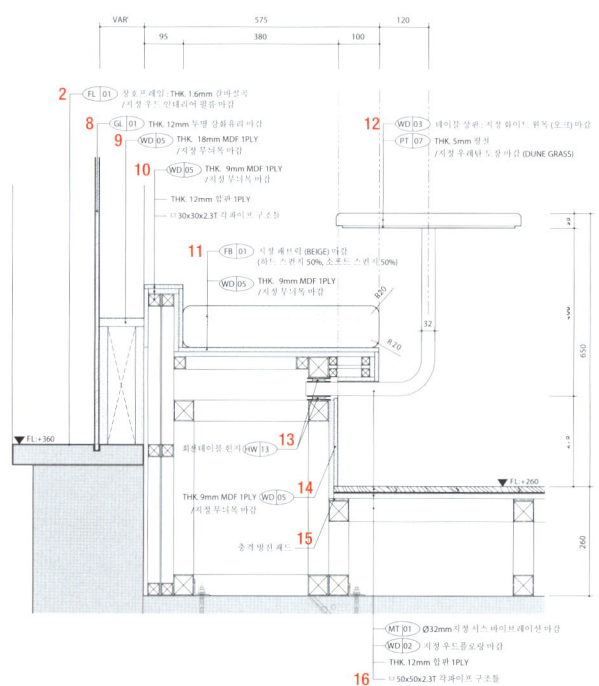

소파 단면 M2 / sofa section M2

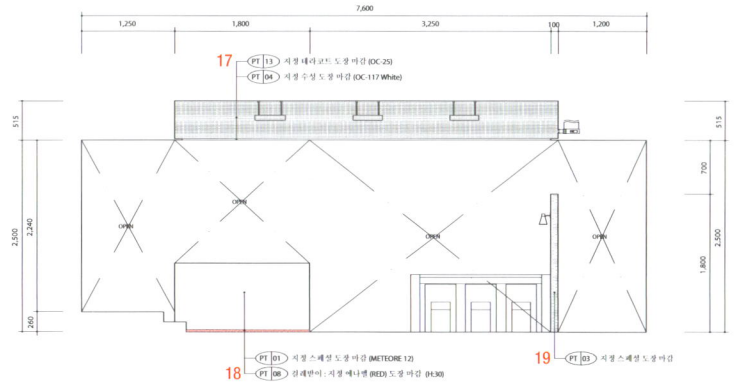

2층 홀 입면 N / 2F hall elevation N

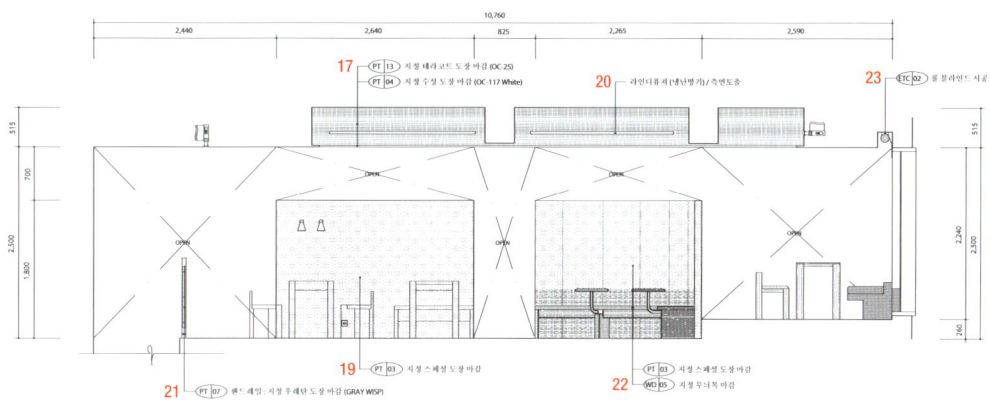

2층 홀 입면 O / 2F hall elevation O

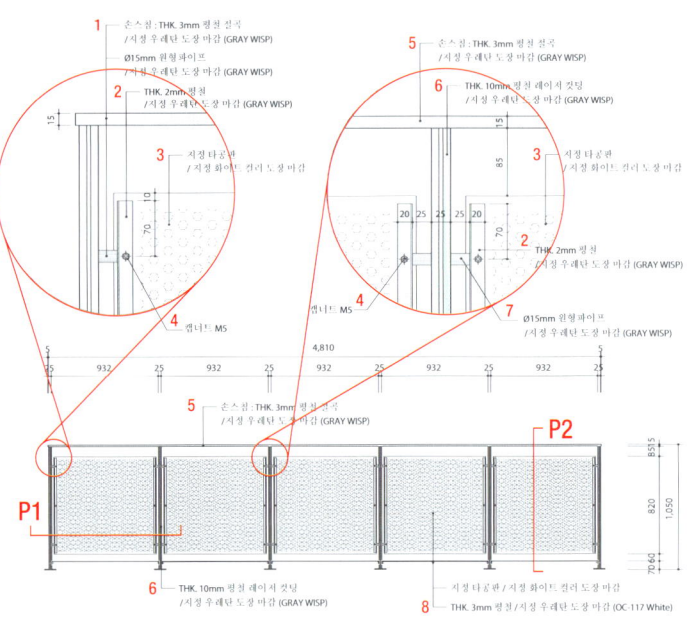

핸드레일 입면 P / handrail elevation P

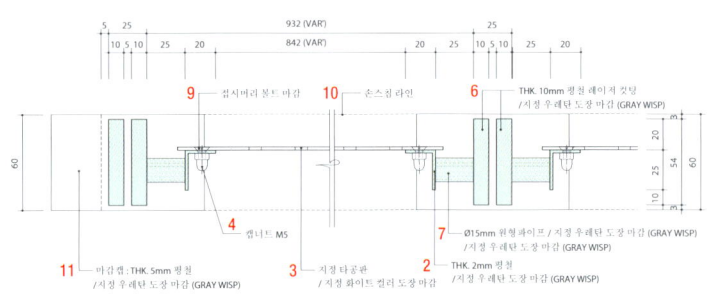

핸드레일 단면 P1 / handrail section P1

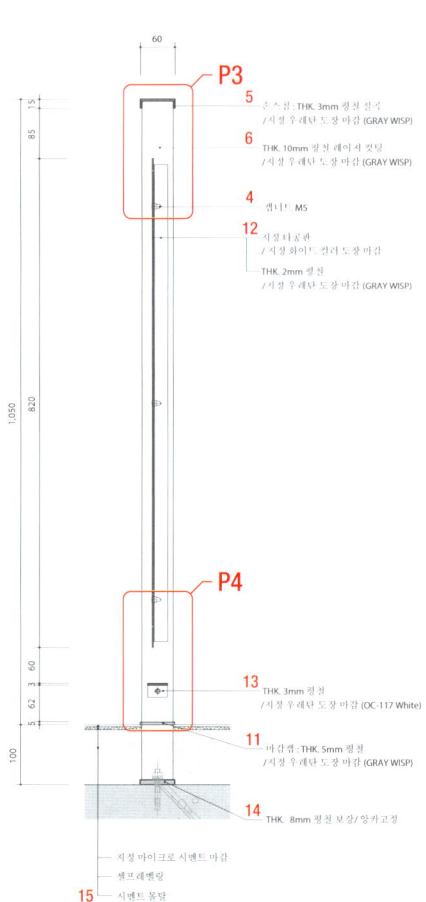

핸드레일 단면 P2 / handrail section P2

1 손스침 : T3 평철 절곡 위 지정 우레탄 도장 / Ø15 원형 파이프 위 지정 우레탄 도장   2 T2 평철 위 지정 우레탄 도장   3 지정 타공판 위 지정 흰색 도장   4 캡너트 M5   5 손스침 : T3 평철 절곡 위 지정 우레탄 도장   6 T10 평철 레이저 커팅 위 지정 우레탄 도장   7 Ø15 원형 파이프 위 지정 우레탄 도장   8 지정 타공판 위 지정 흰색 도장 / T3 평철 위 지정 우레탄 도장   9 접시머리 볼트   10 손스침 라인   11 마감캡 : T5 평철 위 지정 우레탄 도장   12 지정 타공판 위 지정 흰색 도장 / T2 평철 위 지정 우레탄 도장   13 T3 평철 위 지정 우레탄 도장   14 T8 평철 보강, 앵커 고정   15 지정 마이크로시멘트 / 셀프레벨링 / 시멘트 모르타르   16 접시머리 볼트 / 캡너트 M5   17 T10 평철 레이저 커팅 위 지정 우레탄 도장 / 지정 타공판 위 지정 흰색 도장 / T2 평철 위 지정 우레탄 도장   18 손스침 : T3 평철 절곡 위 지정 우레탄 도장 / 지정 타공판 위 지정 흰색 도장   19 T10 평철 레이저 커팅 위 지정 우레탄 도장 / Ø15 원형 파이프 위 지정 우레탄 도장 / 캡너트 M5   20 스트링거 : 지정 우레탄 도장

1 Handrail : App. urethane painting on T3 flat steel bending / App. urethane painting on Ø15 round pipe   2 App. urethane painting on T2 flat steel   3 App. white painting on perforated sheet   4 Cap nut M5   5 Handrail : App. urethane painting on T3 flat steel   6 App. urethane painting on T10 flat steel laser cutting   7 App. urethane painting on Ø15 round pipe   8 App. white painting on perforated sheet / App. urethane painting on T3 flat steel   9 Flat head bolt   10 Handrail line   11 End cap : App. urethane painting on T3 flat steel   12 App. white painting on perforated sheet / App. urethane painting on T2 flat steel   13 App. urethane painting on T3 flat steel   14 T8 flat steel reinforcement, Anchor joint   15 App. microcement / Self-leveling / Cement mortar   16 Flat head bolt / Cap nut M5   17 App. urethane painting on T10 flat steel laser cutting / App. white painting on perforated sheet / App. urethane painting on T2 flat steel   18 Handrail : App. urethane painting on T3 flat steel / App. white painting on perforated sheet   19 App. urethane painting on T10 flat steel laser cutting / App. urethane painting on Ø15 round pipe / Cap nut M5   20 Stringer : App. urethane painting

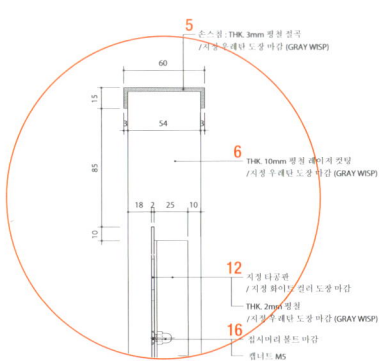

핸드레일 상세 P3 / handrail detail P3

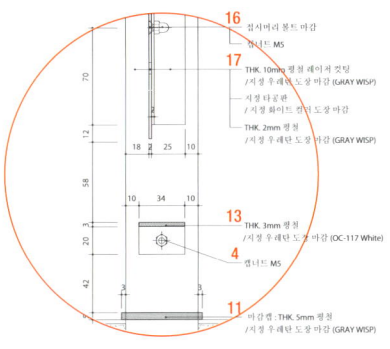

핸드레일 상세 P4 / handrail detail P4

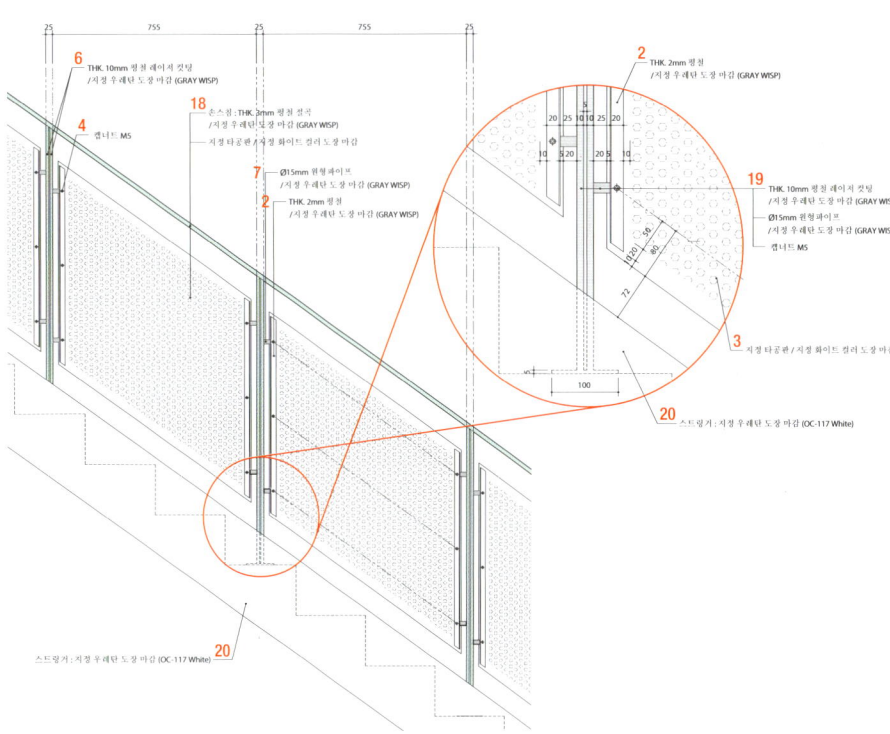

핸드레일 입면 Q / handrail elevation Q

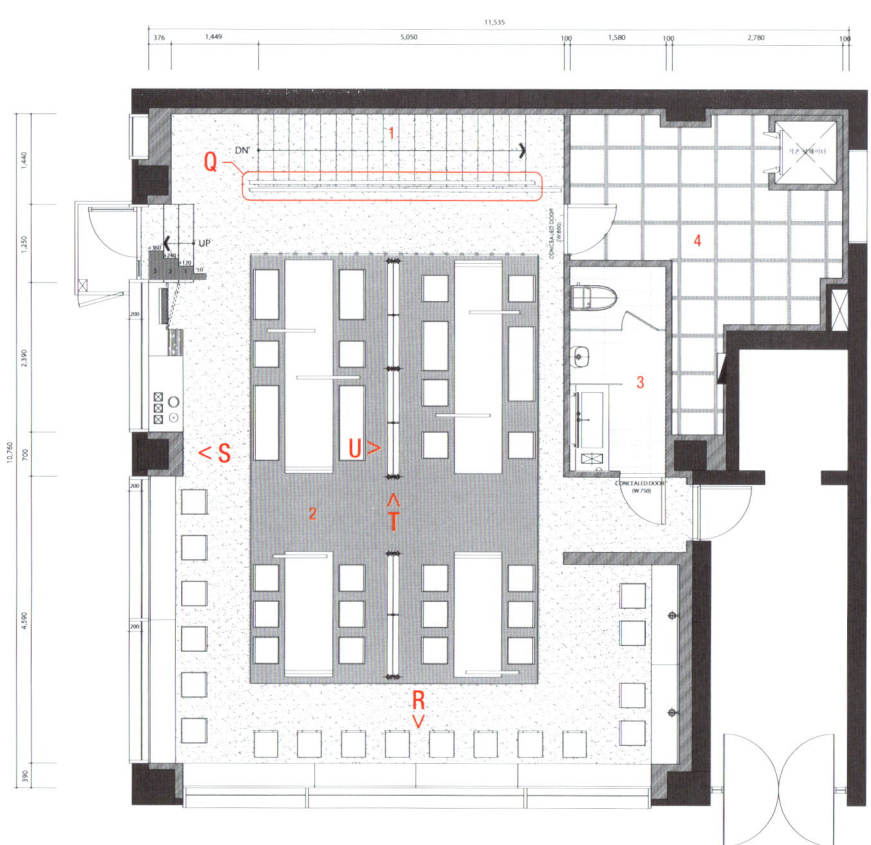

3층 평면도 / 3rd floor plan

1 계단 **2** 홀 **3** 화장실 **4** 백룸

1 Stairs **2** Hall **3** Restroom **4** Back room

1 T12 투명 강화유리 2 지정 우드 인테리어 필름 3 지정 스페셜 페인트 / 걸레받이 : 지정 에나멜 도장 4 롤 블라인드 5 제작 스피커 6 지정 손잡이 7 LED 바 8 지정 무늬목 9 핸드레일 : 지정 우레탄 도장 10 지정 마이크로시멘트

1 T12 clear tempered glass 2 App. wood interior film 3 App. special paint / Baseboard : App. enamel painting 4 Roll blind 5 Custom-made blind 6 App. handle 7 LED bar 8 App. wood veneer 9 Handrail : App. urethane painting 10 App. microcement

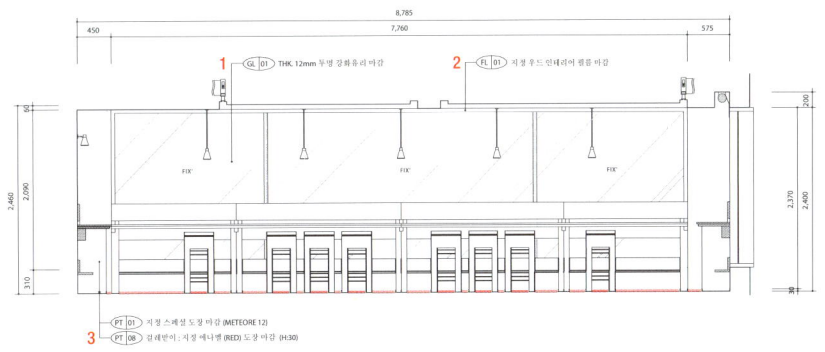

3층 홀 입면 R / 3F hall elevation R

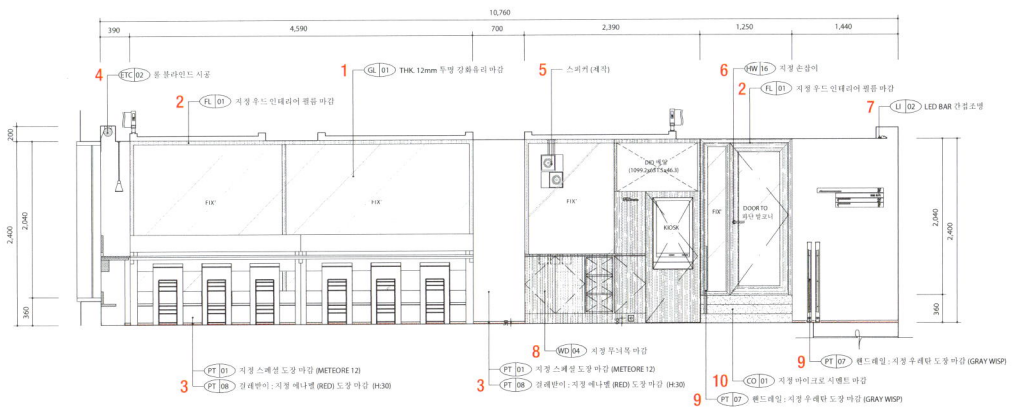

3층 홀 입면 S / 3F hall elevation S

1 롤 블라인드 2 지정 스페셜 페인트 / 걸레받이 : 지정 에나멜 도장 3 기둥 : 지정 우레탄 도장 4 지정 손잡이 / 지정 스페셜 페인트 / 걸레받이 : 지정 에나멜 도장 5 라인디퓨저 6 T10 평철 위 지정 우레탄 도장 7 선반 : T3 평철 레이저 가공 위 우레탄 도장 / 지정 타공판 위 지정 도장 8 타공판 커버 : T1.6 갈바륨 절곡 위 지정 우레탄 도장 9 지정 타공판 위 지정색 도장 10 레이저 가공 후 끼워넣기 11 T3 평철 위 지정 우레탄 도장 12 브래킷 : T5 평철 위 지정 우레탄 도장 13 T1.6 갈바륨 절곡 위 지정 우레탄 도장 14 □20X20X2.3T 각파이프 보강 15 캡너트 M5 16 지정 우레탄 도장 17 브래킷 : T5 평철 위 지정 우레탄 도장 / Ø15 원형 파이프 위 지정 우레탄 도장 18 선반 : T3 평철 레이저 가공 위 우레탄 도장 / 타공판 커버 : T1.6 갈바륨 절곡 위 지정 우레탄 도장 / 지정 타공판 위 지정색 도장 19 Ø15 원형 파이프 위 지정 우레탄 도장 20 제작 우드플로링 / 셀프레벨링 / 시멘트 모르타르 21 고정용 파이프 천장 매입 / □30X30X2.3T 각파이프 보강 22 T9 합판 보강 / T9.5 석고보드 위 지정 수성도장 23 Ø15 원형 파이프 위 지정 에나멜 도장 24 마감캡 : T2 평철 위 지정 에나멜 도장 25 선반 : T3 평철 레이저 가공 위 우레탄 도장 / 브래킷 : T5 평철 위 지정 우레탄 도장 26 T1.6 갈바륨 절곡 위 지정 우레탄 도장 / 지정 타공판 위 지정색 도장 27 □20X20X2.3T 각파이프 보강 위 지정 우레탄 도장 / T1.6 갈바륨 절곡 위 지정 우레탄 도장

1 Roll blind 2 App. special paint / Baseboard : App. enamel painting 3 Pole : App. urethane painting 4 App. handle / App. special paint / Baseboard : App. enamel painting 5 Line diffuser 6 App. urethane painting on T10 flat steel 7 Shelf : App. urethane painting on T3 flat steel laser processing / App. painting on perforated sheet 8 Perforated sheet cover : App. urethane painting on T1.6 galvalume bending 9 App. color painting on perforated sheet 10 Putting in after laser processing 11 App. urethane painting on T3 flat steel 12 Bracket : App. urethane painting on T5 flat steel 13 App. urethane painting on T1.6 galvalume bending 14 □20X20X2.3T square pipe reinforcement 15 Cap nut M5 16 App. urethane painting 17 Bracket : App. urethane painting on T5 flat steel / App. urethane painting on Ø15 round pipe 18 Shelf : App. urethane painting on T3 flat steel laser processing / Perforated sheet cover : App. urethane painting on T1.6 galvalume bending / App. color painting on perforated sheet 19 App. urethane painting on Ø15 round pipe 20 Custom-made wood flooring / Self-leveling / Cement mortar 21 Embedding pipe in ceiling / □30X30X2.3T square pipe reinforcement 22 T9 plywood reinforcement / App. water paint on T9.5 gypsum board 23 App. urethane painting on Ø15 round pipe on enamel painting 24 End cap : App. enamel painting on T2 flat steel 25 Shelf : App. urethane painting on T3 flat steel laser processing / Bracket : App. urethane painting on T5 flat steel 26 App. urethane painting on T1.6 galvalume bending / App. color painting on perforated sheet 27 App. urethane painting on □20X20X2.3T square pipe reinforcement / App. urethane painting on T1.6 galvalume bending

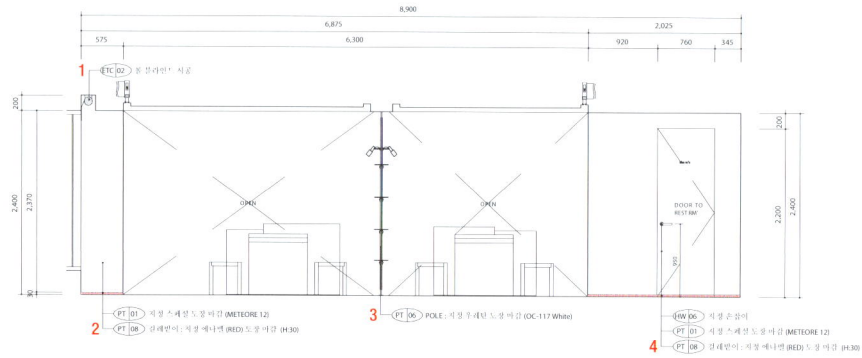

3층 홀 입면 T / 3F hall elevation T

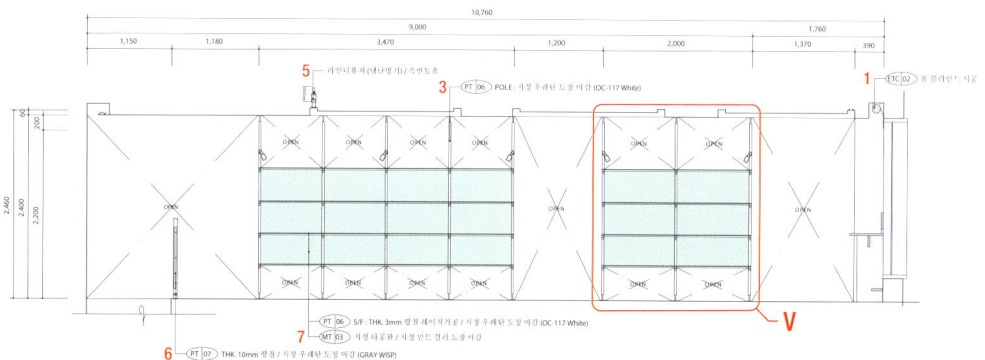

3층 홀 입면 U / 3F hall elevation U

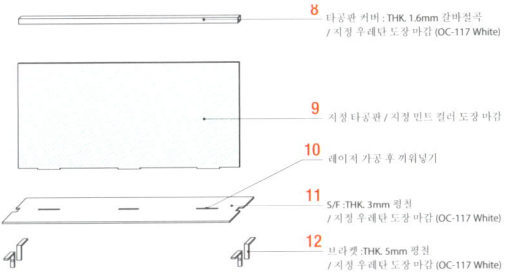

파티션 상세 / partition detail

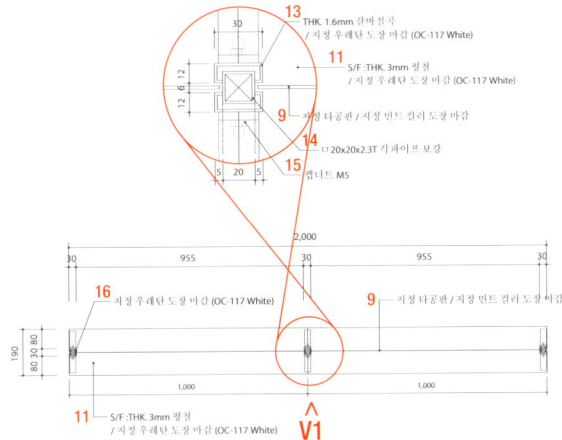

파티션 평면 V / partition top view V

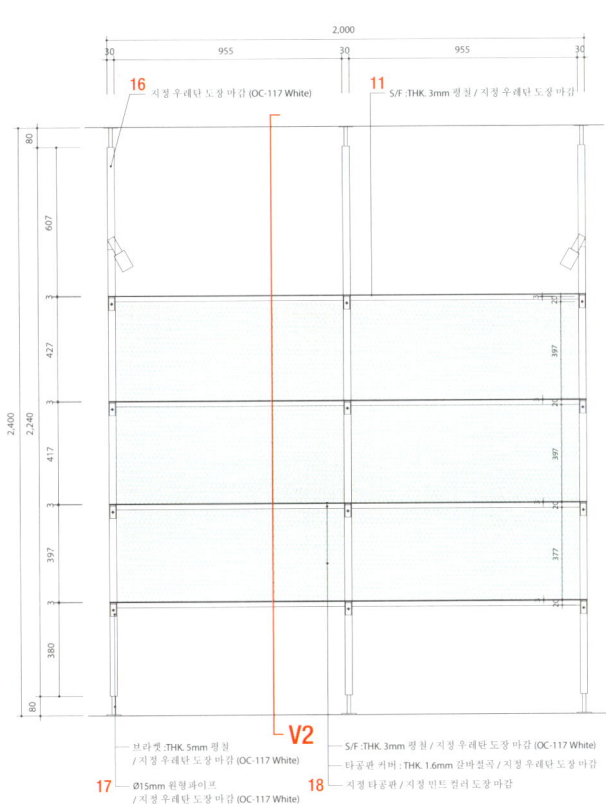

파티션 정면 V1 / partition front view V1

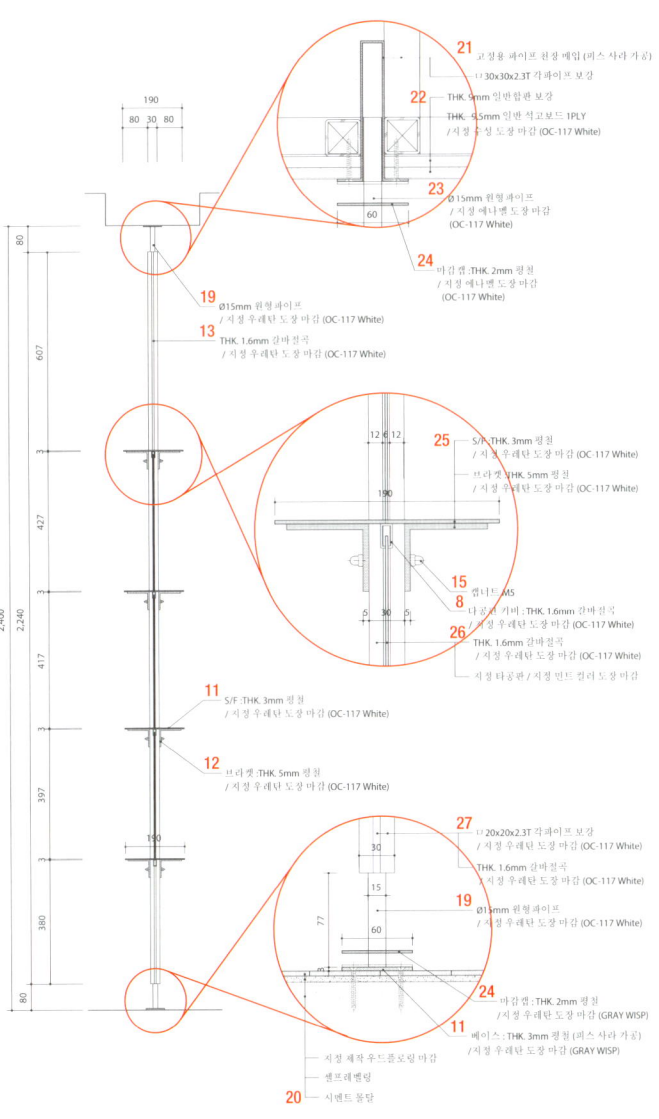

파티션 단면 V2 / partition section V2

SOOSOO COFFEE

PROJECT MARK | Jaehong Son, Jeeseung Yang, Jihyung Song

수수커피 여의도점은 번화한 도심 속에 꾸밈없이 단정한 분위기가 주는 편안한 느낌의 공간으로 디자인되었다. 이곳은 여의도의 중심에 위치하여 IFC몰, 더 현대, 한강과 가까워 데이트하는 연인부터 바쁜 직장인들까지 다양한 사람들이 모이는 장소다. 내부의 높은 천장은 개방적인 분위기를 조성한다. 천장 전체를 목모보드로 마감하여 음향이 울리지 않도록 고려했고, 계절감이 공간 내부로 들어올 수 있도록 4면을 창으로 활용했다. 문이나 수납장 디자인도 생동감 있는 느낌을 주도록 설계되었다. 자주 사용하지 않는 장비들은 수납장 속에 숨겨 깔끔하게 가려주고 자주 사용하는 장비와 물건들은 반투명한 FRP 소재를 통해 표현했다. 이 보일듯 말듯한 뿌연 면을 가진 거대한 미닫이문은 공간 사용자들의 앞에서 이 공간의 꾸밈없고 라이브함을 대변한다. 가구는 각 테이블별로 사용자의 목적을 고려하여 설계되었다. 작업을 위한 큰 테이블, 미팅을 위한 4인용 테이블, 잠시 머물고 갈 사람을 위한 바 테이블, 소통을 즐기는 사람들을 위한 창가 테이블 등 다양한 용도에 맞는 테이블이 각 형태에 맞게 디자인 되었다. 본연의 성질을 가지고 있는 소재들을 그대로 노출하여 자연스러운 분위기를 연출했고, 같은 소재를 연마 방식을 다르게 하여 거칠면서도 단정한 소재들의 조화가 이루어지도록 했다. 이러한 다양한 소재들이 각자의 특성을 살려 수수커피의 꾸밈없이 단정한 분위기를 만들어 냈다.

Soosoo Coffee Yeouido Branch is an unpretentious and neat space in the bustling downtown area. Located in the heart of Yeouido and close to IFC Mall, The Hyundai, and the Hangang River, it attracts diverse visitors from dating couples to busy office workers. The high ceiling creates an open atmosphere. The entire ceiling is finished with wood wool board to prevent sound reverberation, and windows are installed on all four sides to capture seasonal changes inside the space. The doors and storage units are designed to add vibrancy to the space. Rarely used equipment is neatly tucked away in storage cabinets, while frequently used equipment and items are covered with translucent FRP material. The large sliding door with its subtly frosted surface highlights the space's unpretentious and lively charm. The furniture is designed considering each table's intended use. Various types of tables are installed: large tables for work, four-person tables for meetings, bar tables for short stays, and window-side tables for those who enjoy socializing. Materials are left exposed to maintain their natural properties, creating an organic feel. The same materials are polished in different ways to create a harmonious blend of rough and neat textures. Consequently, the unique physical properties of various materials come together to complete the brand's unique modest and tidy atmosphere.

디자인 손재홍, 양지승, 송지형 / 프로젝트 마크
위치 서울특별시 영등포구 국제금융로2길 24
용도 카페
면적 178.8m²
마감 콘크리트, 무늬목, 스테인리스 스틸, 유리섬유
완공 2024. 1
디자인팀 김어진, 팽종인
사진 조동현

Location 24 Gukjaegeumyung-ro 2-gil, Yeongdeungpo-gu, Seoul
Use Cafe
Area 178.8m²
Finishing Concrete, Wood veneer, Stainless steel, Glass fiber
Completion 2024. 1
Photographer Donghyun Cho

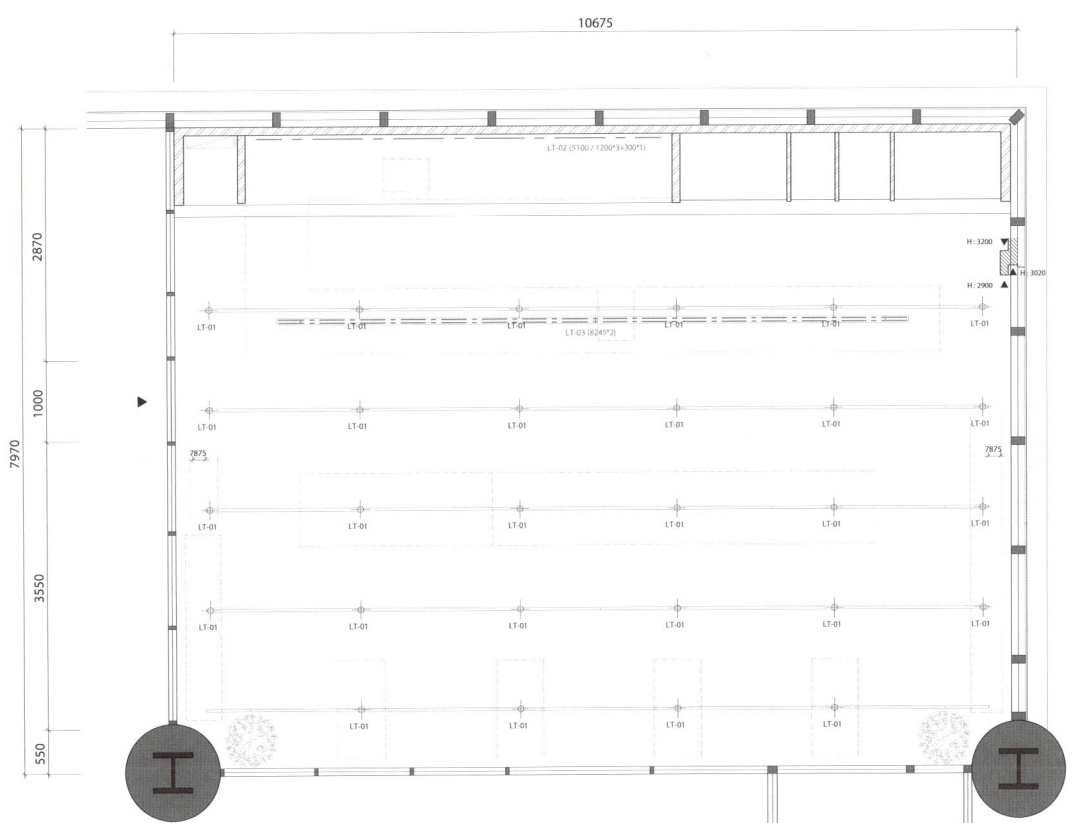

천장도 / ceiling plan

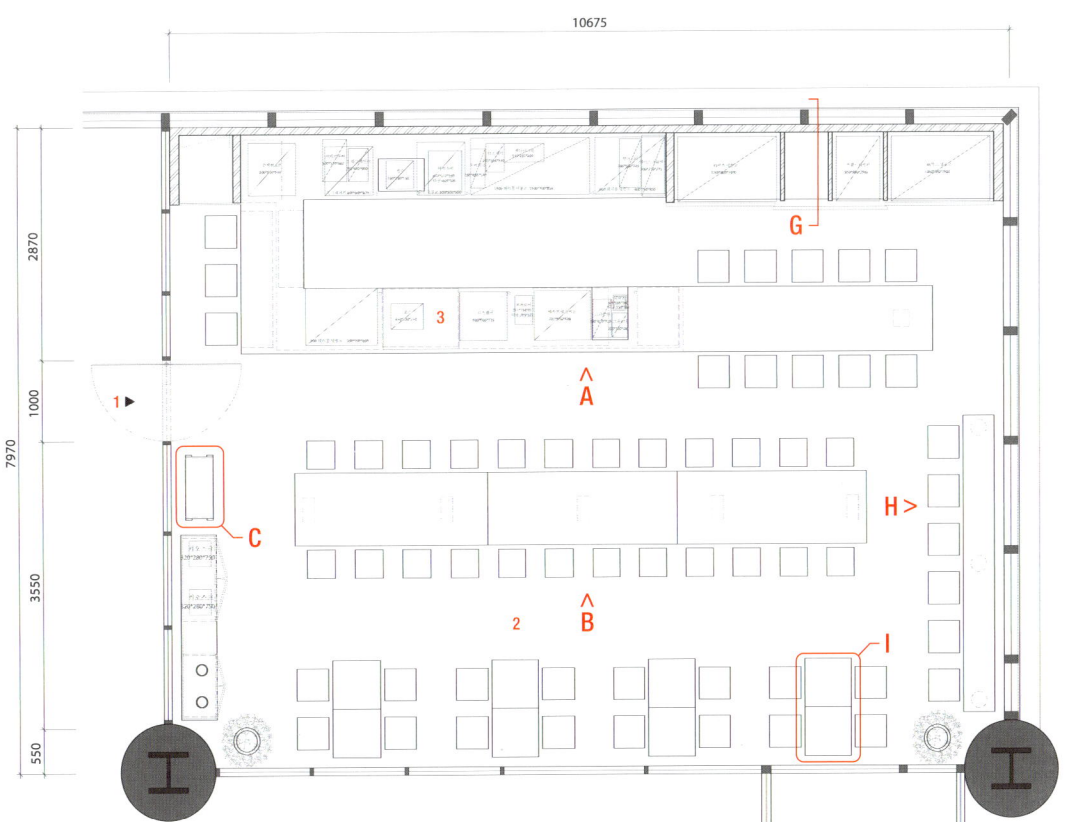

평면도 / floor plan

1 입구 2 홀 3 커피 바

1 Entrance 2 Hall 3 Coffee bar

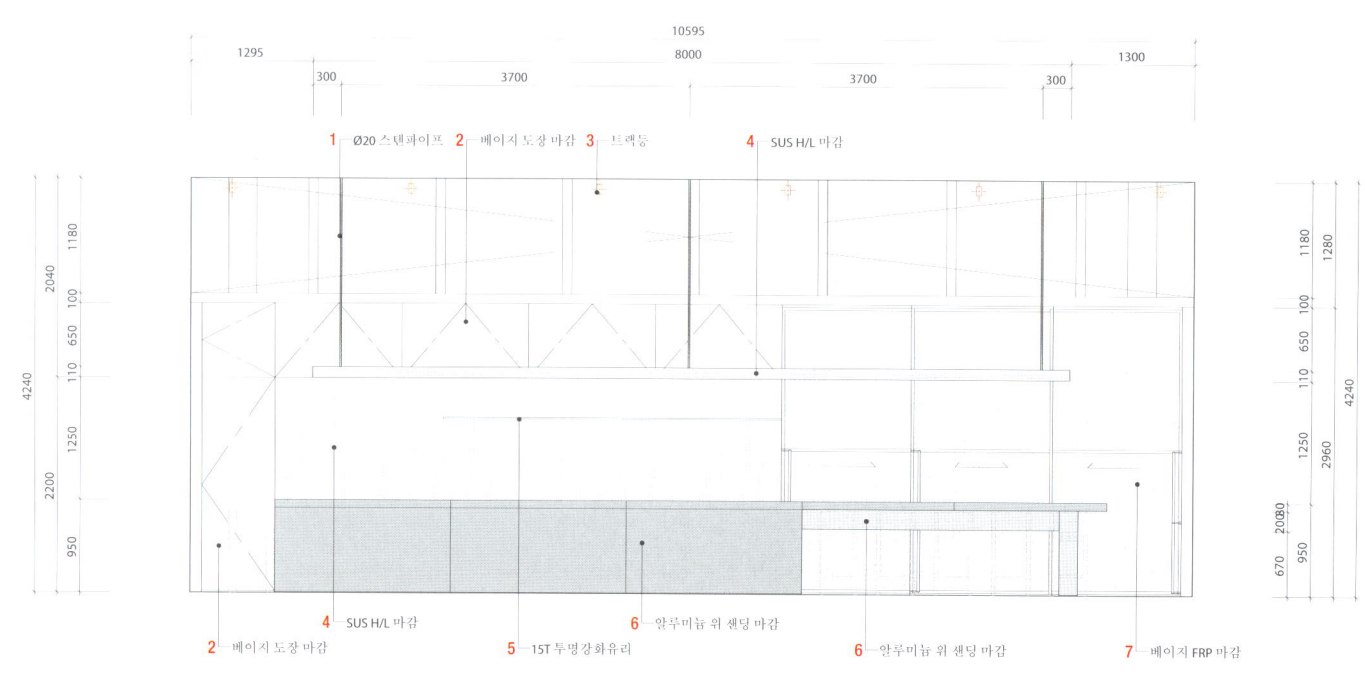

홀 입면 A / hall elevation A

1 Ø20 스테인리스 스틸 파이프 2 베이지 도장 3 트랙조명 4 SUS 헤어라인 5 T15 투명 강화유리 6 알루미늄 위 샌딩 마감 7 베이지 FRP 8 서비스 스테이션 9 검은색 도장 10 상판 : 검은색 FRP 11 T10 프로스트 유리 12 SUS 바이브레이션

1 Ø20 stainless steel pipe 2 Beige painting 3 Track lighting 4 SUS hairline 5 T15 clear tempered glass 6 Sanding aluminum 7 Beige FRP 8 Service station 9 Black painting 10 Top : Black FRP 11 T10 frost glass 12 SUS vibration

홀 입면 B / hall elevation B

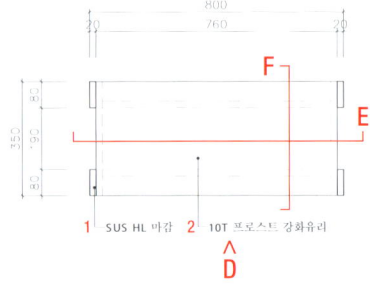

선반 평면 C / shelf top view C

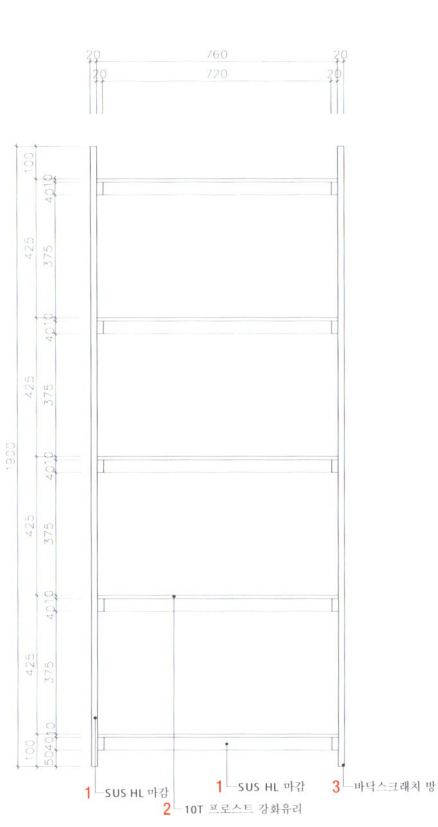

선반 정면 D / shelf front view D

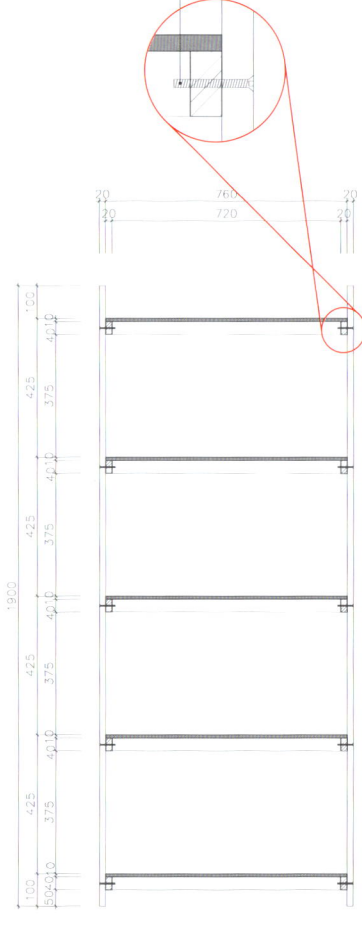

선반 단면 E / shelf section E

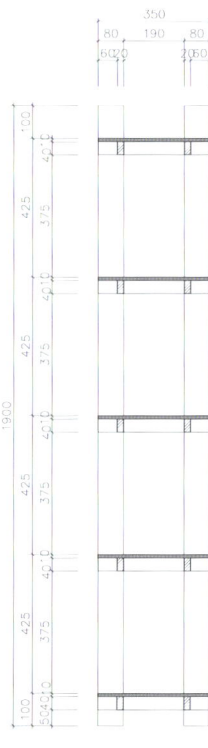

선반 단면 F / shelf section F

1 SUS 헤어라인   2 T10 프로스트 강화유리   3 바닥 스크래치 방지 발   4 접시머리 렌치볼트 M5   5 T9 석고보드   6 T9 합판   7 30X30 각재   8 가구장   9 기존 창틀에 상부 구조 고정   10 디밍 조명   11 T15 투명유리   12 지정 금속 시트   13 T9 합판 2겹   14 50X30 각파이프 구조

1 SUS hairline   2 T10 frost tempered glass   3 Floor protector foot   4 Flat head cap screw M5   5 T9 gypsum board   6 T9 plywood   7 30X30 timber   8 Cabinet installation   9 Fixing upper structure to the existing window frame   10 Dimmable lighting   11 T15 clear glass   12 App. metal sheet   13 T9 plywood 2ply   14 50X30 square pipe structure

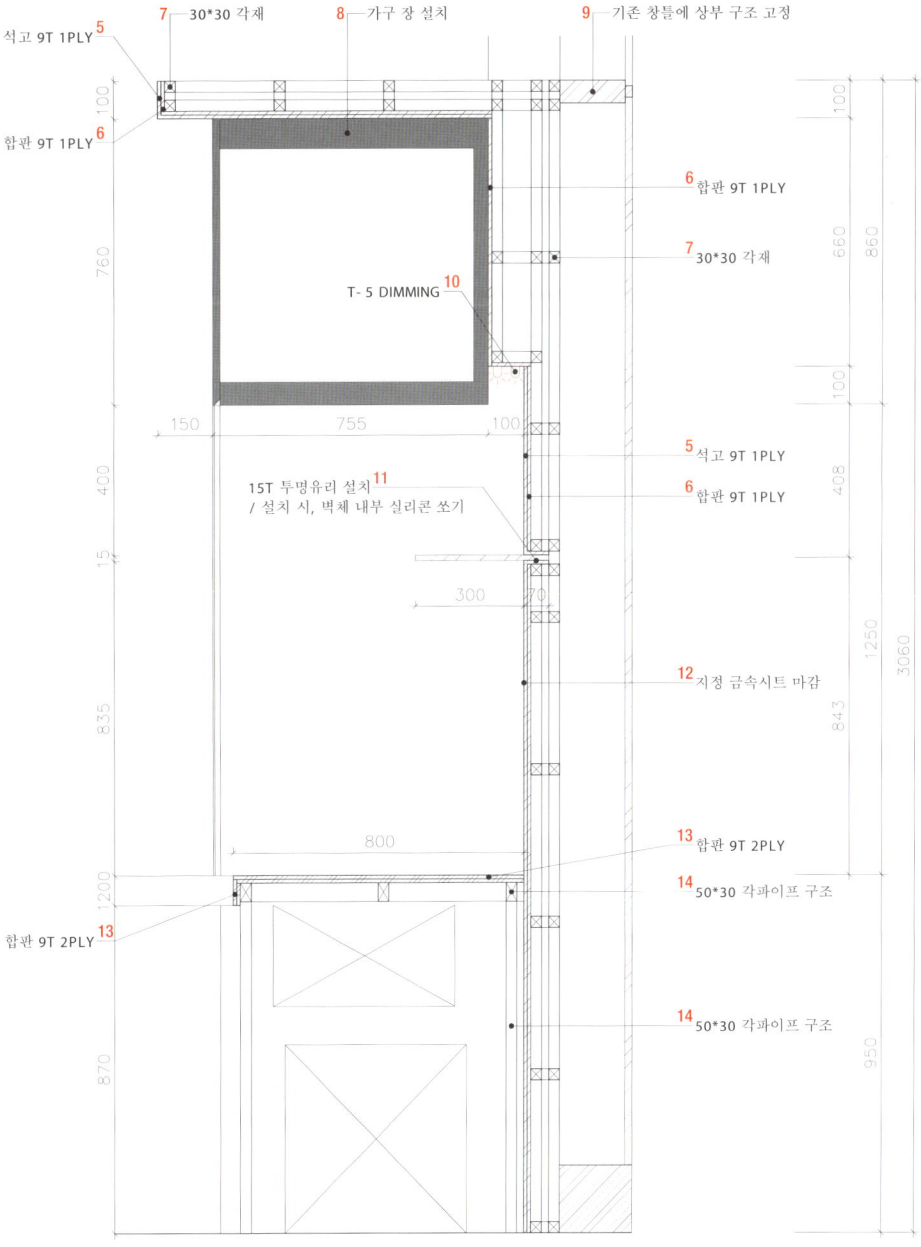

주방 수납장 단면 G / kitchen cabinet section G

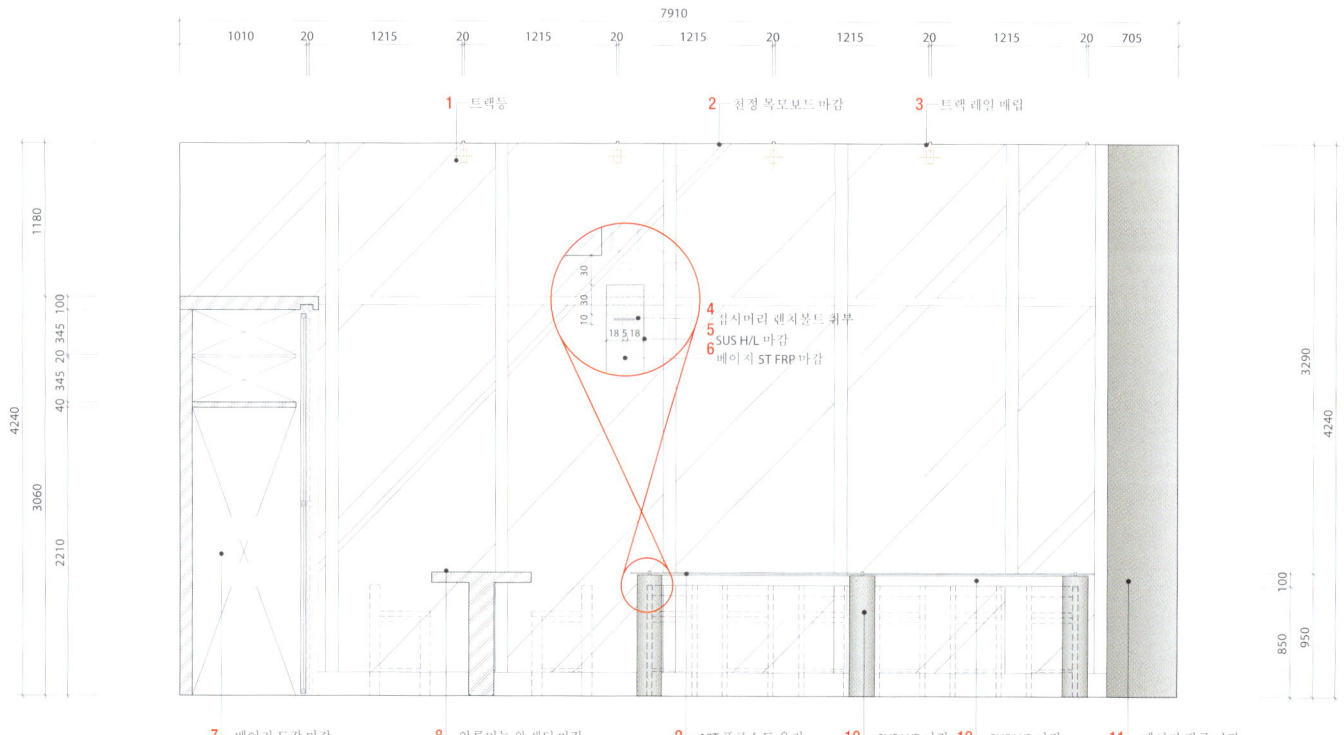

홀 입면 H / hall elevation H

1 트랙조명 2 천장 목모보드 3 트랙 레일 매립 4 접시머리 렌치볼트 5 SUS 헤어라인 6 T5 베이지 FRP 7 베이지 도장 8 알루미늄 위 샌딩 마감 9 T10 프로스트 유리 10 SUS 바이브레이션 11 베이지 필름
12 지정 유리 볼트 13 T10 프로스트 강화유리 14 Ø60 SUS 헤어라인 15 Ø20 SUS 환봉 헤어라인 16 20X40 SUS 헤어라인 17 바닥 스크래치 방지 발

1 Track lightiing 2 Ceiling wood wool acoustic panel 3 Embedding track rail 4 Flat head cap screw 5 SUS hairline 6 T5 beige FRP 7 Beige painting 8 Sanding aluminum 9 T10 frost glass 10 SUS vibration 11 Beige film
12 App. glass standoff bolt 13 T10 frost tempered glass 14 Ø60 SUS hairline 15 Ø20 SUS round pipe hairline 16 20X40 SUS hairline 17 Floor protector foot

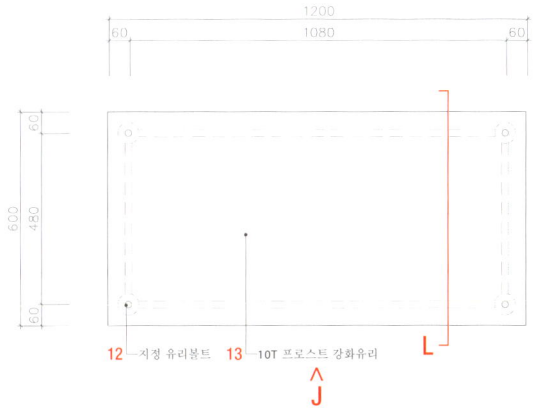

테이블 평면 I / table top view I

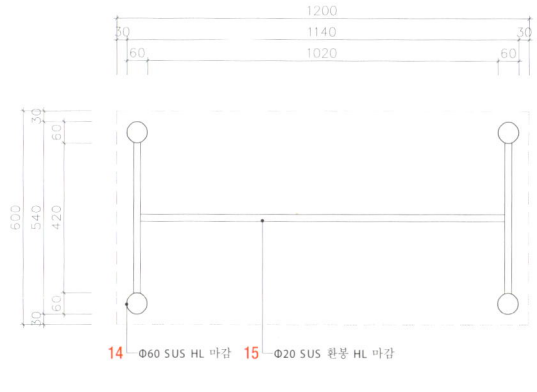

테이블 단면 K / table section K

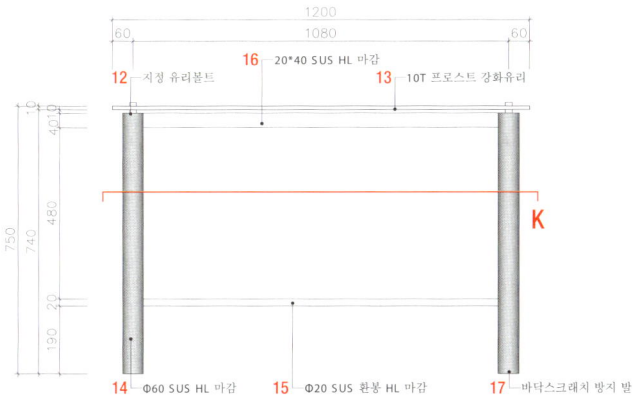

테이블 정면 J / table front view J

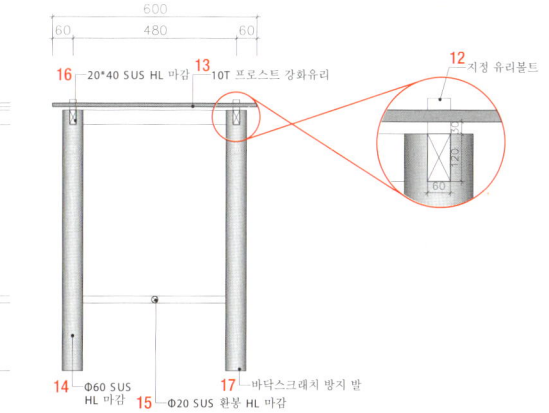

테이블 단면 L / table section L

2025 ANNUAL INTER DETAI

CLINIC · RELIGION

216 OGANACELL DERMATOLOGY CLINIC
오가나셀 피부과 의원 잠실점

224 CELLIN CLINIC
셀린의원 홍대점

236 JEILSOMANG CHURCH
제일소망교회

OGANACELL DERMATOLOGY CLINIC

INTOEX | Yunjun Yang

오가나셀은 국내에서 가장 높은 마천루인 롯데월드타워에 위치한 프리미엄 피부과 의원이다. 일반 높이의 건물에서는 느낄 수 없는 압도적인 전망과 화사한 채광을 살리는 일에 초점을 두었으며 병원이 갖춰야 하는 기능적 공간을 모두 갖추면서도 주변을 한 눈에 조망할 수 있도록 대기공간에 임팩트를 주었다. 자연으로부터 들어오는 빛과 자연스러운 조망을 그대로 받아들이는 공간이기에 전체적인 색감도 따뜻한 우드와 절제된 톤의 스타코를 사용해 내추럴하고 부드러운 이미지를 드러냈다. 목재 특유의 질감과 이미지를 고조하기 위해 그 외의 부분은 안정적인 아이보리, 그레이로 지지함으로써 부담스럽지 않은 컬러와 자연스러운 소재가 외부의 빛과 조망으로 어우러짐을 의도했다. 대기 공간은 창가로 향할수록 부채꼴처럼 넓게 펼쳐지는 형태로 구획했다. 채광을 최대한 살리고자 창가 쪽으로 대기 공간을 넓게 열어주고 입구 부분의 좁아지는 매스는 전면부에 여유로운 공간감을 형성한다. 복도로 들어서면 유선형이라는 건물의 특징이 연결된 곡선과 직선이 조화롭게 어우러지며 공간을 풍성하게 채워준다. 파사드부터 이어진 우드벽체는 곡선 라인을 통해 차례로 천장과 복도로 확장된다. 덕분에 벽체, 천장 등의 요소가 서로 나뉘지지 않고 하나의 큰 공간으로 인식되게 의도했으며 부드러운 선의 흐름이 자연스러운 동선을 유도한다. 대기 공간이 채광을 최대한 끌어올려 밝고 화사한 느낌이었다면 복도부터 이어지는 실들은 상대적으로 차분하다. 더욱 진중하게 치료에만 집중하도록 색감도 한 톤을 낮춘 그레이로 맞추고 조명도 최소화했다. 어둑한 공간에 간접조명을 이어 동선을 유도했으며 타인의 시선으로부터 방해받지 않도록 중간 대기 공간도 개별적으로 나뉘 앉을 수 있게 배치했다.

Oganacell is a premium dermatology clinic located in Lotte World Tower, the tallest skyscraper in Korea. In this context, designer's accent was on maintaining the awe-inspiring views and rich natural light, which are unmatched by typical buildings of regular height. While ensuring the clinic meets all necessary functional standards, they placed special emphasis on crafting a striking waiting area that offers panoramic vistas of the surroundings. Oganacell Dermatology Clinic Jamsil Branch embraces serene views and natural light, which are reflected through warm wood tones and subtle Stucco hues, offering a soothing ambiance. By boldly utilizing transparent glass instead of separate signage, the exterior view is seamlessly invited in. To maximize natural light, the waiting area extends generously towards the windows, while the narrowing structures at the entrance increase a feeling of openness upfront. On entering the corridor, the fusion of curved and straight architectural elements consisting of streamed lines enriches the space with abundance. Wood walls, extending from the entrance, trace the curved lines, gradually expanding to encompass the ceiling and corridor. This integration creates a sense of unity, where walls and ceilings blend into one expansive space, facilitating a fluid circulation guided by natural lines. While the waiting area exudes a bright and lively atmosphere by enhancing natural light, the connecting corridors adopt a more subdued ambiance. To foster an environment conducive to treatment, the color palette is toned down to a lighter shade of gray, reducing the dependence on artificial lighting. Indirect lighting is employed to navigate through the dimmer spaces, and the middle waiting area is designed to offer individual privacy, shielding patients from external distractions.

디자인 양윤준 / 인투익스
위치 서울특별시 송파구 올림픽로 300, 11층
용도 피부과 의원
면적 313㎡
마감 바닥 – 포세린 타일 / 벽 – 페인트, 은경, 우드 필름, 패브릭 / 천장 – 비닐 페인트
완공 2023. 9
디자인팀 유예영, 강신영
시공팀 구광현
사진 강명국

Location 11F, 300, Olympic-ro, Songpa-gu, Seoul
Use Dermatology clinic
Area 313㎡
Finishing Floor - Porcelain tile / Wall - Paint, Mirror, Wood film, Fabric / Ceiling - Vinyl paint
Completion 2023. 9
Design team Yeyoung You, Shinyoung Kang
Construction team Gwanghyeon Ku
Photographer Myungguk Kang

평면도 / floor plan

1 입구 2 인포메이션 데스크 3 대기실 4 상담실 5 진료실 6 사진실 7 파우더룸 8 직원실 9 관리준비실 10 VIP 관리실 11 피부관리실 12 레이저실 13 시술실

1 Entrance 2 Information desk 3 Waiting room 4 Consultation room 5 Doctor's office 6 Photo room 7 Powder room 8 Staff room 9 Preparation room 10 VIP skincare room 11 Skincare room 12 Laser room 13 Procedure room

217

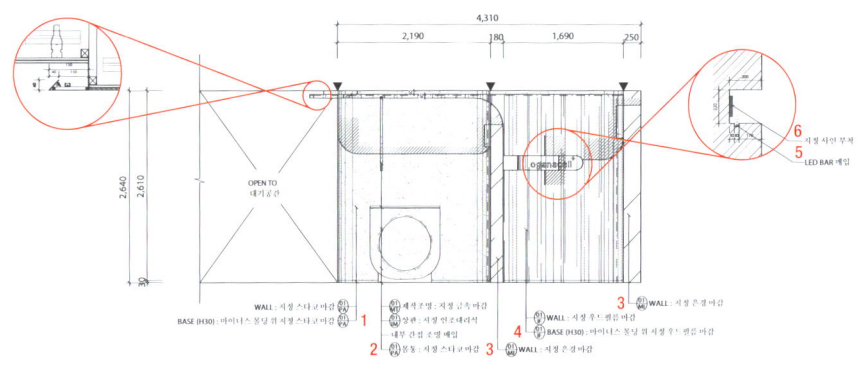

리셉션 입면 A / reception elevation A

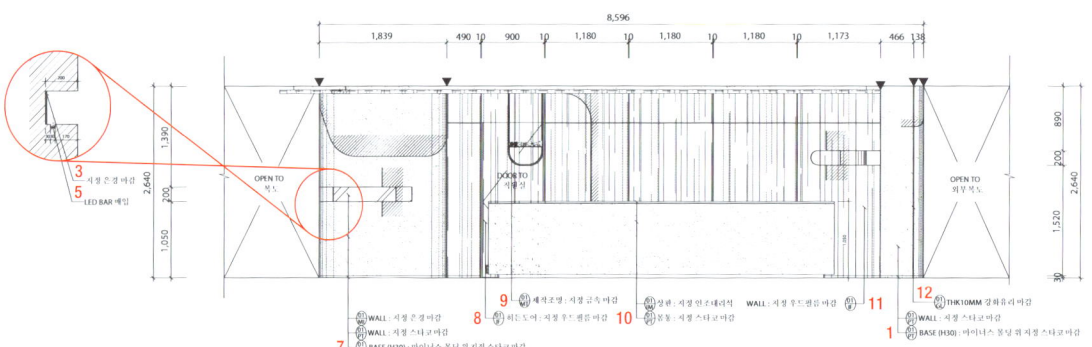

리셉션 입면 B / reception elevation B

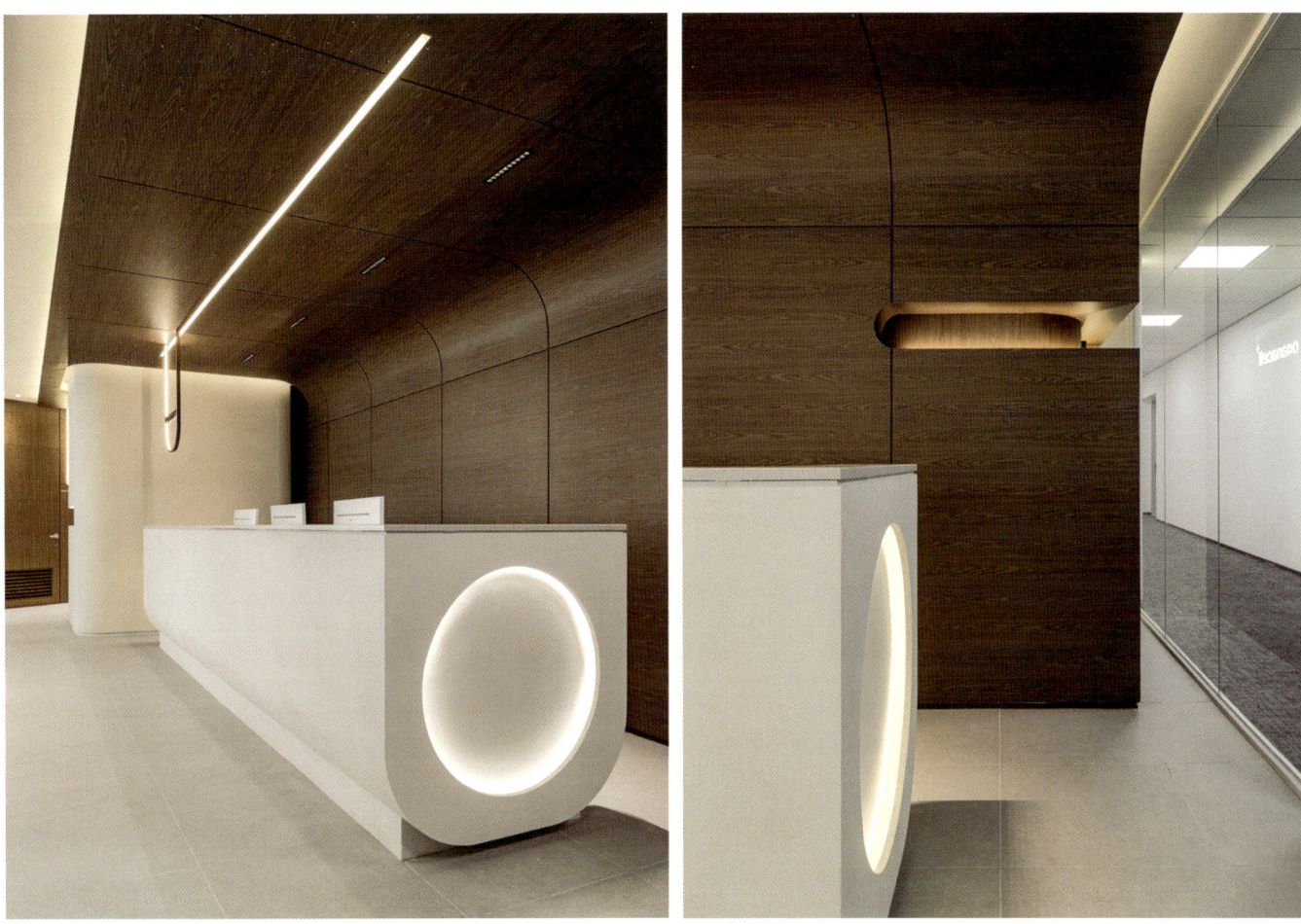

1 벽 : 지정 스타코 / 걸레받이 : 마이너스 몰딩 위 지정 스타코 2 제작 조명 : 지정 금속 / 상판 : 지정 인조대리석 / 내부 간접조명 / 몸통 : 지정 스타코 3 지정 은경 4 벽 : 지정 우드 필름 / 걸레받이 : 마이너스 몰딩 위 지정 우드 필름 5 LED 바 6 지정 사인 7 벽 : 지정 은경 / 벽 : 지정 스타코 / 걸레받이 : 마이너스 몰딩 위 지정 스타코 8 히든 도어 : 지정 우드 필름 9 제작 조명 : 지정 금속 10 상판 : 지정 인조대리석 / 몸통 : 지정 스타코 11 벽 : 지정 우드 필름 12 T10 강화유리 13 문 : T10 강화유리 / 문 프레임 : 지정 도장

1 Wall : App. stucco / Base : App. stucco on minus moulding 2 Custom-made light : App. metal / Top : App. engineered marble / Inside indirect lighting / Base : App. stucco 3 App. mirror 4 Wall : App. wood film / Base : App. wood film on minus moulding 5 LED bar 6 App. sign 7 Wall : App. mirror / Wall : App. stucco / Base : App. stucco on minus moulding 8 Hidden door : App. wood film 9 Custom-made light : App. metal 10 Top : App. engineered marble / Base : App. stucco 11 Wall : App. wood film 12 T10 tempered glass 13 Door : T10 tempered glass / Door frame : App. painting

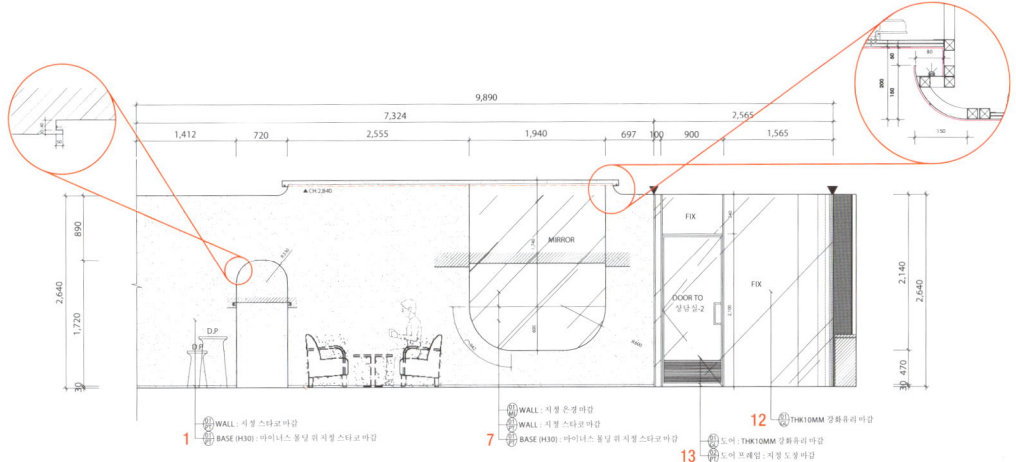

대기공간 입면 C / waiting area elevation C

219

1 상판 : T12 테라조 2 데스크 상판 : 지정 인테리어 필름 / T5 강화유리 3 상판 : T12 테라조 / 몸통 : 지정 스타코 도장, 투명 무광 코팅 / 하부 : 지정 스타코 도장, 투명 무광 코팅 4 데스크 상판 : 지정 인테리어 필름, T5 강화유리 5 T10 강화유리 위 안전 필름 6 F.C.U 지정 도장 7 벽 : 지정 스타코 도장 / 걸레받이 : 마이너스 몰딩 위 지정 스타코 도장 8 벽 : 지정 우드 필름 9 상부 간접조명 / 선반 하부 간접조명 10 상부 간접조명 / 벽 : 지정 스타코 도장 / 걸레받이 : 마이너스 몰딩 위 지정 스타코 도장

1 Top : T12 terazzo 2 Desk top : App. interior film / T5 tempered glass 3 Top : T12 terazzo / Base : App. stucco painting, Clear matt coating / Lower : App. stucco painting, Clear matt coating 4 Desk top : App. interior film, T5 tempered glass 5 Safety film on T10 tempered glass 6 F.C.U : App. painting 7 Wall : App. stucco painting / Base : App. stucco painting on minus moulding 8 Wall : App. wood film 9 Upper indirect lighting / Indirect lighting below shelf 10 Upper indirect lighting / Wall : App. studdo painting / Base : App. stucco painting on minus moulding

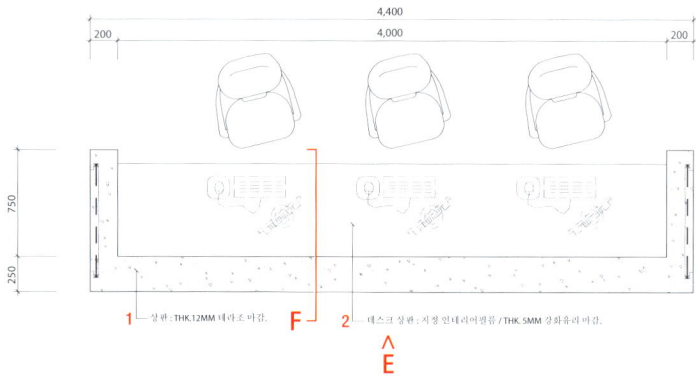

인포메이션 데스크 평면 D / information desk top view D

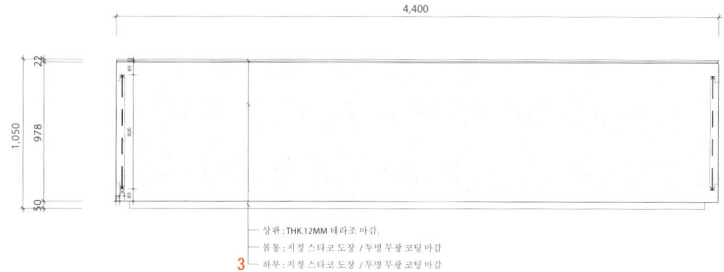

인포메이션 데스크 정면 E / information desk front view E

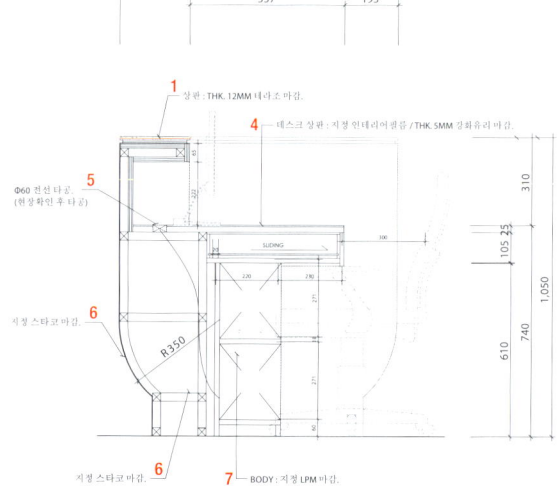

인포메이션 데스크 단면 F / information desk section F

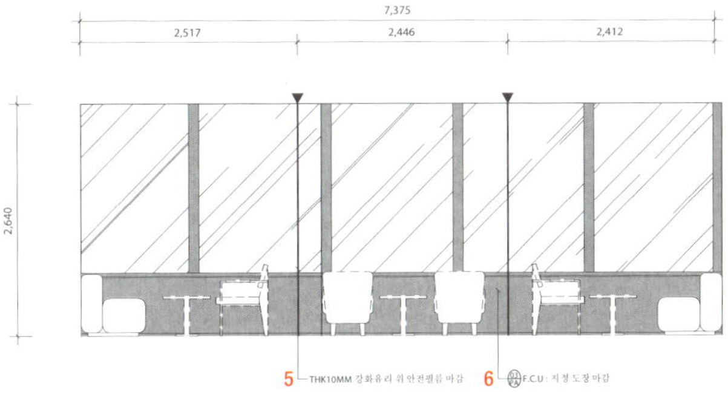

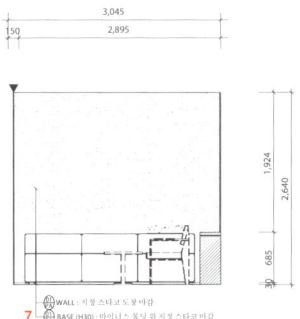

대기공간 입면 G / waiting area elevation G

상담실 입면 H / consultation room elevation H

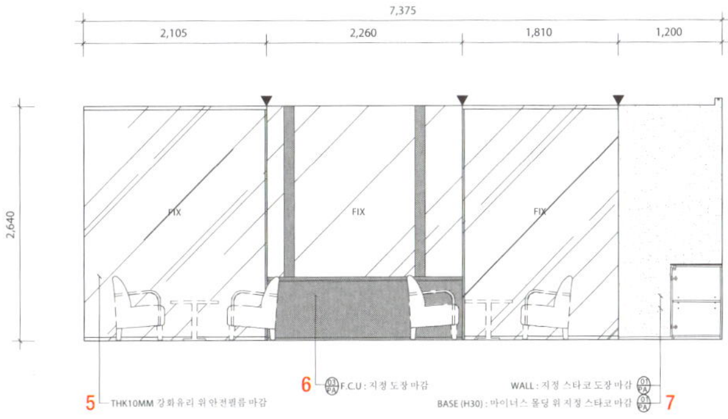

상담실 입면 I / consultation room elevation I

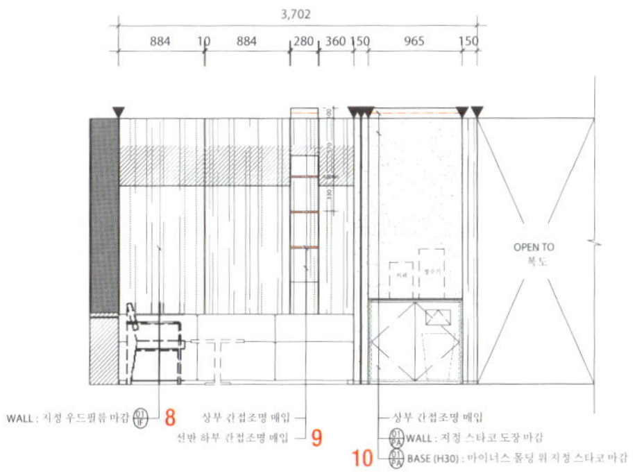

상담실 입면 J / consultation room elevation J

1 철제 그릴 설치   2 지정 LED   3 문 프레임 : 지정 인테리어 필름   4 인방 : 지정 인테리어 필름 / 문 : 지정 인테리어 필름   5 벽 : 지정 우드 필름 / 걸레받이 : 마이너스 몰딩 위 지정 우드 필름   6 프레임 : 지정 도장   7 상부 간접조명 / 벽 : 지정 패브릭 / 좌석 : 지정 패브릭 / 좌석 하부 간접조명   8 벽 : 지정 스타코 / 걸레받이 : 마이너스 몰딩 위 지정 스타코   9 지정 은경   10 벽 : 지정 타일 / 지정 인조대리석 / 지정 LPM   11 게이트 : 지정 도장   12 벽 : 지정 도장 / 걸레받이 : 마이너스 몰딩 위 지정 스타코 도장   13 프레임 : 지정 필름   14 문 : 지정 필름 / 철제 그릴   15 지정 커튼   16 프레임 위 지정 인테리어 필름   17 벽 : 지정 벽지 / 걸레받이 : 몰딩 위 지정 필름

1 Steel grill installation   2 App. LED   3 Door frame : App. interior film   4 Lintel : App. interior film / Door : App. interior film   5 Wall : App. wood film / Base : App. wood film on minus moulding   6 Frame : App. painting   7 Upper indirect lighting / Wall : App. fabric / Seat : App. fabric / Indirect lighting below seat   8 Wall : App. stucco / Base : App. stucco on minus moulding   9 App. mirror   10 Wall : App. tile / App. engineered marble / App. LPM   11 Gate : App. painting   12 Wall : App. painting / Base : App. stucco painting on minus moulding   13 Frame : App. film   14 Door : App. film / Steel grill   15 App. curtain   16 App. interior film on frame   17 Wall : App. wallpaper / Base : App. film on moulding

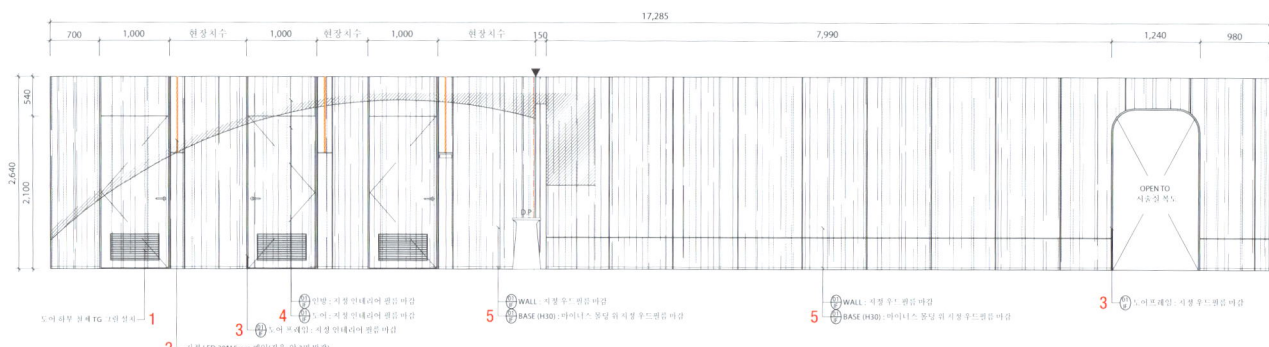

복도 입면 K / corridor elevation K

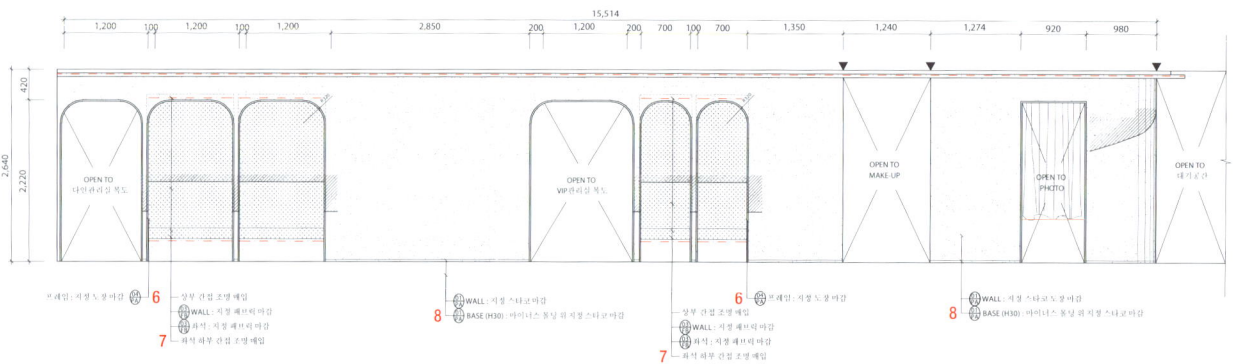

복도 입면 L / corridor elevation L

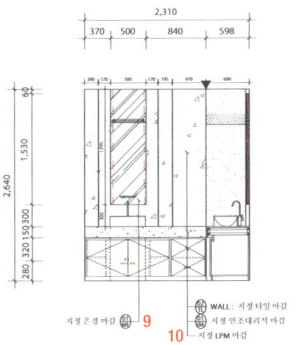

파우더룸 입면 M1 /
powder room elevation M1

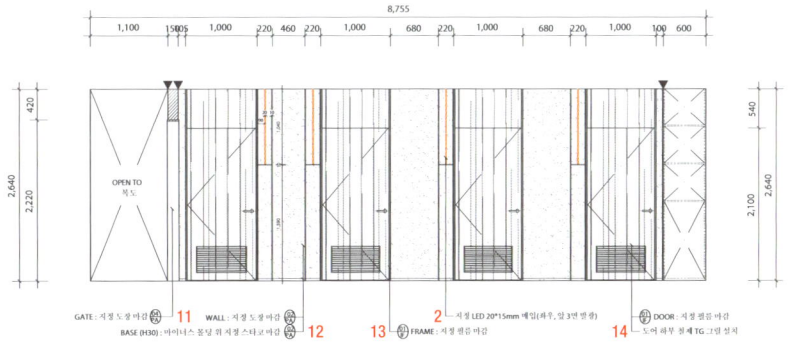

복도 입면 N / corridor elevation N

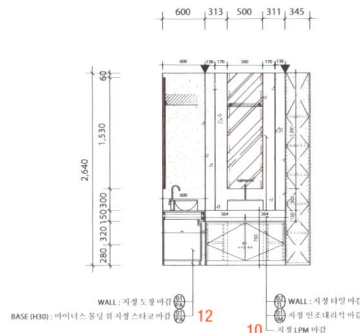

파우더룸 입면 M2 /
powder room elevation M2

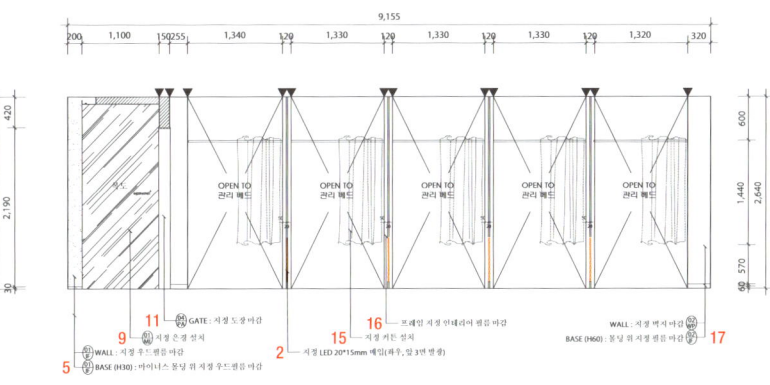

복도 입면 O / corridor elevation O

CELLIN CLINIC

INTOEX | Yunjun Yang

디자이너는 기존의 셀린 지점들과의 비교적 유사한 공간 분위기를 느끼되 홍대점만의 공간 장점을 극대화하여 셀린의원에 새로운 바람을 넣어주고 싶었다. 이 공간의 강점은 약 5m의 높은 천장고와 대로를 바라볼 수 있는 스카이 라운지 타입의 탁 트인 빌딩의 뷰를 가지고 있다는 것이었다. 클라이언트는 무엇보다 이 두 가지의 강점을 적극 반영하길 원하였고, 디자이너는 공간 안에 뷰의 극대화, 높은 천고의 쾌적한 환경 조성, 정해져 있는 금액 안에서 공간 구성과 표현, 이 세 가지에 가장 큰 중점을 두고 디자인을 시작하였다. 메인 입구의 파사드는 군더더기 없이 부드러운 선의 형태를 복도에서부터 고객을 맞이하는 공간까지 끌어 들어오도록 하였으며, 그 끝과 주변에 곡선과 소재를 동일하게 반복적으로 사용하여 공간을 입면적으로 풍부하고 자연스럽게 표현하였다. 메인 공간을 지나쳐 조금은 어스름한 양방향의 복도를 지나면 강렬한 컬러의 다크 레드의 벽과 천장을 마주한다. 파우더룸과 시술 전의 고객이 머무르는 공간을 하나로 묶어 홍대 셀린만의 유니크한 공간을 만들어 주고자 하였다. 로비를 다양한 높낮이, 둥근 선들로 공간을 강조하였다면, 중간 허리 공간은 강한 매스감을 가진 하나의 형태감, 강렬하지만 간접적인 빛들로 공간을 강조하고자 하였다.

Designer's goal was to create an atmosphere similar to the existing CELLIN Branches, while also maximizing the unique spatial advantages of the Hongdae location, bringing a fresh touch to the CELLIN Clinic. Upon entering the area, one immediately notices the Hongdae Branch's prominent spatial advantages, highlighted by the Sky Lounge-style space featuring a lofty 5-meter-high ceiling and a commanding view of the big avenue below. The client expressed a strong desire to fully incorporate these two strengths into the design. Therefore, they focused on three aspects for the design: maximizing the view within the space, creating a pleasant environment with high ceilings, ensuring optimal spatial layout and expression within the given budget. They designed the main entrance façade to feature seamless lines and smooth shapes, creating a welcoming transition from the corridor to the reception area. By consistently repeating rounded lines and materials at the end and around the space, they enhanced the area with elevations that express these elements naturally. After traversing a slightly dim, two-way corridor beyond the main area, clients are met with striking dark red walls and ceilings. Their goal was to craft a distinctive space for clients to stay, merging both the powder room and pre-procedure room into a unified and characteristic space representative of the CELLIN Clinic Hongdae Branch. While they accentuated the space in the lobby with diverse heights and curved lines, their intention for the midsection area was to create a solid and cohesive form of mass, enhanced by intense yet indirect lighting.

디자인 양윤준 / 인투익스
위치 서울특별시 마포구 양화로 140, 1층
용도 피부과 의원
면적 394.09㎡
마감 바닥 – 포세린 타일, 콩자갈, 데코타일 / 벽 – 스타코, 유리, 은경, 목망, 벽지 / 천장 – 비닐페인트, 벽지
완공 2023. 12
디자인팀 구한나, 강신영
사진 강명국

Location 1F, 140, Yanghwa-ro, Mapo-gu, Seoul
Use Dermatoloty clinic
Area 394.09㎡
Finishing Floor - Porcelain tile, Pebbles, Deco tile / Wall - Stucco, Glass, Mirror, Wooden lattice, Wallpaper / Ceiling - Vinyl paint, Wallpaper
Completion 2023. 12
Design team Hanna Ku, Shinyoung Kang
Photographer Myungguk Kang

평면도 / floor plan

1 입구 2 상담실 3 인포메이션 데스크 4 대기실 5 탈의실 6 사진실 7 진료실 8 라커룸/파우더룸 9 시술실 10 시술준비실 11 레이저실 12 VIP룸 13 관리준비실 14 사무실 15 제모실 16 직원실

1 Entrance 2 Consultation room 3 Information desk 4 Waiting room 5 Changing room 6 Photo room 7 Doctor's office 8 Locker room / Powder room 9 Procedure room 10 Preparation room for procedure 11 Laser room 12 VIP room 13 Preparation room for skincare 14 Office 15 Hair removal room 16 Staff room

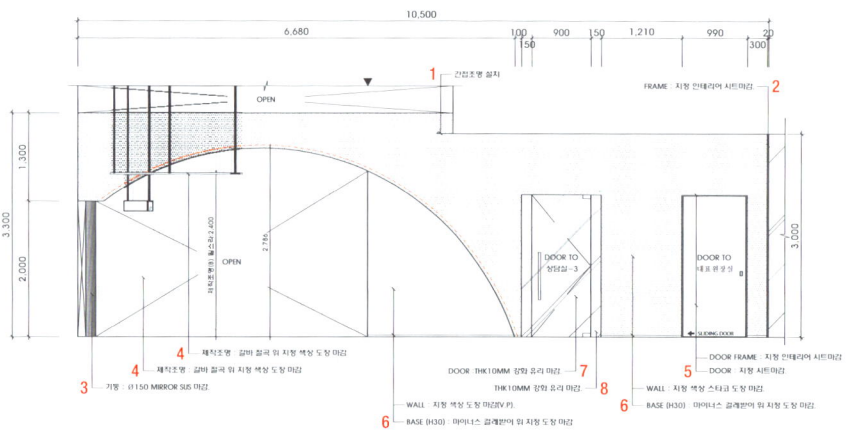

대기실 입면 A / waiting room elevation A

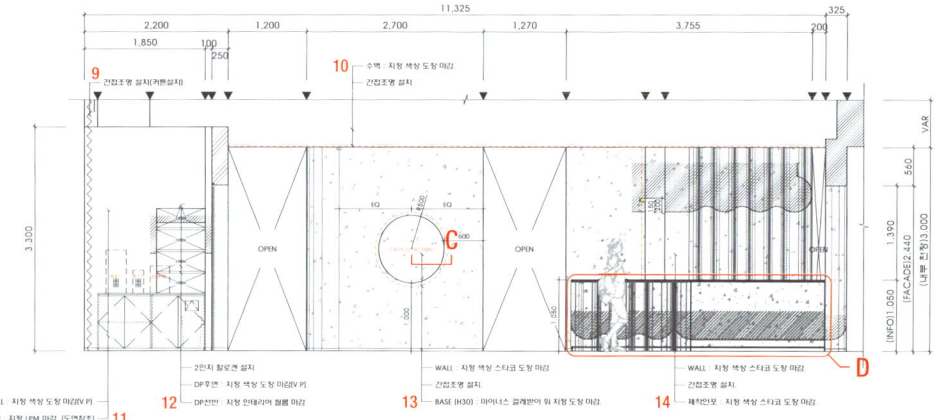

리셉션 입면 B / reception elevation B

벽 단면 상세 C / wall section detail C

1 간접조명 2 프레임 : 지정 인테리어 시트 3 기둥 : Ø150 SUS 미러 4 제작 조명 : 갈바륨 절곡 위 지정색 도장 5 문 프레임 : 지정 인테리어 시트 / 문 : 지정 시트 6 벽 : 지정색 비닐페인트 / 걸레받이 : 마이너스 걸레받이 위 지정 도장 7 문 : T10 강화유리 8 T10 강화유리 9 간접조명(커튼 설치) 10 수벽 : 지정색 도장 / 간접조명 11 벽 : 지정색 비닐페인트 / 제작 가구 : 지정 LPM 12 2" 할로겐 / 후면 : 지정색 비닐페인트 / 선반 : 지정 인테리어 필름 13 벽 : 지정색 스타코 도장 / 간접조명 / 걸레받이 : 마이너스 걸레받이 위 지정 도장 14 벽 : 지정색 스타코 도장 / 간접조명 / 지정색 스타코 도장 15 T9.5 석고보드 2겹 위 지정 벽지 / KS 65 스터드 / T9.5 석고보드 2겹 위 지정색 스타코 도장 / T9.5 석고보드 2겹 위 지정색 스타코 도장 16 간접조명 / T1.2 갈바륨 절곡 위 지정색 스타코 도장 17 지정 LPM / T12 지정 인조대리석 18 상판 타공 / 인포메이션 하부 마감 라인 19 상판 : T12 지정 인조대리석 / 몸통 : 지정 스타코 도장 / 걸레받이 : 지정 스타코 도장 20 T12 지정 인조대리석 21 지정 LPM, T5 강화유리 22 T9.5 석고보드 / 몸통 : T1.2 갈바륨 절곡 위 지정 스타코 도장

1 Indirect lighting 2 Frame : App. interior sheet 3 Column : Ø150 SUS mirror 4 Custom-made light fixture : App. color painting on galvalume bending 5 Door frame : App. interior sheet / Door : App. sheet 6 Wall : App. color vinyl paint / Base : App. painting on minus baseboard 7 Door : T10 tempered glass 8 T10 tempered glass 9 Indirect lighting (curtain installation) 10 Reveal : App. color painting / Indriect lighting 11 Wall : App. color vinyl paint / Custom-made furniture : App. LPM 12 2" Hallogen / Back : App. color vinyl paint / Shelf : App. interior film 13 Wall : App. color stucco painting / Indirect lighting / Base : App. painting on minus baseboard 14 Wall : App. color stucco painting / Indirect lighting / App. color stucco painting 15 App. wallpaper on T9.5 gypsum board 2ply / KS 65 stud / App. color stucco painting on T9.5 gypsum board 2ply / App. color stucco painting on T9.5 gypsum board 2ply 16 Indirect lighting / App. color stucco painting on T1.2 galvalume bending 17 App. LPM / T12 app. engineered marble 18 Top perforation / Lower information desk finishing line 19 Top : T12 engineered marble / Body : App. stucco painting / Base : App. stucco painting 20 T12 app. engineered marble 21 App. LPM, T5 tempered glass 22 T9.5 gypsum board / Body : App. stucco painting on T1.2 galvalume

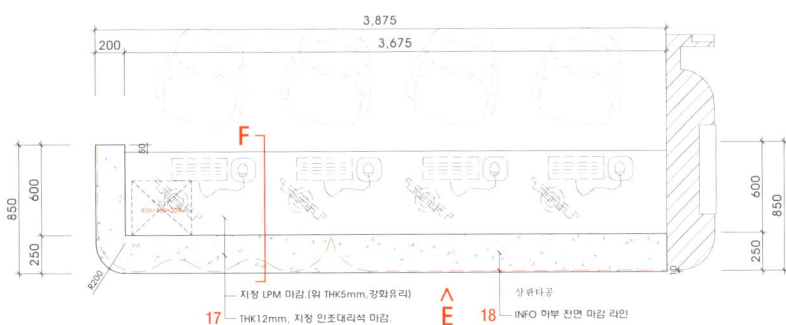

리셉션 데스크 평면 D / reception desk top view D

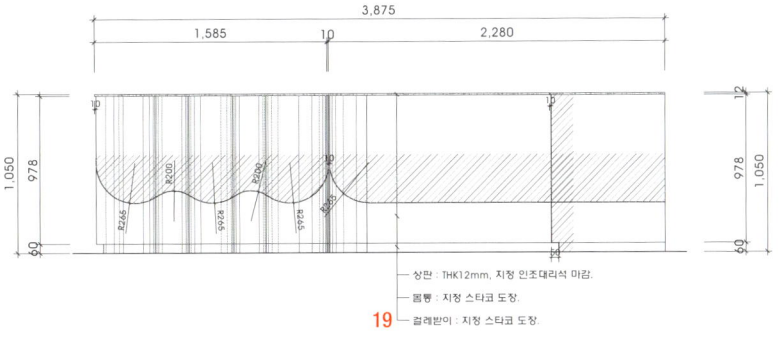

리셉션 데스크 정면 E / reception desk front view E

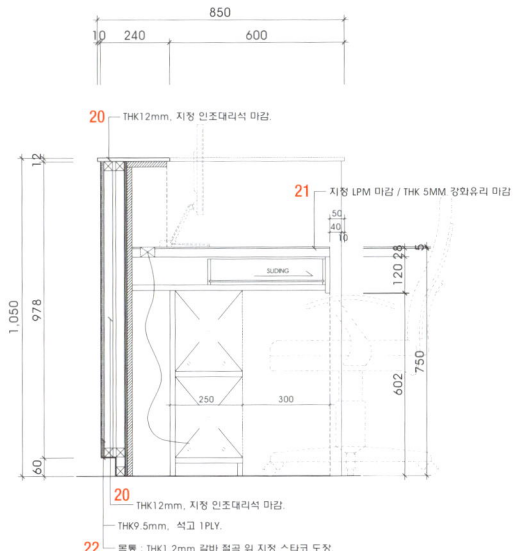

리셉션 데스크 단면 F / reception desk section F

1 간접조명 2 제작 조명 : 갈바륨 절곡 위 지정색 도장 3 수벽 : 지정색 도장 4 벽 : 지정색 스타코 도장 / 걸레받이 : 마이너스 걸레받이 위 지정 도장 5 자동문 : T10 강화유리 / 바닥 : 슬로프 형성 6 T10 강화유리 7 프레임 : 지정 인테리어 시트 8 점검구 가구 : 지정색 인테리어 시트 / 벽 : 지정색 스타코 도장 / 걸레받이 : 마이너스 걸레받이 위 지정 도장 9 T1.2 갈바륨 절곡 위 지정색 도장 / T1.2 갈바륨 절곡 위 지정색 도장 10 상담실 천장 : 지정색 스타코 도장 / 간접조명 11 수벽 : 지정색 도장 / 간접조명 12 T10 강화유리 / 문 : T10 강화유리 13 상담실 벽 : 지정색 스타코 도장 14 지정 커튼 15 기둥 : Ø150 SUS 미러 16 기둥 : 은경 설치 17 코너 몰딩 : SUS 헤어라인 18 KS 65 스터드 / T9.5 석고보드 / T9 합판 / T5 은경 19 벽 : 지정색 스페셜 도장

1 Indirect lighting 2 Custom-made ligh fixture : App. color painting on galvalume bending 3 Reveal : App. color painting 4 Wall : App. color stucco painting / Base : App. painting on minus baseboard 5 Automatic door : T10 tempered glass / Floor : Slope formation 6 T10 tempered glass 7 Frame : App. interior sheet 8 Access door : App. color interior sheet / Wall : App. color stucco painting / Base : App. painting on minus baseboard 9 App. color painting on T1.2 galvalume bending / App. color painting on T1.2 galvalume bending 10 Consultation room ceiling : App. color stucco painting / Indirect lighting 11 Reveal : App. color painting / Indirect lighting 12 T10 tempered glass / Door : T10 tempered glass 13 Consultation room wall : App. color stucco painting 14 App. curtain 15 Column : Ø150 SUS mirror 16 Column : Mirror installation 17 Corner moulding : SUS hairline 18 KS 65 stud / T9.5 gypsum board / T9 plywood / T5 mirror 19 Wall : App. color special painting

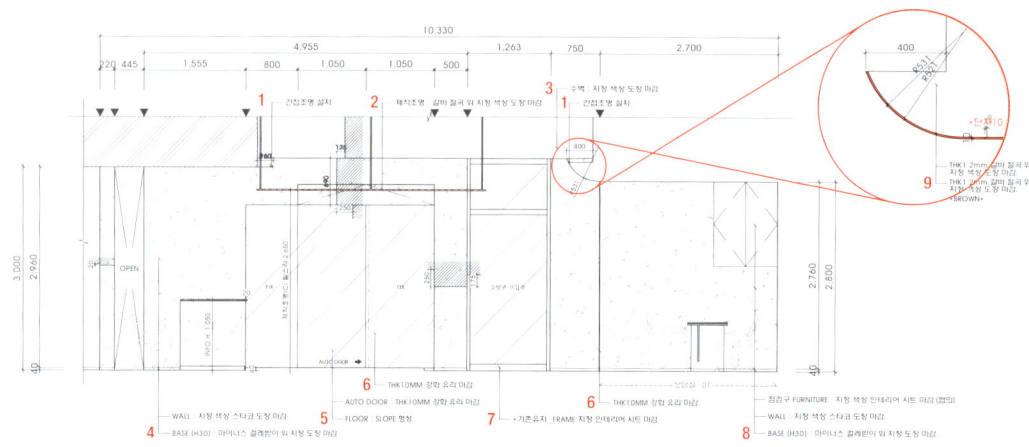

대기실 입면 G / waiting room elevation G

대기실 입면 H / waiting room elevation H

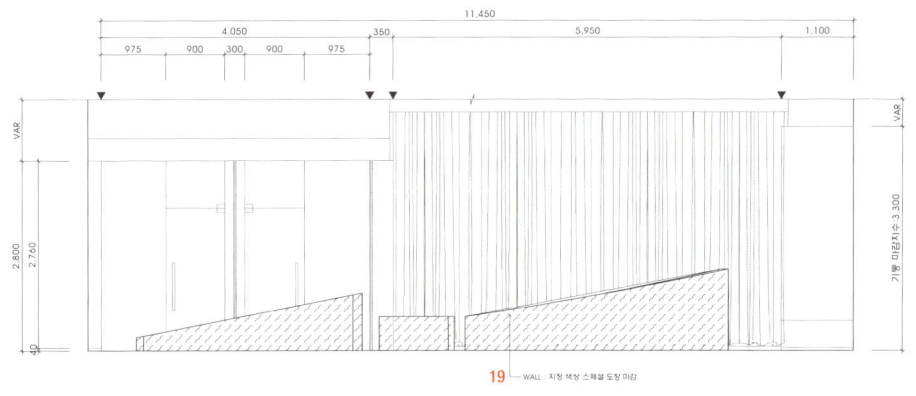

대기실 입면 I / waiting room elevation I

1 프레임 : 지정 인테리어 시트 2 T10 강화유리 3 문 : T10 강화유리 4 프레임 : 지정 인테리어 시트 / 지정 커튼 5 벽 : 지정색 비닐페인트 도장 / 걸레받이 : 마이너스 걸레받이 위 지정 도장 6 게이트 : 지정색 도장 / 벽부등 7 T9.5 석고보드 2겹 위 지정 벽지 / KS 65 스터드 / T9.5 석고보드 2겹 위 지정 스타코 도장 8 간접조명 9 제작 펜던트 조명 10 수벽 : 지정색 도장 11 금속 절곡 위 지정색 도장 12 벽 : 지정색 스타코 도장 / 걸레받이 : 마이너스 걸레받이 위 지정 도장 13 금속 절곡 위 지정 인테리어 시트 14 상판 : T1.2 SUS 헤어라인 절곡 / 몸통 : 지정색 스페셜 도장 / 제작 소파 : 지정 패브릭 15 게이트 : 지정색 도장 16 지정 펜던트 17 벽 : 은경 18 재료분리대 : 금속 절곡 위 지정색 도장 19 목망 위 지정색 도장 / 걸레받이 : 마이너스 걸레받이 위 지정색 도장 20 T1.2 갈바륨 절곡 위 인테리어 시트 21 □30X30 각파이프 보강대 / T9.5 석고보드 / T9 합판 / T5 거울 22 □30X30 각파이프 보강대 / T9.5 석고보드 2겹 위 지정색 도장 / 목망 위 지정색 도장 23 T1.2 갈바륨 절곡 위 지정색 도장 24 T1.2 갈바륨 절곡 위 지정색 스타코 도장

1 Frame : App. interior sheet 2 T10 tempered glass 3 Door : T10 tempered glass 4 Frame : App. interior sheet / App. curtain 5 Wall : App. color vinyl paint / Base : App. painting on minus baseboard 6 Gate : App. color painting / Wall lamp 7 App. wallpaper on T9.5 gypsum board 2ply / KS 65 stud / App. stucco painting on T9.5 gypsum board 2ply 8 Indirect lighting 9 Custom-made pendant light 10 Reveal : App. color painting 11 App. color painting on metal bending 12 Wall : App. color stucco painting / Base : App. painting on minus baseboard 13 App. interior sheet on metal bending 14 Top : T1.2 SUS hairline bending / Body : App. special painting / Custom-made sofa : App. fabric 15 Gate : App. color painting 16 App. pendant light 17 Wall : Mirror 18 Material separator : App. color painting on metal bending 19 App. color painting on wooden lattice / Base : App. color painting on minus baseboard 20 Interior sheet on T1.2 galvalume bending 21 □30X30 square pipe / T9.5 gypsum board / T9 plywood / T5 mirror 22 □30X30 square pipe / App. color painting on T9.5 gypsum board 2ply / App. color painting on wooden lattice 23 App. color painting on T1.2 galvalume bending 24 App. color stucco painting on T1.2 galvalume bending

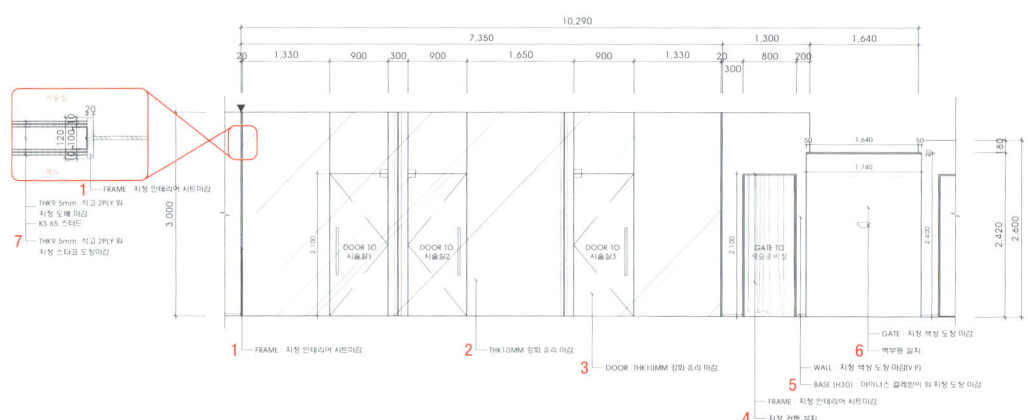

복도 입면 J / corridor elevation J

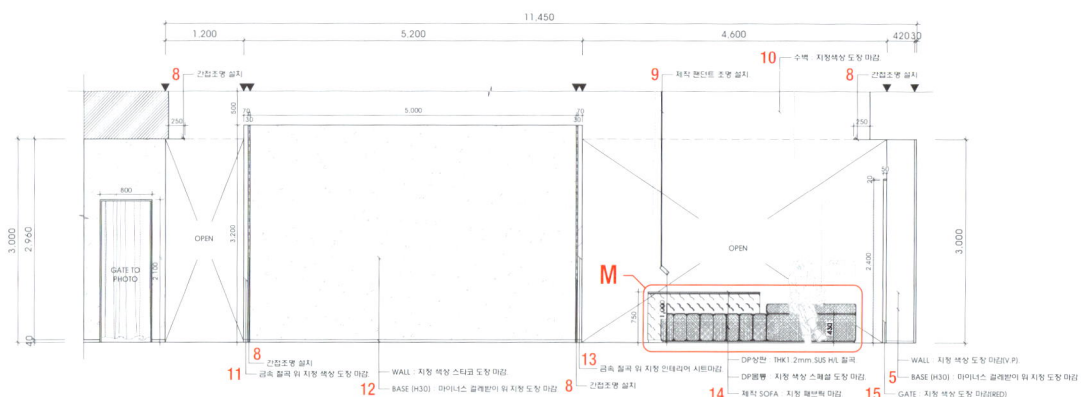

대기실 입면 K / waiting room elevation K

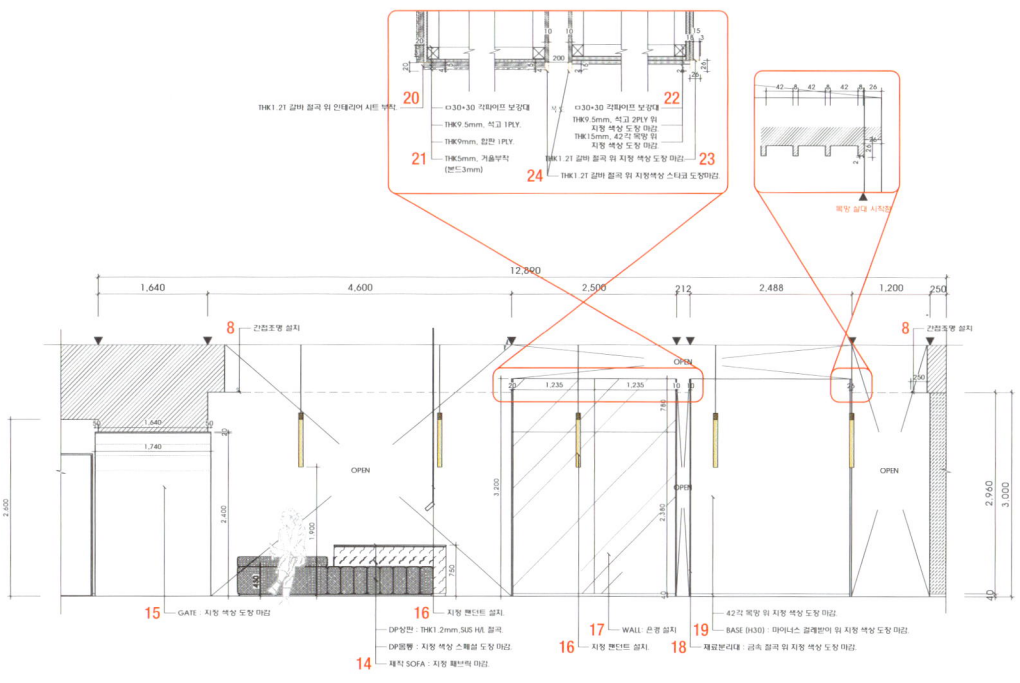

대기실 입면 L / waiting room elevation L

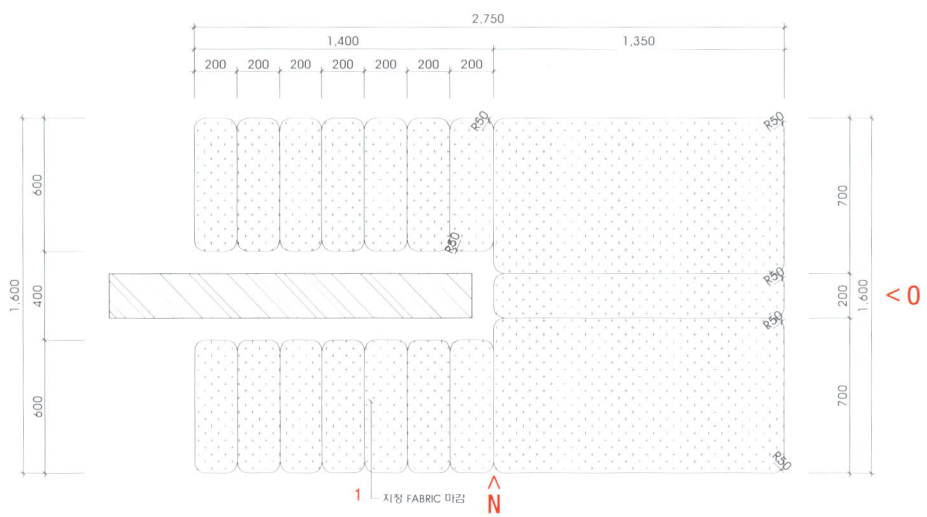

소파 평면 M / sofa top view M

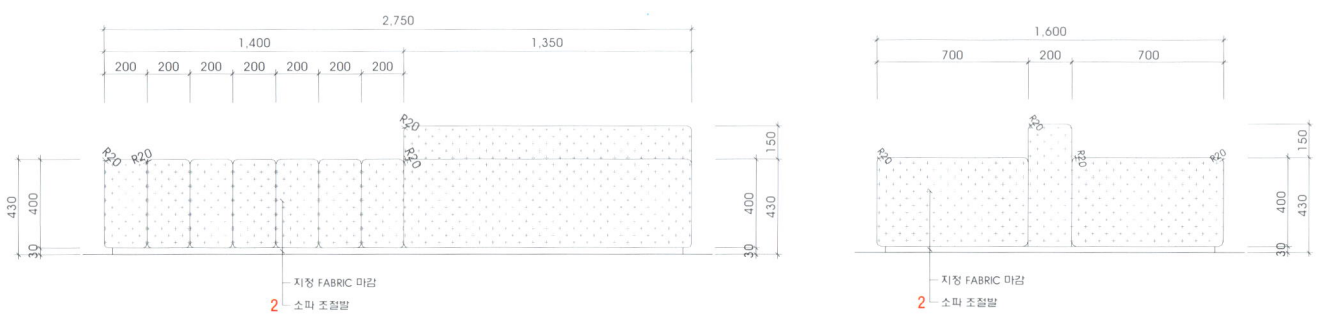

소파 정면 N / sofa front view N

소파 측면 O / sofa side view O

1 지정 패브릭 2 지정 패브릭 / 소파 조절발 3 프레임 : 지정 인테리어 시트 4 벽 : 지정색 스타코 도장 / 걸레받이 : 마이너스 걸레받이 위 지정 도장 5 제작 가구 : 지정 LPM 6 프레임 : T1.2 SUS 헤어라인 절곡 7 상판 & 물턱 : T12 인조대리석 / 제작가구 : 지정 LPM 8 T5 은경(후면 간접조명) 9 벽 : 지정색 스타코 도장 / 벽부등 / 콘센트 10 T9 합판 / T5 은경 11 T40 / 간접조명

1 App. fabric 2 App. fabric / Sofa adjustable feet 3 Frame : App. interior sheet 4 Wall : App. color stucco painting / Base : App. painting on minus baseboard 5 Custom-made furniture : App. LPM 6 Frame : T1.2 SUS hairline bending 7 Top & Water sill : T12 engineered marble / Custom-made furniture : App. LPM 8 T5 mirror (back indirect lighting) 9 Wall : App. color stucco painting / Wall lamp / Outlet 10 T9 plywood / T5 Mirror 11 T40 / Indirect lighting

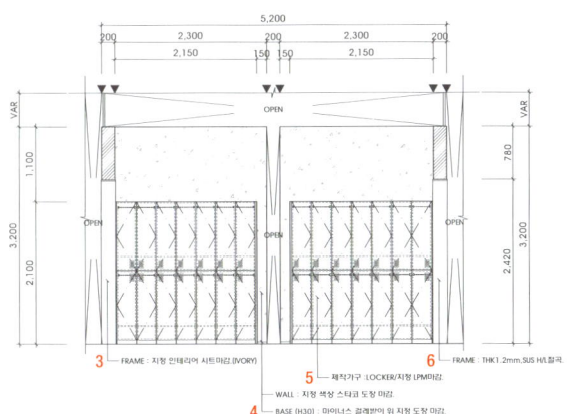

파우더룸 입면 P1 / powder room elevation P1

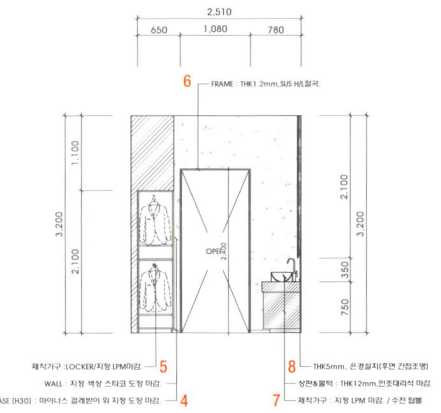

파우더룸 입면 P2 / powder room elevation P2

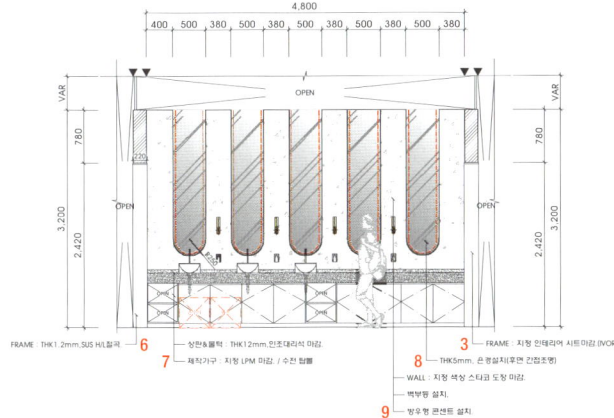

파우더룸 입면 P3 / powder room elevation P3

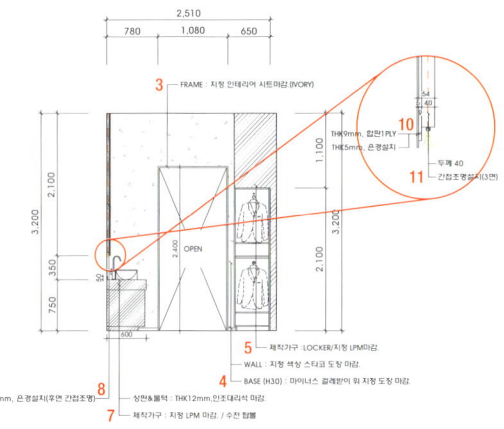

파우더룸 입면 P4 / powder room elevation P4

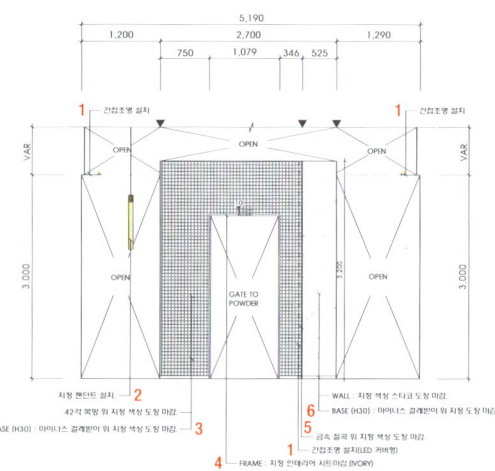

대기실 입면 Q / waiting room elevation Q

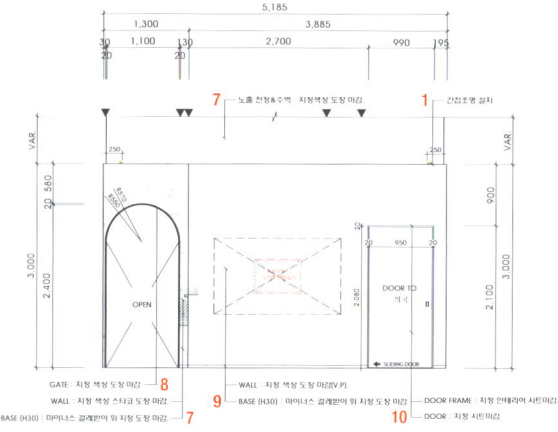

대기실 입면 R / waiting room elevation R

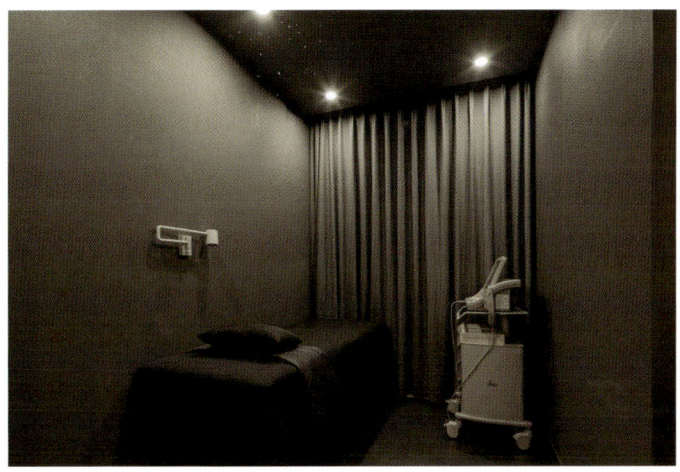 

1 간접조명 2 지정 펜던트 조명 3 목망 위 지정색 도장 / 걸레받이 : 마이너스 걸레받이 위 지정색 도장 4 프레임 : 지정 인테리어 시트 5 금속 절곡 위 지정색 도장 6 벽 : 지정색 스타코 도장 / 걸레받이 : 마이너스 걸레받이 위 지정 도장 7 노출천장 & 수벽 : 지정색 도장 8 게이트 : 지정색 도장 9 벽 : 지정색 비닐페인트 / 걸레받이 : 마이너스 걸레받이 위 지정색 도장 10 문 프레임 : 지정 인테리어 시트 / 문 : 지정 시트 11 게이트 : 지정 시트 12 벽 : 지정 벽지 / 걸레받이 : 평몰딩 위 지정 시트 13 파티션 : 지정 시트 / 조명 : 2인치 매입등 14 조명 설치 15 □50X50 각파이프 보강대 / T9.5 석고보드 2겹 위 지정 도배 16 2인치 할로겐 매입 / T9 합판 / □30X30 각파이프 보강대 / T9 합판 위 지정 시트 17 파티션 상단 : T9 MDF 위 지정 인테리어 시트 / 벽 : 지정 벽지 / 걸레받이 : 평몰딩 위 지정 시트

1 Indirect lighting 2 App. pendant light 3 App. color painting on wooden lattice / Base : App. painting on minus baseboard 4 Frame : App. interior sheet 5 App. color painting on metal bending 6 Wall : App. color stucco painting / Base : App. painting on minus baseboard 7 Exposed ceiling & Reveal : App. color painting 8 Gate : App. color painting 9 Wall : App. color vinyl paint / Base : App. painting on minus baseboard 10 Door frame : App. interior sheet / Door : App. sheet 11 Gate : App. sheet 12 Wall : App. wallpaper / Base : App. sheet on flat panel moulding 13 Partition : App. sheet / Lighting : 2" recessed lighting 14 Lighting 15 □50X50 square pipe / App. wallpaper on T9.5 gypsum board 16 Embedding 2" hallogen / T9 plywood / □30X30 square pipe / App. sheet on T9 plywood 17 Partition top : App. interior sheet on T9 MDF / Wall : App. wallpaper / Base : App. sheet on flat panel moulding

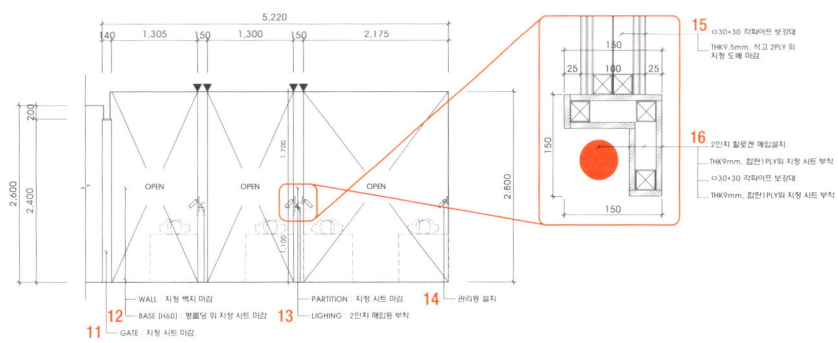

피부관리실 입면 S / skincare room elevation S

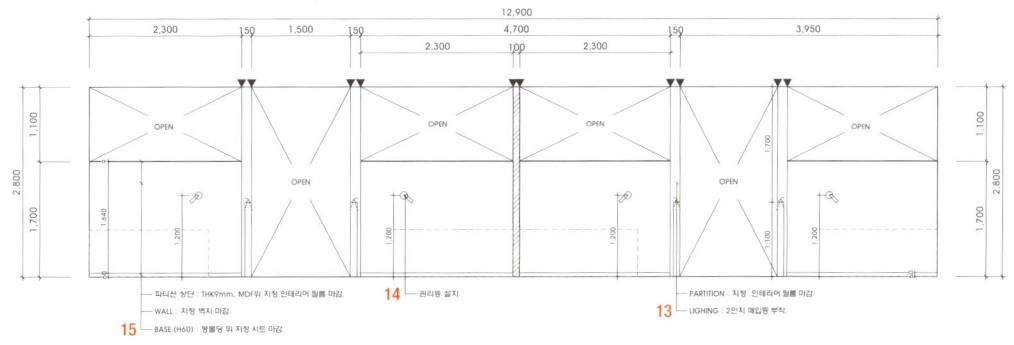

피부관리실 입면 T / skincare room elevation T

JEILSOMANG CHURCH

Danaham Associate | Seohyeon Yun

Emotional Connection
제일소망교회는 종교 시설을 넘어 지역 커뮤니티로써 많은 소통과 유대를 전하는 곳이다. 디자이너는 교회를 표현하는 수많은 키워드 중 공간화할 수 있는 요소를 고민하며 특별한 공간 시나리오를 기획하였다. 어린이부터 어르신까지, 다양한 연령대 수용은 물론 새로 발돋움할 젊은 가족까지 고려한 커뮤니티 공간이 될 것이다.

환대, 열림, 집중, 소통
1층은 환대를 공간화하며 책장의 바운더리를 낮추고 통창 디자인으로 설계되었다. 외부에서 내부를 쉽게 엿볼 수 있고, 더 나아가 부담없이 공간에 들어올 수 있는 열린 액션을 의도한 것이다. 대예배당은 오롯한 집중이 필요한 공간이기 때문에 성스럽고 아름다운 무드를 동시에 깃들이며, 메인 디자인 요소로 십자가에 주목했다. LED 십자가와 천장의 곡선 형태를 결합하여 만들어낸 디자인은 신성하고 몰입적인 분위기를 선사하며 시선이 머무는 디자인이자 공간을 아우르는 코어 디테일이다. 카페테리아는 식사만을 위한 장소가 아닌, 자유로운 소통의 공간이 된다.

층을 오르며 점점 깊어지고 짙어지는 마음들
1층부터 5층까지 일관된 톤앤매너이지만 공간에 올라갈수록 고조되는 감정을 디자인적으로 설계했다. 공간 시나리오에 따라 감각이 발아될 수 있도록 세심한 디테일을 심어두었다. 지역 커뮤니티로써 사람들이 오래 머물고 소통할 안양 제일소망교회, 앞으로 더 많이 교감하고 연결되길 바란다.

Emotional Connection
Jeilsomang Church is both a religious facility and a local community hub where meaningful connections and bonds are formed. Among many keywords that represent a church, the designer selected elements that could be translated into a spatial language to develop a spatial narrative. The goal was to create a community space that welcomes people of all ages, from children to the elderly, while also embracing young families opening a new chapter of life.

Hospitality, Openness, Focus, and Communication
The first floor embodies the concept of hospitality with lowered bookshelves and floor-to-ceiling windows. The aim was to create an open space where people can easily see inside from the outside and casually walk in. The main worship hall, requiring complete focus, is designed to present a sacred and beautiful setting, with the cross as its key design element. A LED cross combines with curved ceiling structures to create a sacred and immersive atmosphere while serving as the visual centerpiece of the space. The cafeteria serves not just as a dining area but as a space for casual communication.

Emotions that Intensify with Every Floor
While maintaining consistent tones and manners from the first to fifth floors, the design expresses intensifying emotions as you go up. Small design elements are carefully applied to increase the emotional density along the spatial narrative. As a local community hub where people stay and interact comfortably, Anyang Jeilsomang Church will establish itself as a place for communication and bonding.

디자인 윤서현 / 다나함 어소시에이트
위치 경기도 안양시 동안구 귀인로 80
용도 종교시설
면적 1,773㎡
마감 바닥 - 타일, LVT 타일 / 벽 - 스페셜 페인트, 타일, 시트, 브라운경 / 천장 - 바리솔, 필름, 페인트
완공 2024. 4
디자인팀 유지연
공사팀 한재균, 김지혁, 강준원, 장동환
사진 제트 스튜디오

Location 80, Guin-ro, Dongan-gu, Anyang-si, Gyeonggi-do
Use Religious facility
Area 1,773㎡
Finishing Floor - Tile, LVT tile / Wall - Special paint, Tile, Sheet, Brown mirror / Ceiling - Barrisol, Film, Paint
Completion 2024. 4
Photographs Z Studio

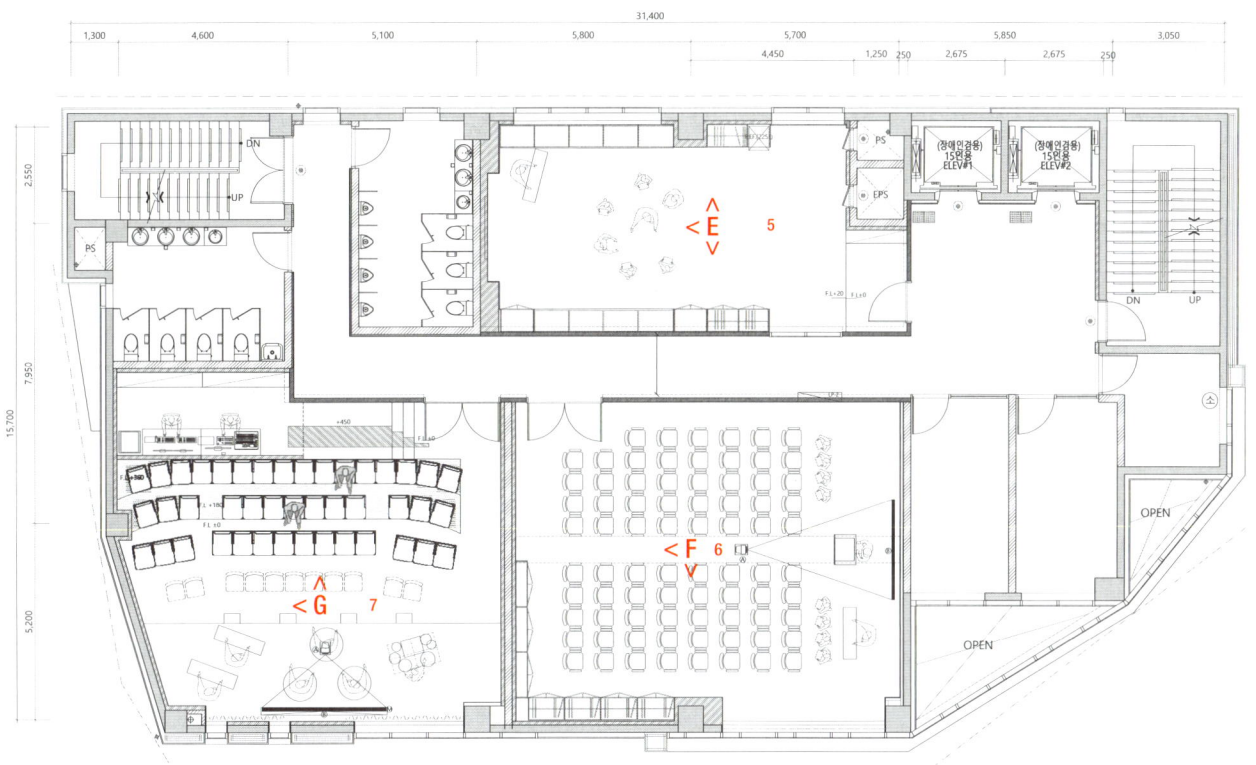

2층 평면도 / 2nd floor plan

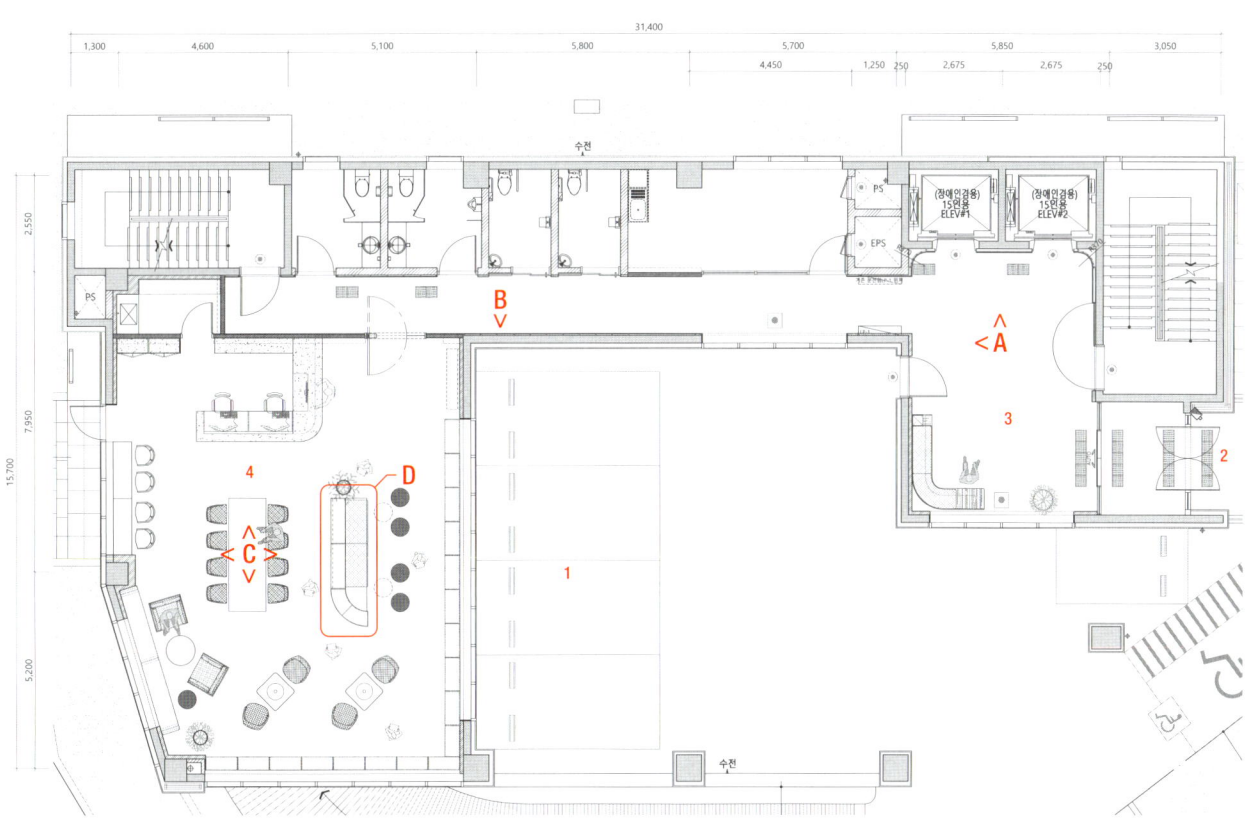

1층 평면도 / 1st floor plan

1 주차장 **2** 입구 **3** 로비 **4** 북카페 **5** 유치부실 **6** 세미나실 **7** 소극장

1 Parking lot **2** Entrance **3** Lobby **4** Book cafe **5** Kinderten **6** Seminar room **7** Theater

1 지정 도장 2 지정 타일 / 10mm 바닥 줄눈 3 사인 4 바리솔 / 지정 타일 5 지정 필름 6 지정 도장 / 랩핑 걸레받이 7 지정 타일 / 지정 필름 / 랩핑 걸레받이 8 지정 도장 / 지정 도장 9 지정 도장 / 지정 타일 10 마이너스 걸레받이 11 블라인드 12 지정 도장 / 마이너스 걸레받이 13 지정 필름 14 무지주 선반 / 지정 도장

1 App. painting 2 App. tile / 10mm floor joint reveal 3 Sign 4 Barrisol / App. tile 5 App. film 6 App. painting / Vinyl wrap baseboard 7 App. tile / App. film / Vinyl wrap baseboard 8 App. painting / App. painting 9 App. painting / App. tile 10 Minus baseboard 11 Blind 12 App. painting / Minus baseboard 13 App. film 14 Floating shelf / App. painting

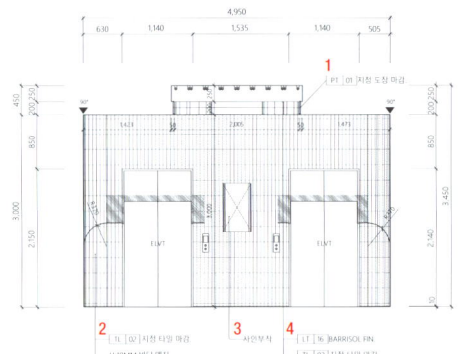

1층 홀 입면 A1 / 1F hall elevation A1

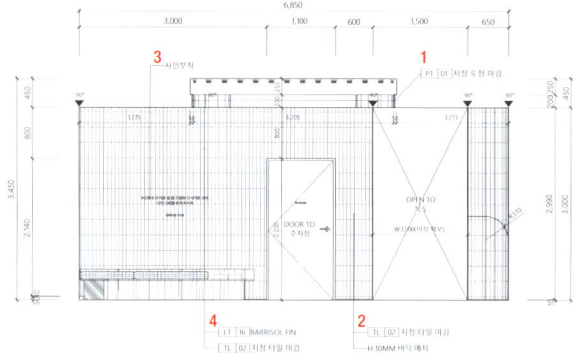

1층 홀 입면 A2 / 1F hall elevation A2

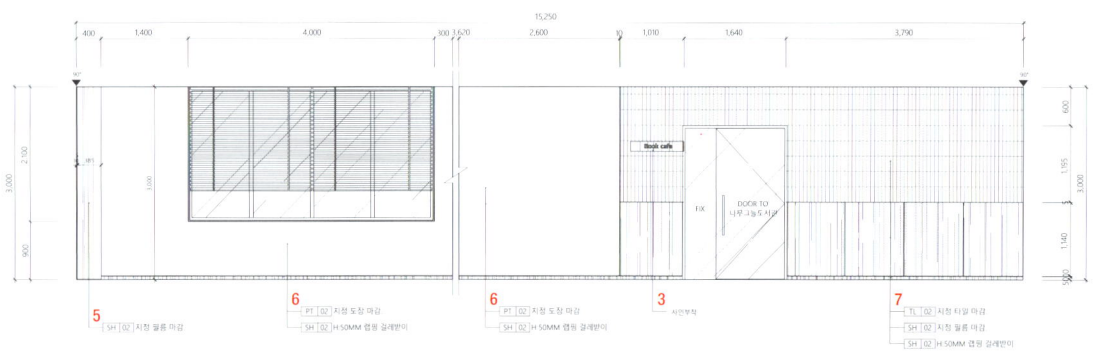

1층 복도 입면 B / 1F corridor elevation B

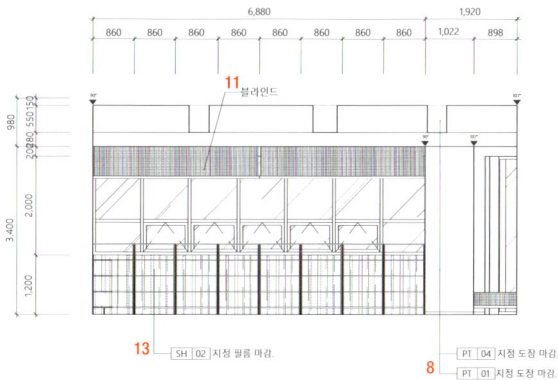

1층 북카페 입면 C1 / 1F book cafe elevation C1

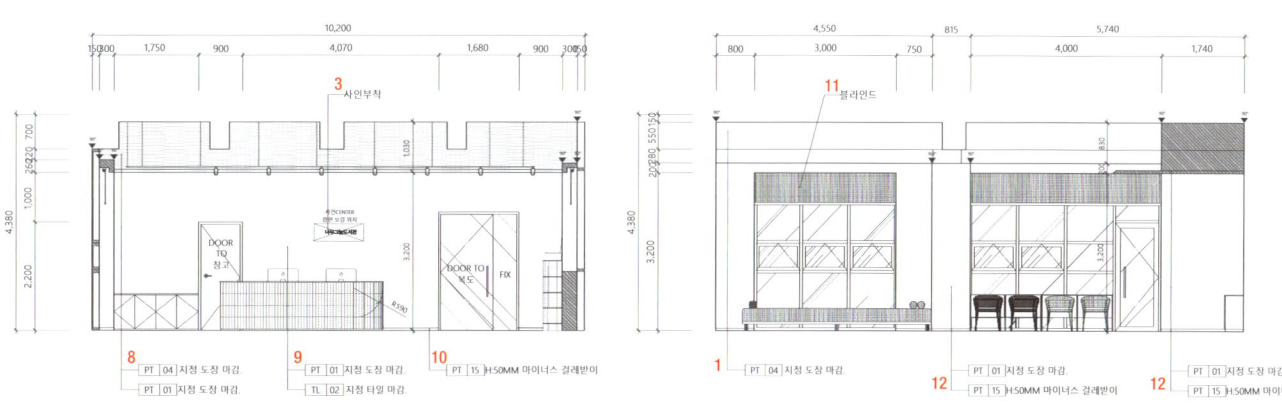

1층 북카페 입면 C2 / 1F book cafe elevation C2

1층 북카페 입면 C3 / 1F book cafe elevation C3

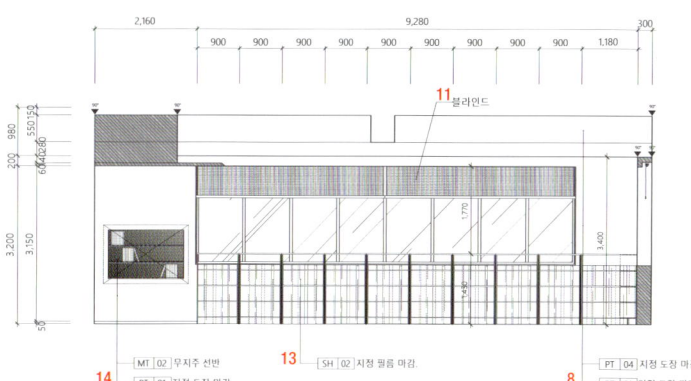

1층 북카페 입면 C4 / 1F book cafe elevation C4

1 지정 도장 2 지정 패브릭 3 지정 패브릭 / 지정 도장 4 T50 패브릭 5 T150 패브릭 6 T9 MDF 위 지정 도장 7 T9 MDF 위 지정 도장 / T9 합판 / □30X30 철제 파이프 / □20X20 철제 파이프 / T9 합판 / T9 MDF 위 지정 도장 8 벨크로 9 지정 커튼 10 지정 쿠션 / 지정 패브릭 / 지정 시트 11 지정 시트 12 지정 쿠션 / 5mm 바닥 줄눈 13 지정 도장 / 마이너스 걸레받이 14 마이너스 걸레받이 15 지정 패브릭 / 지정 패브릭 / 지정 시트 16 지정 백페인트 글래스 17 지정 시트 / 지정 쿠션

1 App. painting 2 App. fabric 3 App. fabric / App. painting 4 T50 fabric 5 T150 fabric 6 App. painting on T9 MDF 7 App. painting on T9 MDF / T9 plywood / □30X30 steel pipe / □20X20 steel pipe / T9 plywood / App. painting on T9 MDF 8 Velcro 9 App. curtain 10 App. cushion / App. fabric / App. sheet 11 App. sheet 12 App. cushion / 5mm floor joint reveal 13 App. painting / Minus baseboard 14 Minus baseboard 15 App. fabric / App. fabric / App. sheet 16 App. back painted glass 17 App. sheet / App. cushion

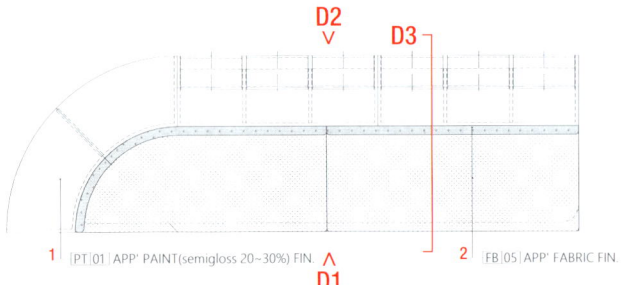

소파 평면 D / sofa top view D

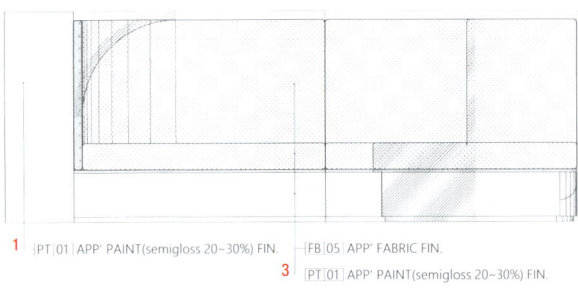

소파 정면 D1 / sofa front view D1

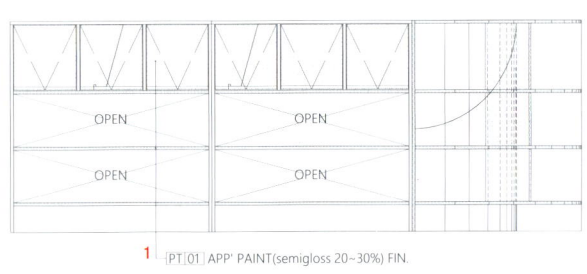

소파 후면 D2 / sofa rear view D2

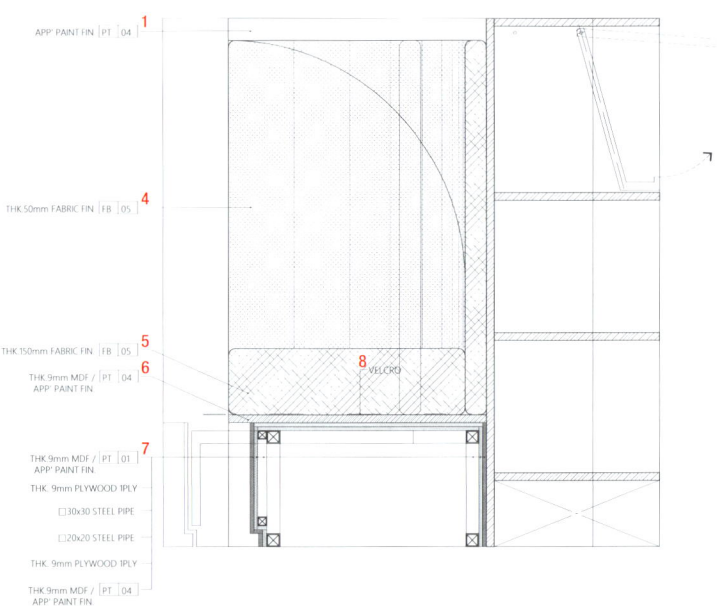

소파 단면 D3 / sofa section D3

2층 유치부실 입면 E1 / 2F kindergarten elevation E1

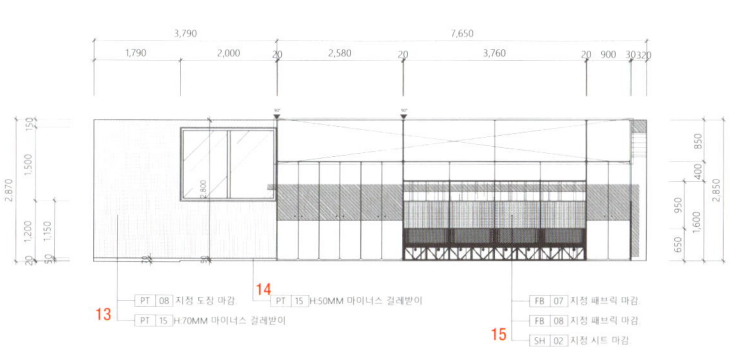

2층 유치부실 입면 E2 / 2F kindergarten elevation E2

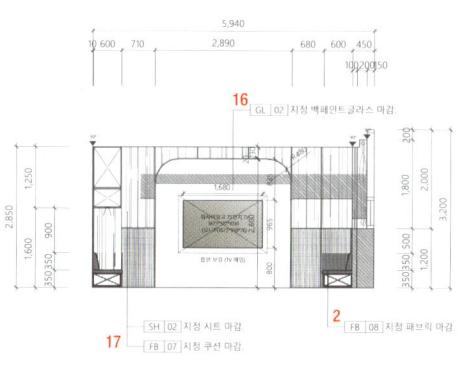

2층 유치부실 입면 E3 / 2F kindergarten elevation E3

1 LED 간접조명 2 블라인드 3 지정 도장 / 랩핑 걸레받이 4 지정 필름 5 사인 부착 / 지정 도장 6 지정 도장 7 지정 타일 / 10mm 줄눈 8 사인 9 지정 필름 / 랩핑 걸레받이 10 메탈 타공판 위 지정 도장
11 아트보드 위 패브릭 / 아트보드 위 패브릭 12 아트보드 위 패브릭 / 방염 흡음타공보드 13 지정 시트 14 랩핑 걸레받이 15 지정색 스페셜 페인트

1 LED indirect lighting 2 Blind 3 App. painting / Vinyl wrap baseboard 4 App. film 5 Sign / App. painting 6 App. painting 7 App. tile / 10mm joint reveal 8 Sign 9 App. film / Vinyl wrap baseboard 10 App. painting on metal perforated sheet 11 Fabric on art board / Fabric on art board 12 Fabric on art board / Fire prevention sound absorbtion panel 13 App. sheet 14 Vinyl wrap baseboard 15 App. color special paint

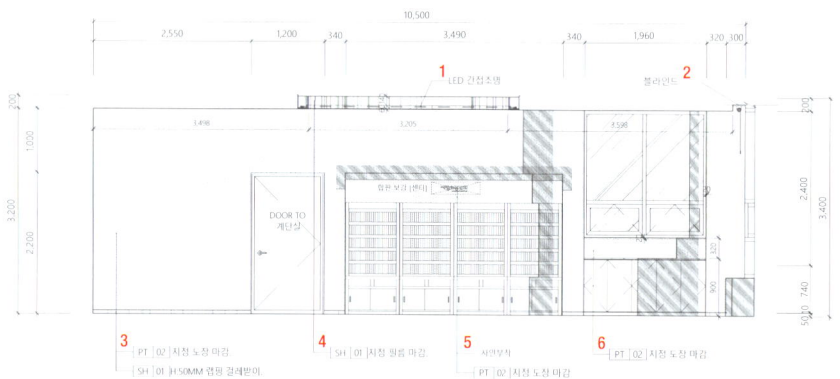

3층 홀 입면 H1 / 3F hall elevation H1

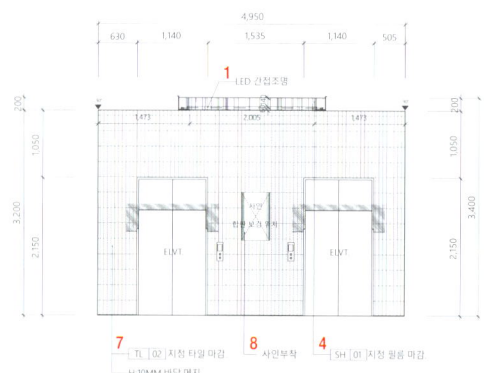

3층 홀 입면 H2 / 3F hall elevation H2

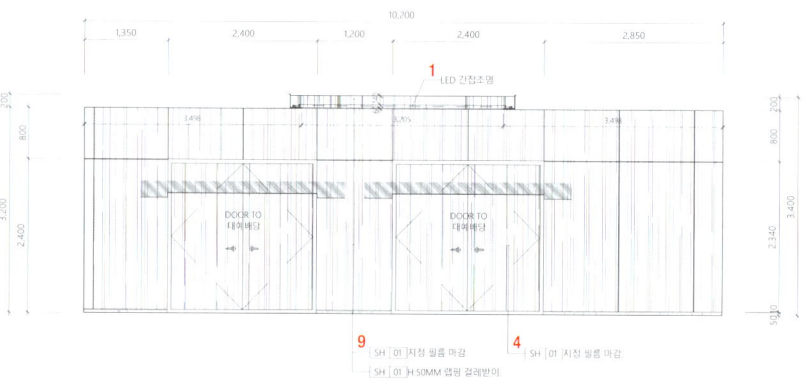

3층 홀 입면 H3 / 3F hall elevation H3

3층 대예배당 입면 I1 / 3F chapel elevation I1

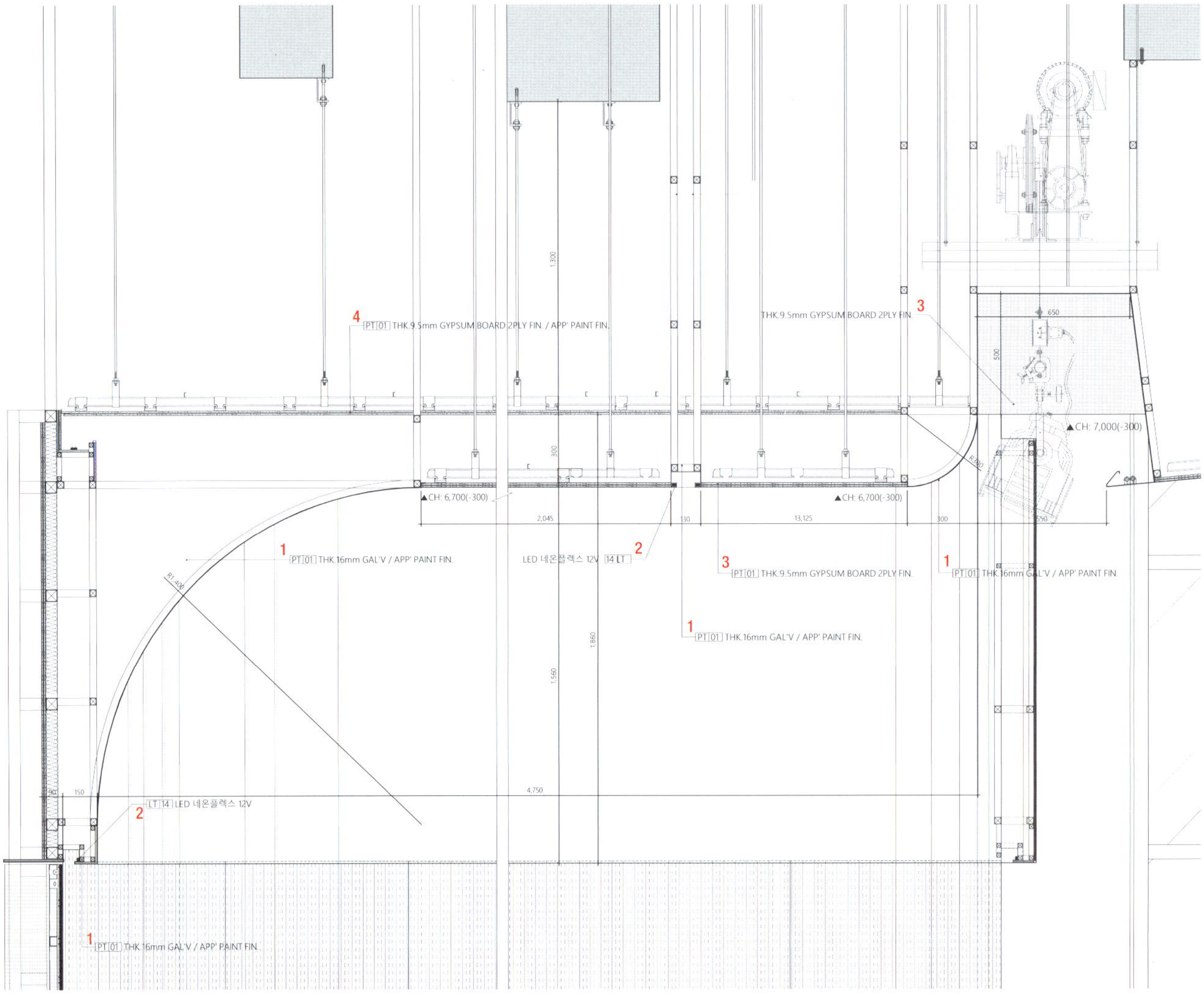

대예배당 천장 단면 상세 J / chapel ceiling section detail J

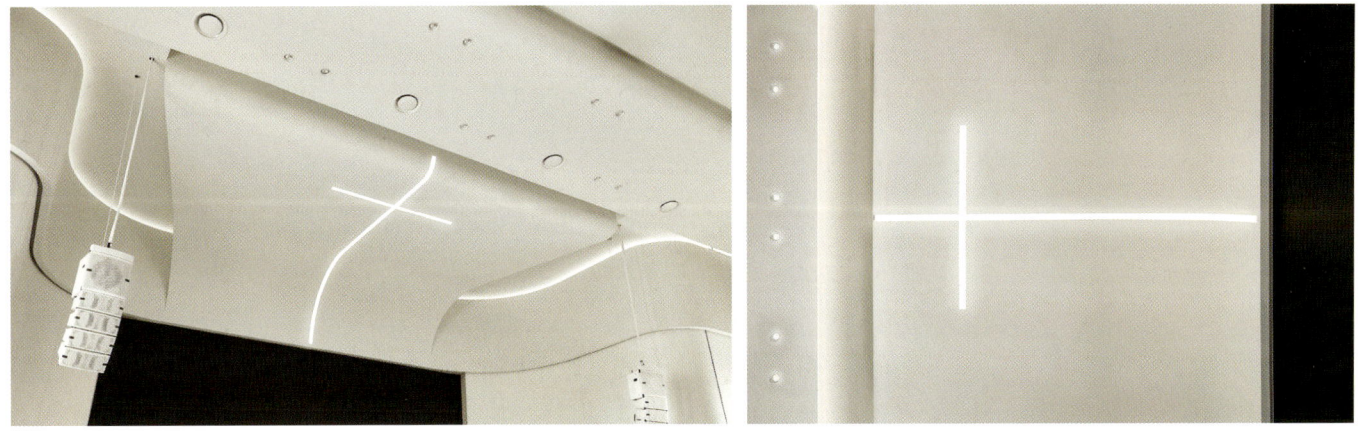

1 T16 갈바륨 위 지정 도장 **2** LED 네온플렉스 12V **3** T9.5 석고보드 2겹 **4** T9.5 석고보드 2겹 위 지정 도장

1 App. painting on T16 galvalume **2** LED Neon Plex 12V **3** T9.5 gypsum board 2ply **4** App. painting on T9.5 gypsum board 2ply

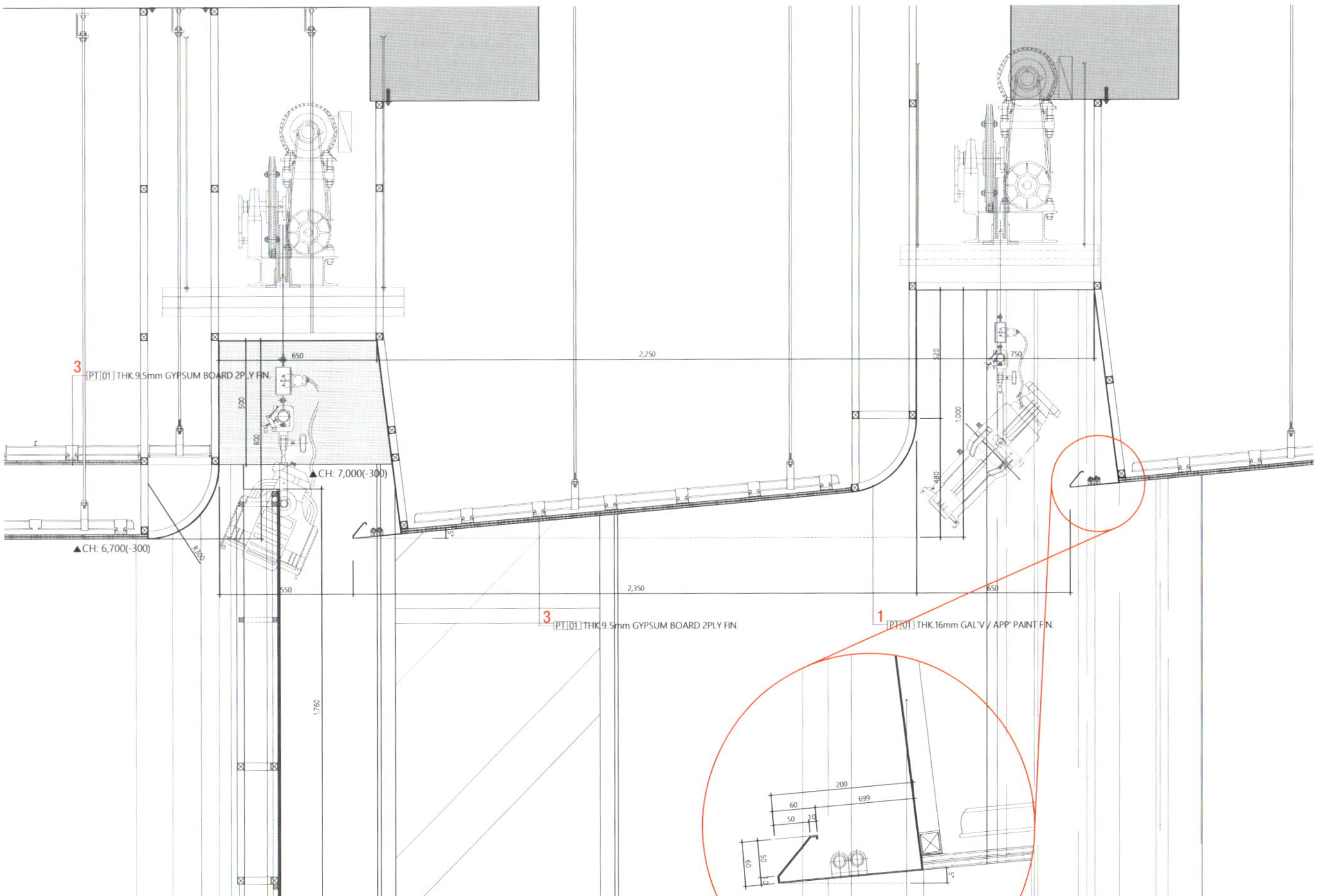

대예배당 천장 단면 상세 K / chapel ceiling section detail K

1 LED 간접조명 **2** 메탈 타공 위 도장 / 아트보드 위 패브릭 **3** 아트보드 위 패브릭 / 마이너스 걸레받이 **4** 금속 위 도장 / 금속 위 도장 **5** 메탈 타공판 위 도장 / 지정 금속 **6** 아트보드 위 패브릭 / 방염 흡음타공보드 **7** 타공유리 **8** 지정 시트

1 LED indirect lighting **2** Painting on metal perforated sheet / Fabric on art board **3** Fabric on art board / Minus baseboard **4** Painting on metal / Painting on metal **5** Painting on metal perforated sheet / App. metal **6** Fabric on art board / Fire prevention sound absorbtion panel **7** Perforated glass **8** App. sheet

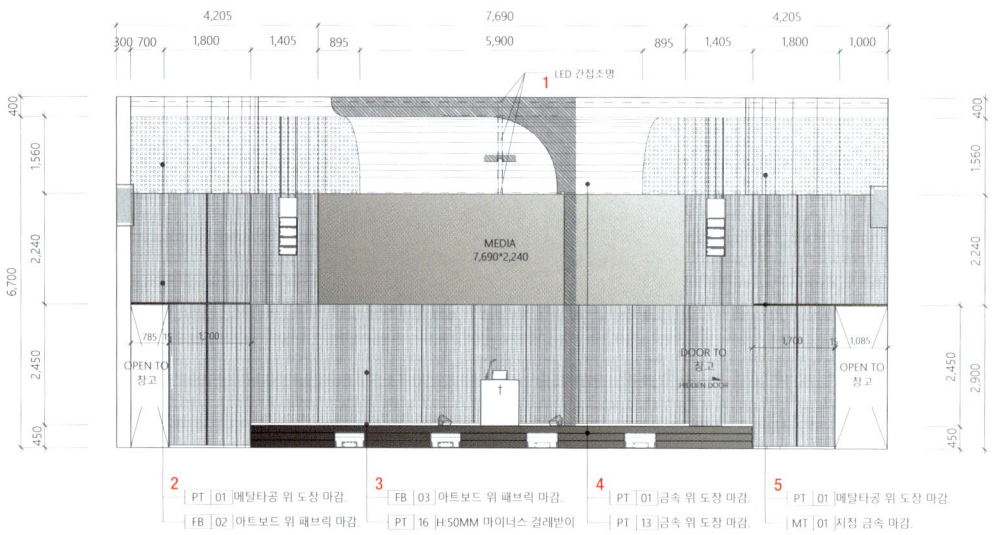

3층 대예배당 입면 I2 / 3F chapel elevation I2

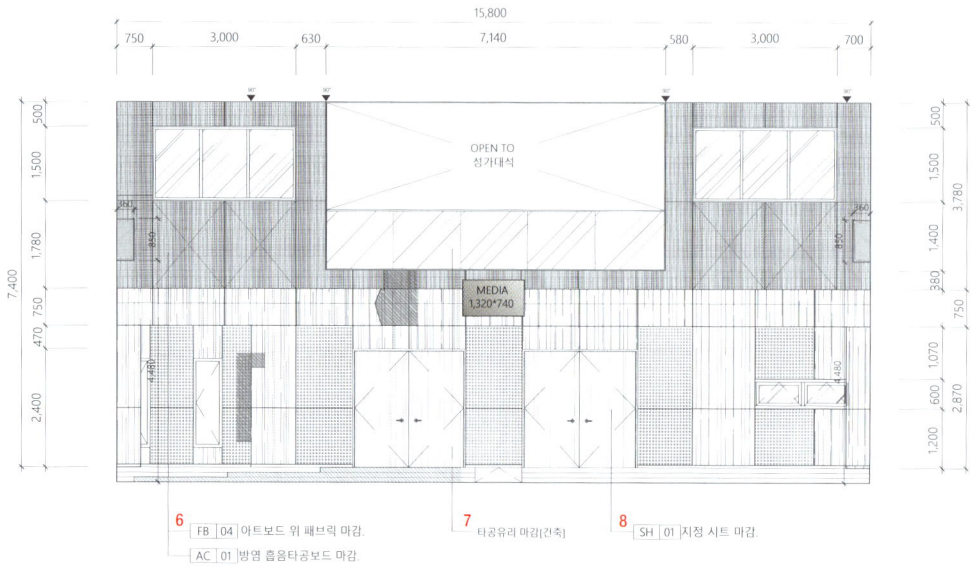

3층 대예배당 입면 I3 / 3F chapel elevation I3

5층 평면도 / 5th floor plan

1 엘리베이터 홀 2 접대실 3 카페테리아 4 교역자 사무실 5 담임목사실

1 Elevator hall 2 Reception room 3 Cafeteria 4 Office 5 Pastor's office

1 블라인드　2 LED 간접조명　3 지정 도장 / 마이너스 걸레받이　4 지정 벽등(합판 보강) / 지정 도장

1 Blind　2 LED indirect lighting　3 App. painting / Minus baseboard　4 App. wall lamp (plywood reinforcement) / App. painting

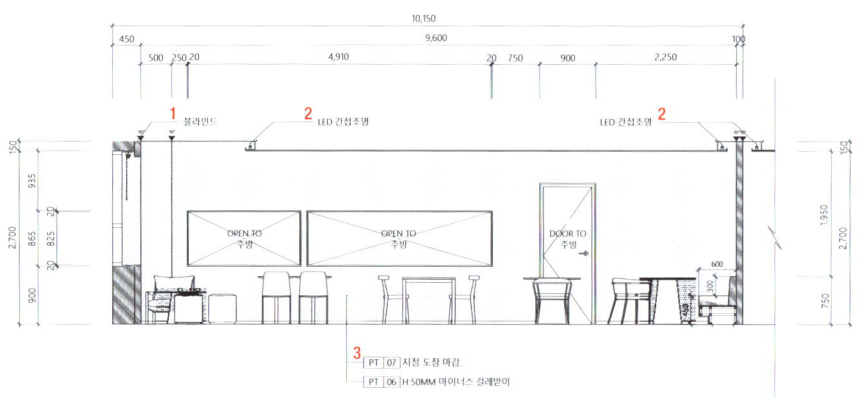

5층 카페테리아 입면 L1 / 5F cafeteria elevation L1

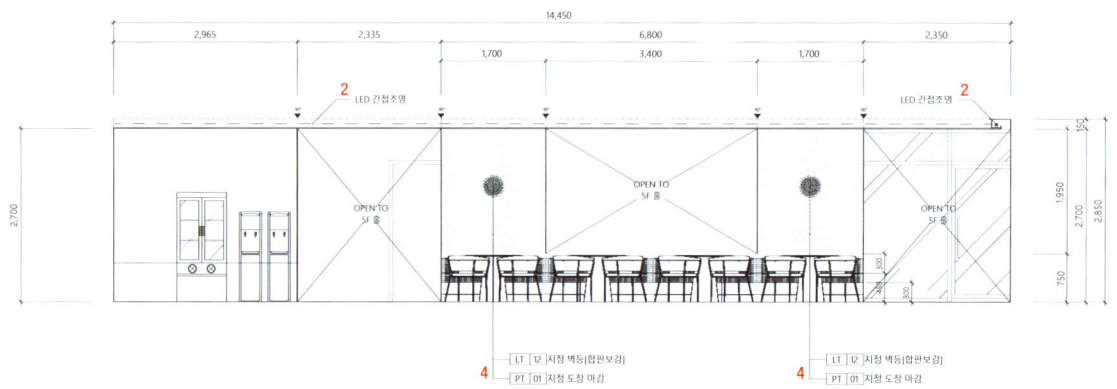

5층 카페테리아 입면 L2 / 5F cafeteria elevation L2

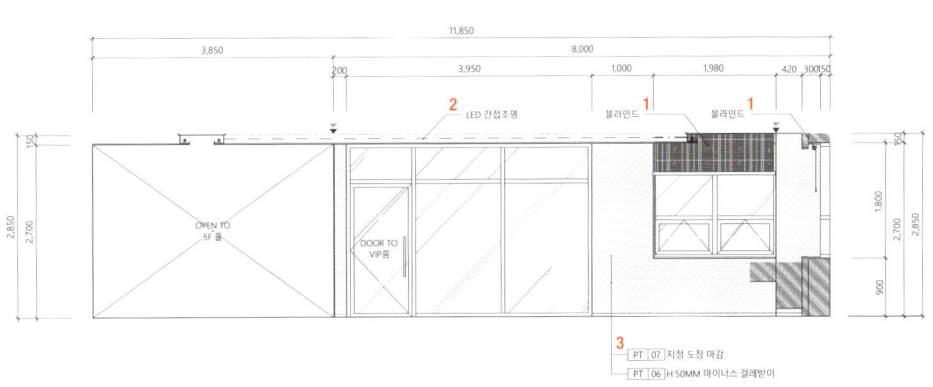

5층 카페테리아 입면 L3 / 5F cafeteria elevation L3

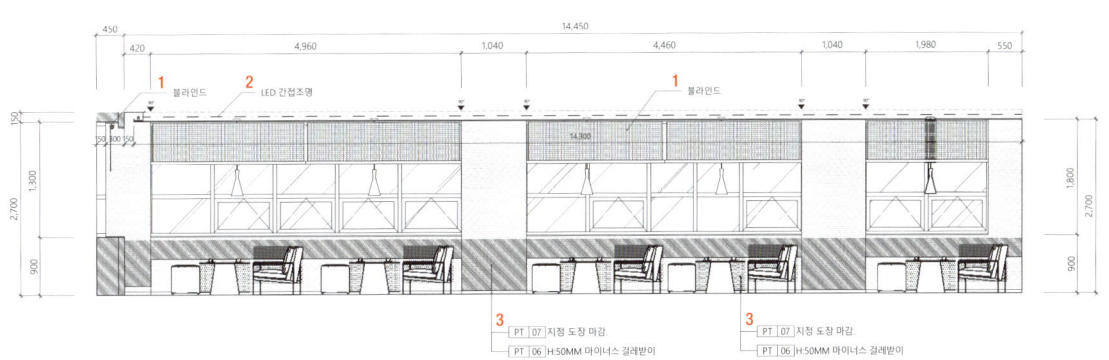

5층 카페테리아 입면 L4 / 5F cafeteria elevation L4

2025 ANNUAL INTERIOR DETAIL

FLAGSHIP STORE

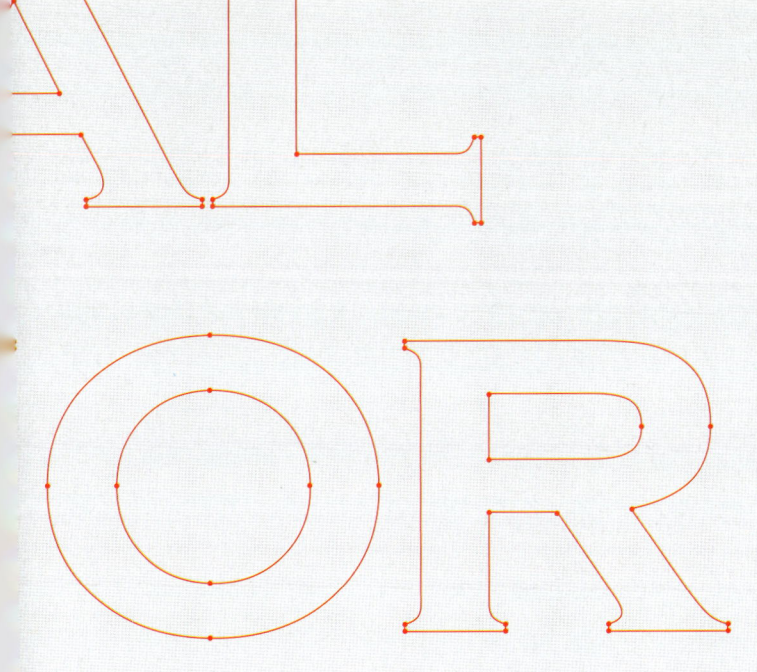

254 INSILENCE SEONGSU
인사일런스 성수

260 PYUNKANG YUL FLAGSHIP STORE
편강 율 플래그십 스토어

268 MEDICUBE FLAGSHIP STORE HONGDAE
메디큐브 플래그십 스토어 홍대

274 JAVIN DE SEOUL
자빈드서울

INSILENCE SEONGSU

design by 83 | Minsuk Kim, Donghyun Nam, Chanun Park

인사일런스 성수 스토어는 한남동에 이은 두 번째 플래그십 스토어이다. '리 빌드'를 주제로 한 성수 스토어는 기존의 틀에서 벗어나 새로운 것을 연결하고 다양한 모양과 방법으로 확장해 나가려는 인사일런스의 새로운 움직임을 담았다. 새롭게 선보이는 성수 스토어는 낡은 것과 새로운 것이 공존하는 지역적 특색을 공간에 접목시켰는데, 기존에 공장으로 사용되어져 오래된 공간에 날 것의 소재, 새로운 소재와 틀에 박히지 않은 형태와 동선으로 브랜드의 무드를 표현하였다. 전체적인 분위기는 오래된 콘크리트 바닥과 벽체, 금속 소재와 PC판, 카펫, 스페셜 페인트 등 여러 물성이 사용되었지만 공간은 하나의 그레이 톤으로 표현하였다. 각 집기들은 매스감 있게 자리잡고 있지만 얇은 마감과 다양한 소재의 사용으로 투박해 보이지 않도록 계획하였다. 각기 다른 소재로 이루어진 공간이 하나의 예술적인 작품으로 보여지기를 연출하였는데, 이는 무한한 가능성이 있는 MZ세대 성지인 성수동에서 브랜드의 무드를 나타낼 수 있도록 계획하였다. 매장 입구에 들어서면 불규칙한 동선을 시작으로 안쪽으로 들어갈수록 작품이 배치된 듯한 비정형적인 가구들로 구성되어 있고, 이동식으로 제작된 가구들은 추후 케이터링 공간으로 활용하도록 계획하였다. 천장의 조명 레일 또한 집기의 배치를 유추하는 듯한 라인으로 구성하였다.

The Insilence Seongsu store, the second flagship store following Hannam-dong, celebrates the theme of 'Rebuild.' This new store showcases Insilence's bold step away from the conventional, weaving together enw connections and expanding in diverse forms and strategies. It reflects a blend of the old and the new, capturing the essence of the Seongsu area. The space, once an industrial factory, now artistically presents the brand's mood using raw and innovative materials in unconventional layouts and pathways. The overall ambiance, with its mix of aged concrete floors and walls, metals, PC panels, carpets, and special paint, showcases a diverse range of textures, yet is cohesively presented in a singular tone of grey. Each fixture is strategically placed, embodying a sense of mass, yet carefully designed to avoid a cumbersome look through the use of thin finishes and diverse materials. The space, crafted from an array of materials, is designed to resemble a unified artistic work, embodying the brand's mood in Seongsu-dong, renowned for the MZ generation's boundless possibilities. As you enter the store, an irregular path unfolds, leading to asymmetrical furniture pieces arranged artfully, giving the impression of being in an art gallery. The furniture, designed to be moile, is intended for future use as a catering area. The lighting rails on the ceiling area intricately designed, featuring lines that subtly hint at layout of the fixtures, contributing to a thoughtful and cohesive aesthetic.

디자인 김민석, 남동현, 박찬언 / 디자인바이팔삼
위치 서울특별시 성동구 연무장17길 5, 2층
용도 플래그십 스토어
면적 123m²
마감 바닥 – 콘크리트 폴리싱, 카펫 / 벽 – 갈바륨, 스테인리스 스틸, 아연도금강판, 폴리카보네이트, OSB 우드, 스페셜 페인트, 메탈 페인트 / 천장 – 스페셜 페인트
디자인팀 조수현, 하윤지
시공팀 홍민욱 / page studio
가구 배지헌 / TT
사진 김동규

Location 2F, 5, Yeonmujang 17-gil, Seongdong-gu, Seoul
Use Flagship store
Area 123m²
Finishing Floor - Concrete polishing, Carpet / Wall - Galvalume, Stainless steel, Zinc steel, OSB wood, Special paint, Metal paint / Ceiling - Special paint
Design team Suhyun Jo, Yunji Ha
Construction team Minwook Hong / page studio
Furniture Jiheon Bae / TT
Photographer Donggyu Kim

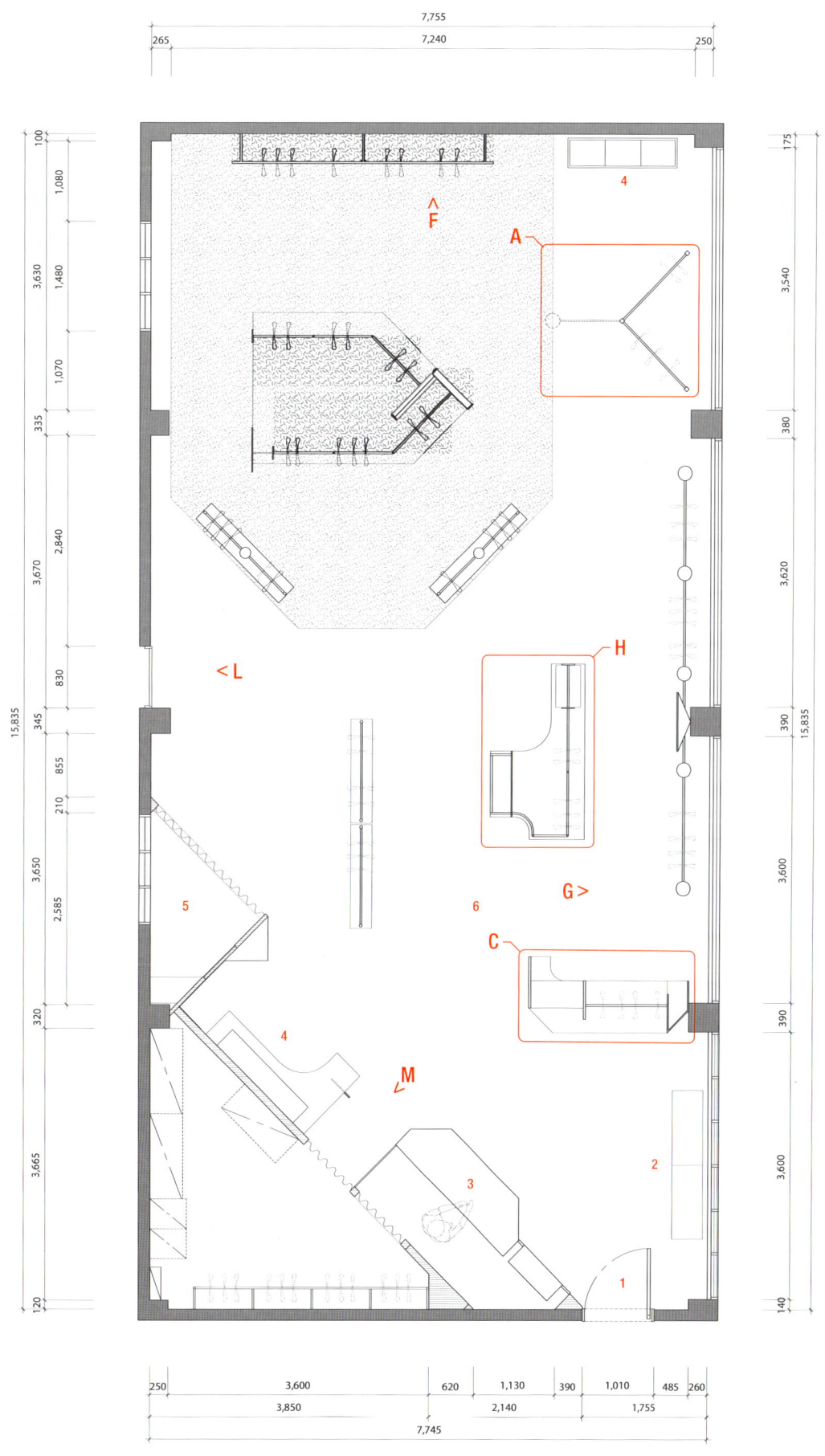

평면도 / floor plan

1 입구 2 대기공간 3 카운터 4 디스플레이 존 5 피팅룸 6 홀

1 Entrance 2 Waiting space 3 Counter 4 Display zone 5 Fitting room 6 Hall

1 □50 지정 아연파이프 2 지정 스테인리스 스틸 미러 3 Ø15 지정 아연파이프 4 Ø25 지정 아연파이프 / □50 지정 아연파이프 5 지정 아연강판 6 □30 지정 아연파이프 / 지정 반투명 폴리카보네이트 / 지정 아연강판 7 지정 반투명 폴리카보네이트 8 Ø25 지정 아연파이프 / 지정 아연강판 9 □30 지정 아연파이프 / 지정 반투명 폴리카보네이트 10 기존 마감 철거 후 샌딩 11 지정 스페셜 페인트 / Ø25 지정 아연파이프 / 지정 스페셜 페인트 12 지정 스틸 그레이팅 13 프레임 : 지정 아연강판 / 지정 스틸 그레이팅 / 지정 아연강판 14 Ø25 지정 아연파이프 15 지정 스페셜 페인트 16 지정 아연강판 / 지정 은경 17 지정 메탈릭 패브릭 / 지정 스페셜 페인트 18 지정 스페셜 페인트 / 지정 아연강판 / 기존 마감 철거 후 샌딩

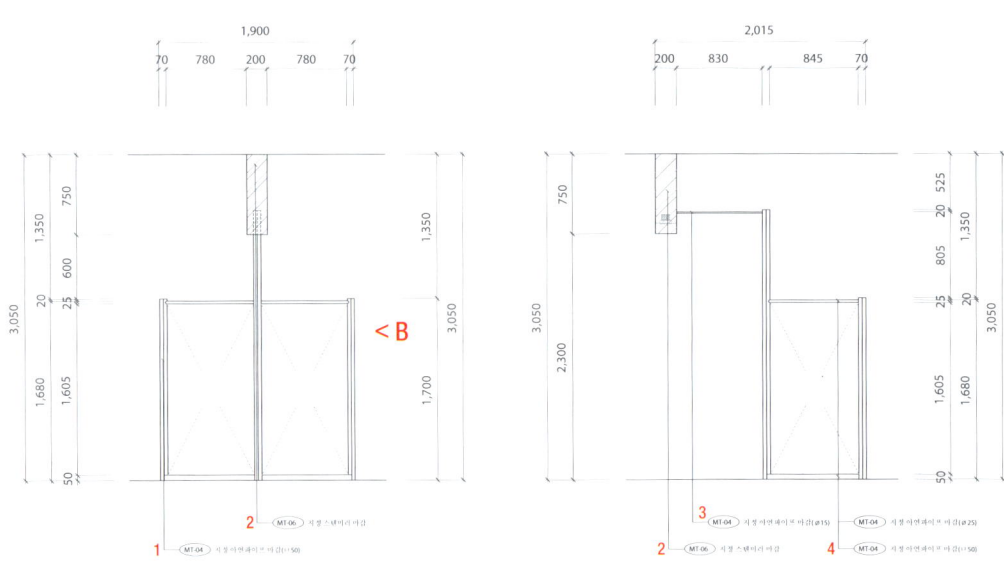

행거 정면 A / hanger front view A   행거 측면 B / hanger side view B

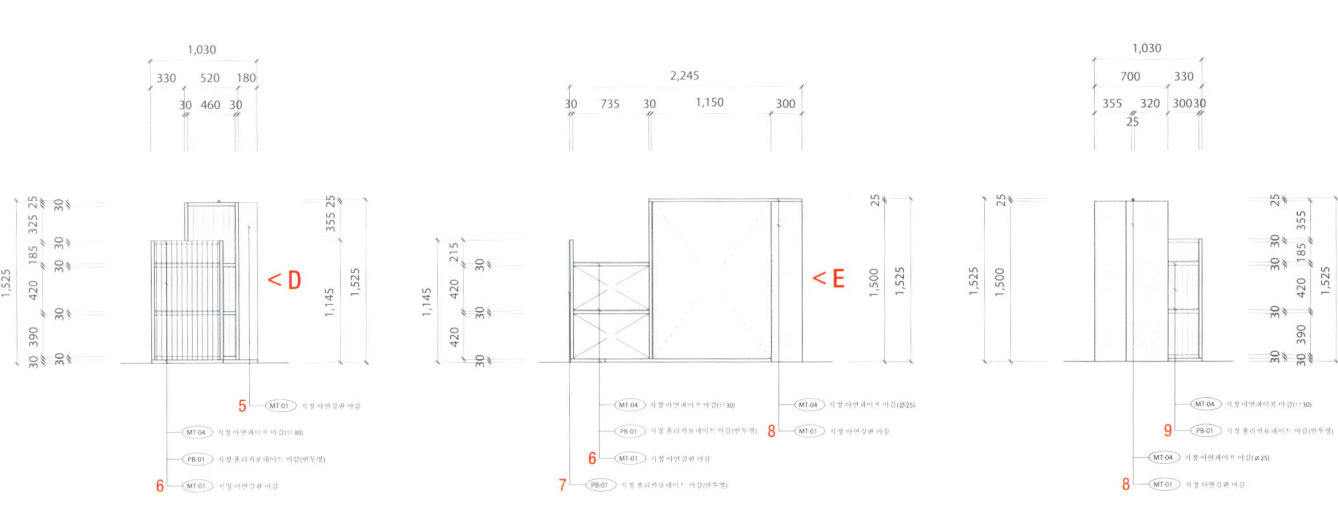

행거 정면 C / hanger front view C   행거 측면 D / hanger side view D   행거 후면 E / hanger rear view E

1 □50 app. galvanized pipe   2 App. stainless steel mirror   3 Ø15 app. galvanized pipe   4 Ø25 app. galvanized pipe / □50 app. galvanized pipe   5 App. galvanized steel plate   6 □30 app. galvanized pipe / App. semi-transparent polycarbonate / App. galvanized steel plate   7 App. semi-transparent polycarbonate   8 Ø25 app. galvanized pipe / App. galvanized steel plate   9 □30 app. galvanized pipe / App. semi-transparent polycarbonate   10 Sanding after remove existing finish   11 App. special paint / Ø25 app. galvanized pipe / App. special paint   12 App. steel grating   13 Frame : App. galvanized steel plate / App. steel grating / App. galvanized steel plate   14 Ø25 app. galvanized pipe   15 App. special paint   16 App. galvanized steel plate / App. mirror   17 App. metallic fabric / App. special paint   18 App. special paint / App. galvanized steel plate / Sanding after remove existing finish

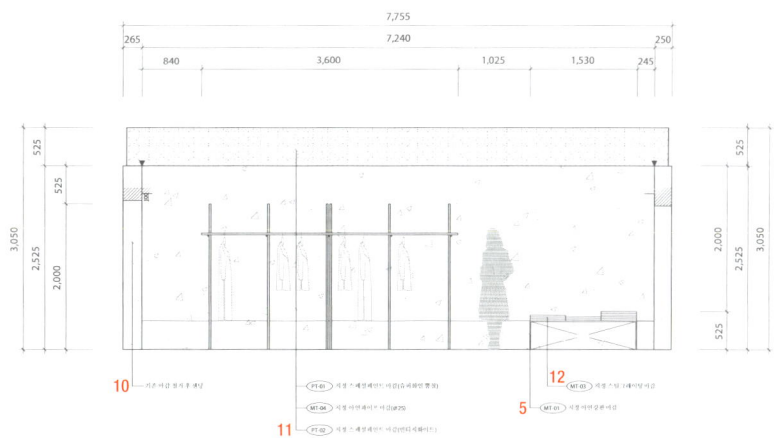

홀 입면 F / hall elevation F

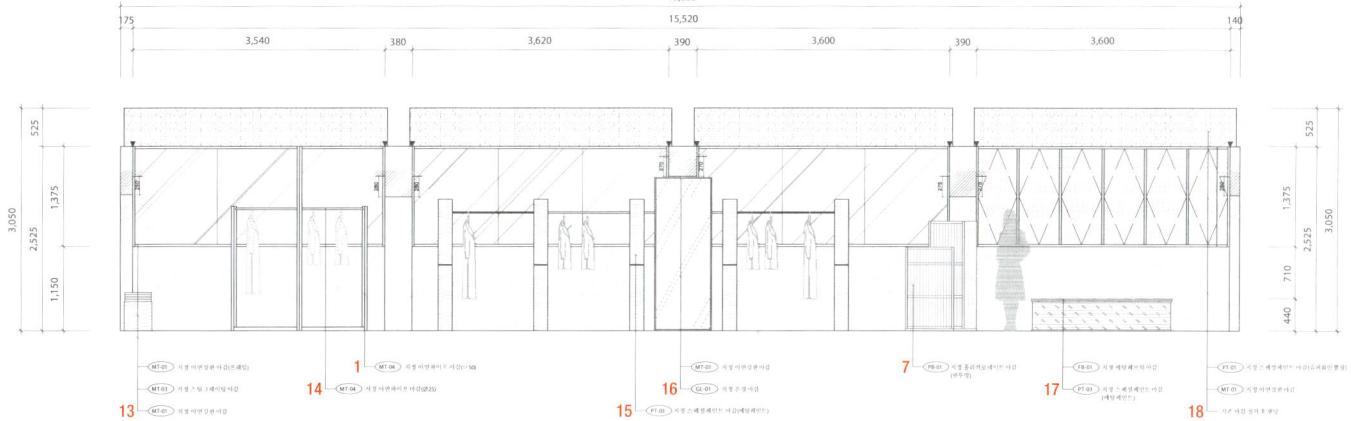

홀 입면 G / hall elevation G

1 Ø25 지정 아연파이프 / 지정 갈바륨 강판　2 T10 지정 투명 강화유리 / 지정 갈바륨 강판　3 Ø25 지정 아연파이프　4 지정 갈바륨 강판　5 지정 은경　6 Ø25 지정 아연파이프 / T10 지정 투명 강화유리 / 지정 갈바륨 강판　7 지정 메탈 커튼　8 지정 아연강판　9 지정 스페셜 페인트 / 기존 마감 철거 후 샌딩　10 지정 반투명 폴리카보네이트 / 지정 아연강판　11 지정 갈바륨 강판 / T8 투명 강화유리　12 지정 아연강판 / T5 지정 스테인리스 스틸 헤어라인

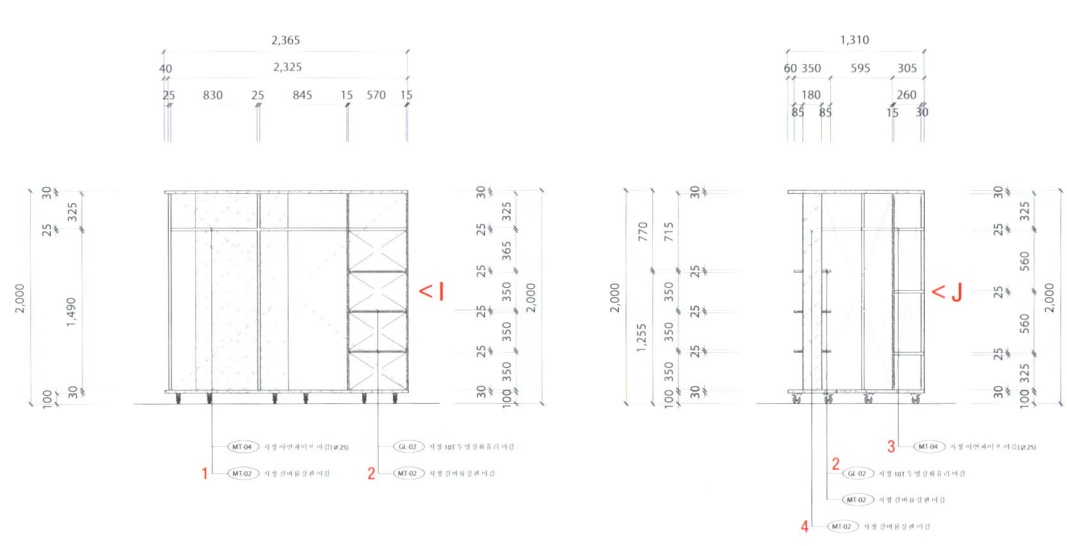

행거 정면 H / hanger front view H   행거 측면 I / hanger side view I

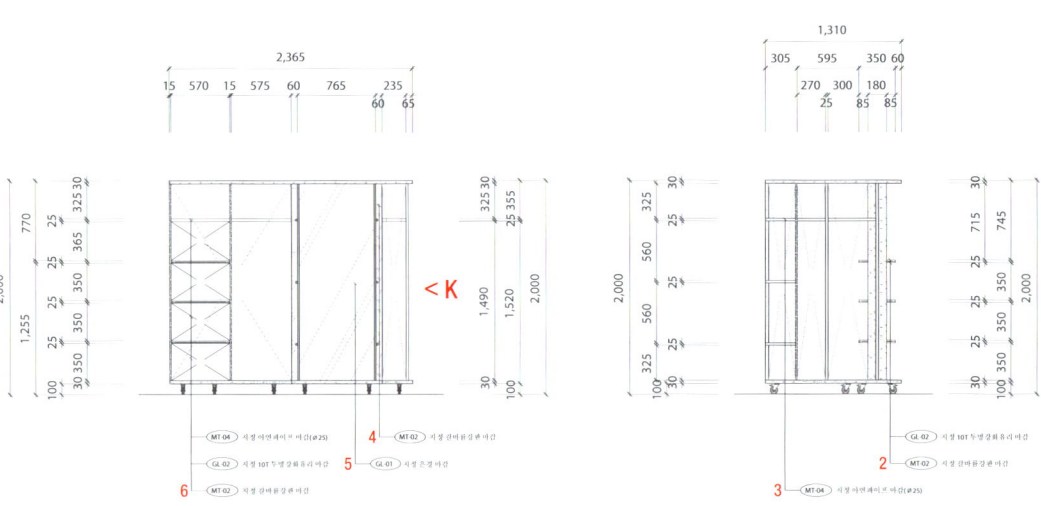

행거 후면 J / hanger rear view J   행거 측면 K / hanger side view K

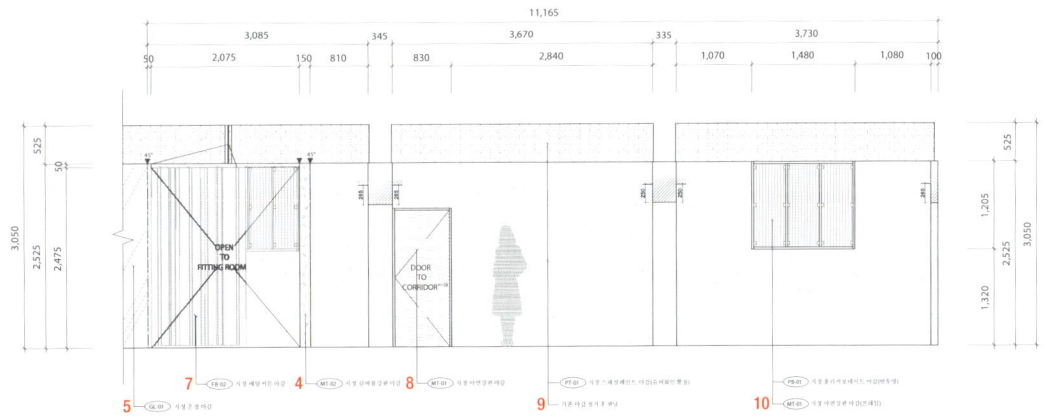

1 Ø25 app. galvanized pipe / App. galvalume steel sheet **2** T10 app. clear tempered glass / App. galvalume steel sheet **3** Ø25 app. galvanized pipe **4** App. galvalume steel sheet **5** App. mirror **6** Ø25 app. galvanized pipe / T10 app. clear tempered glass / App. galvalume steel sheet **7** App. metal curtain **8** App. galvanized steel sheet **9** App. special paint / Sanding after remove existing finish **10** App. semi-transparent polycarbonate / App. galvanized steel sheet **11** App. galvalume steel sheet / T8 clear tempered glass **12** App. galvanized steel sheet / T5 app. stainless steel hairline

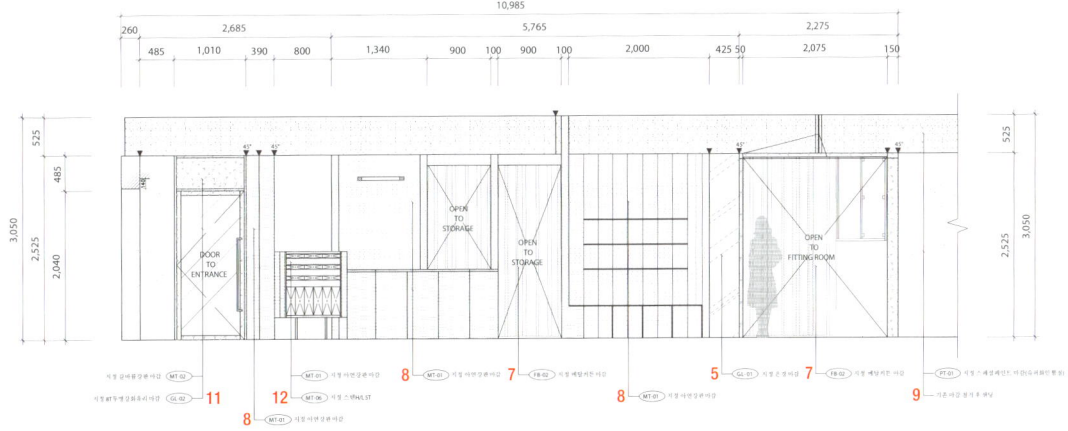

홀 입면 L / hall elevation L

홀 입면 M / hall elevation M

PYUNKANG YUL FLAGSHIP STORE

SHERPA STUDIO | Yeonho Shin, Byeongguk Mo

편강 율 코스메틱은 '비움으로 아름다움을 채우다'라는 슬로건을 가지고, 인위적인 성분을 최소화하고 자연으로부터 온 성분을 가지고 피부 본연의 건강한 아름다움을 찾는 한방 스킨 케어 브랜드이다. 북촌에 위치한 편강 율 플래그십 스토어에서 한국 전통의 요소를 담아 현대적으로 해석해 편강 율의 건강한 아름다움을 전달하고자 했다. 외부 파사드에서는 대청마루에 앉아 처마 밑에서 자연과 공존하면서 편안한 감정을 느낄 수 있도록 경험적 요소 통해 브랜드가 추구하는 가치를 표현한다. 1층 내부로 들어서서 리셉션을 통해 먼저 안내받고, 제품을 자유롭게 사용하고, 전시 공간에 참여할 수 있도록 동선을 기획했다. 눈에 먼저 보이는 천장에 달린 조명은 초롱등의 형태를 차용한 제작 조명이다. 우물 형태를 차용해 만들어진 세면대는 우물반자 형태의 바리솔 조명이 비추고 있다. 그 건너에 보이는 벽면과 리셉션에는 자개가 섞인 천연 흙미장을 했다. 흙미장 안에 제품들이 나열되어 있는데 이는 자연으로부터 온 제품을 간접적으로 전달한다. 2층에서는 현대적으로 브랜드 정체성을 해석했다. 바닥은 우물마루의 형태에서 차용해 패턴을 주고 그대로 이어서 리셉션과 제품 테이블까지 제작했다. 그 뒤의 스테인리스 폴리싱 선반에 제품이 나열되어 있는데 기둥과 선반들이 외부에 있는 자연들을 반사하고, 시각적으로 자연 안에 제품들이 들어와 있을 수 있도록 의도했다. 안쪽 공간에서는 자개를 통해 전시 공간과 체리 원목으로 만든 제작가구 및 전통가구에서도 편강 율의 헤리티지를 경험할 수 있다.

Pyunkang Yul Cosmetics is a Korean skincare brand that promotes the pure, healthy beauty of the skin by minimizing artificial ingredients and actively using natural materials, under the slogan 'Finding the full beauty by Emptying.' The Pyunkang Yul flagship store, located in Bukchon, reinterprets traditional Korean elements in a modern way to express the healthy beauty promoted by the brand. The exterior façade reflects the brand's values through experiential elements, which guides visitors to sit on the traditional Korean-style wooden floor and feel a sense of comfort as they interact with nature under the eaves. On the first floor, the circulation system leads visitors from the entry to the reception, after which they can freely explore the products or browse the exhibition space. The ceiling light, which attracts attention first, is a custom-made fixture inspired by the shape of a traditional Korean lantern. The well-shaped sink is illuminated by Barrisol lights arranged in the form of a coffered ceiling. The walls and reception area on the opposite side are finished with natural clay plastering mixed with mother-of-pearl. Products are displayed within the clay plastering to indirectly convey the characteristics of naturally sourced products. The second floor is a modern interpretation of the brand's identity. The floor pattern, inspired by traditional Korean well-shaped wooden flooring, extends to the reception and product tables. Behind them, polished stainless steel shelves serve as product display stands. The pillars and shelves reflect the natural scenery outside, making the products appear as if they are nestled in nature. In the inner space, the mother-of-pearl details of the exhibition area, cherry wood furnishings and traditional Korean furniture reflect the brand's philosophy.

디자인 신연호, 모병국 / 세르파 스튜디오
위치 서울특별시 종로구 북촌로5가길 35-4
용도 플래그십 스토어
면적 462m²
마감 우드플로링, 알루미늄 주름 강판, 스테인리스 스틸 바이브레이션, 흙미장, 천연석
완공 2024. 2
사진 inandout_studio

Location 35-4, Bukchon-ro 5ga-gil, Jongno-gu, Seoul
Use Flagship store
Area 462m²
Finishing Wood flooring, Aluminum corrugated plate, Stainless steel vibration, Clay plaster, Natural stone
Completion 2024. 2
Photographs inandout_studio

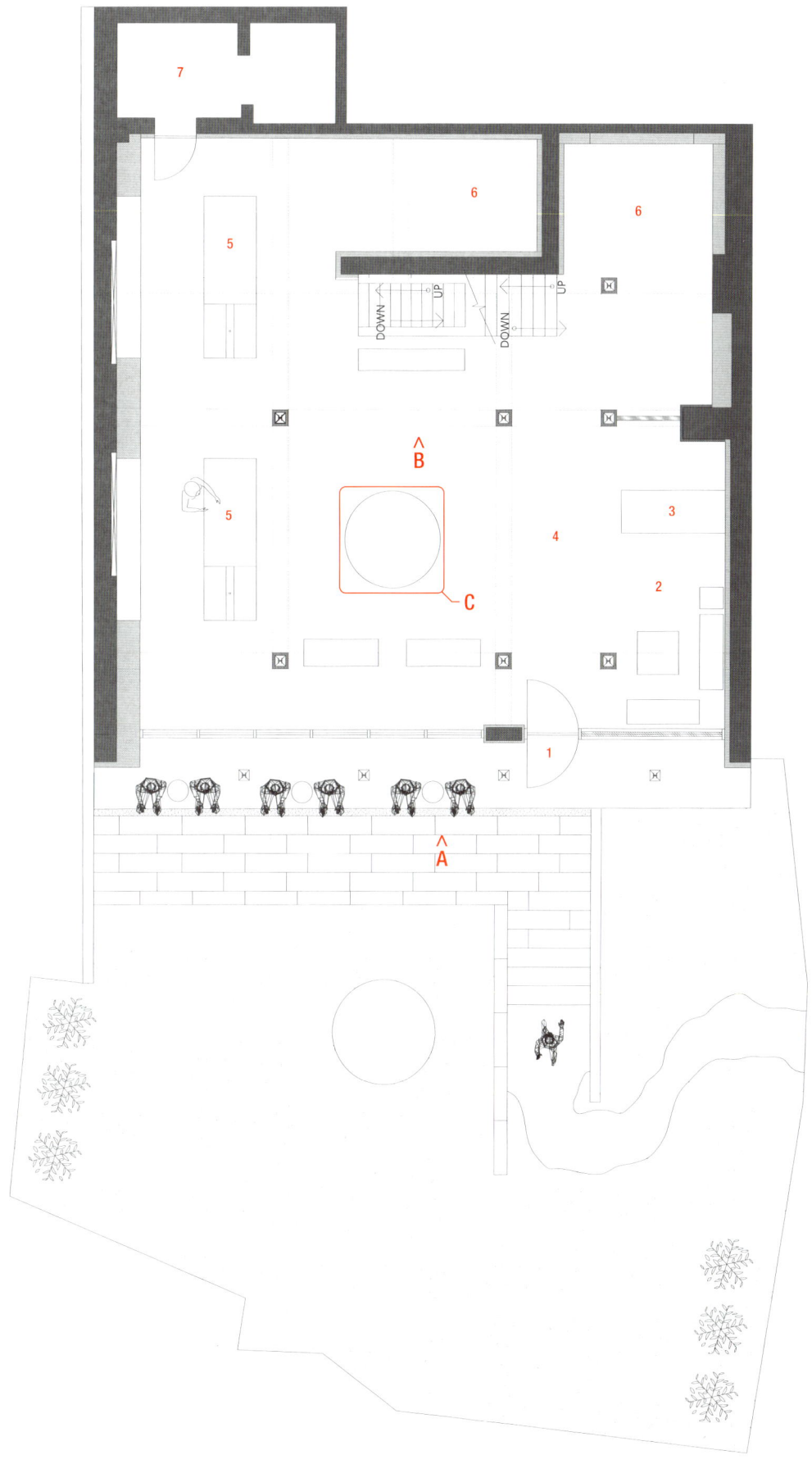

1층 평면도 / 1st floor plan

1 입구 2 라운지 3 리셉션 4 홀 5 제품 체험 존 6 전시 공간 7 창고

1 Entrance 2 Lounge 3 Reception 4 Hall 5 Product testing zone 6 Exhibition space 7 Storage

1 목재 전통문 **2** 우드 플로링(S.P.C) **3** 유리문 **4** 알루미늄 골판 패널 **5** 흙미장 **6** SUS 바이브레이션 **7** 포천석 **8** 리셉션

1 Wooden traditional door **2** Wood flooring (S.P.C) **3** Glass door **4** Aluminium corrugated panel **5** Natural clay plaster **6** SUS vibration **7** Pocheonseok **8** Reception

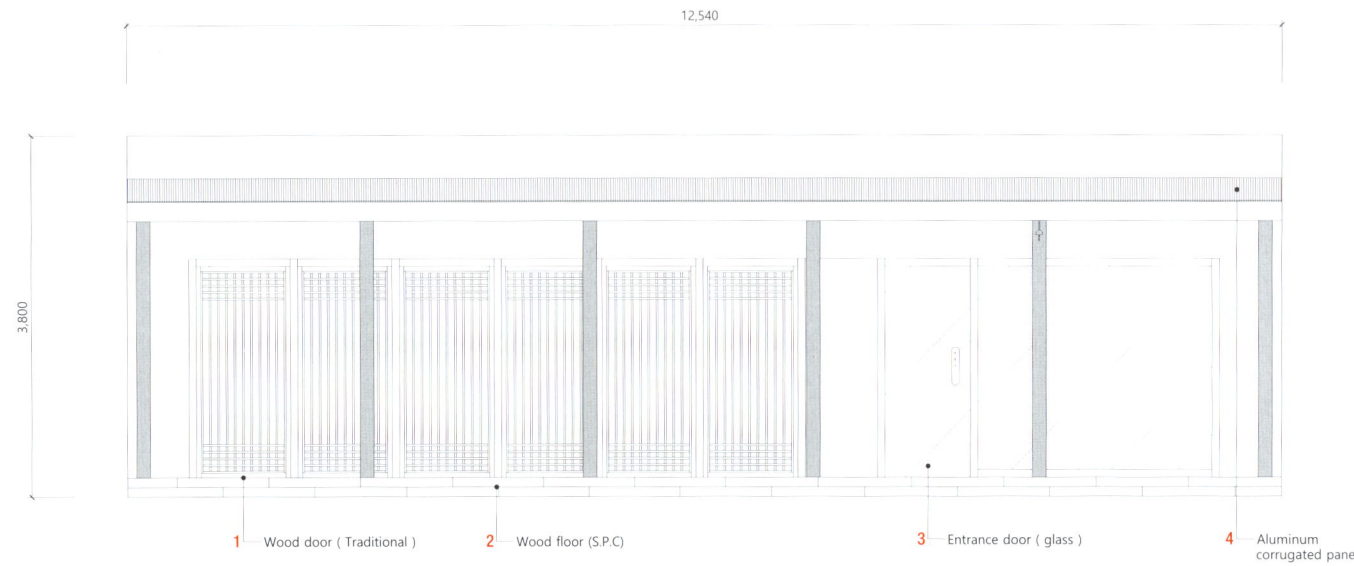

1층 홀 입면 A / 1F hall elevation A

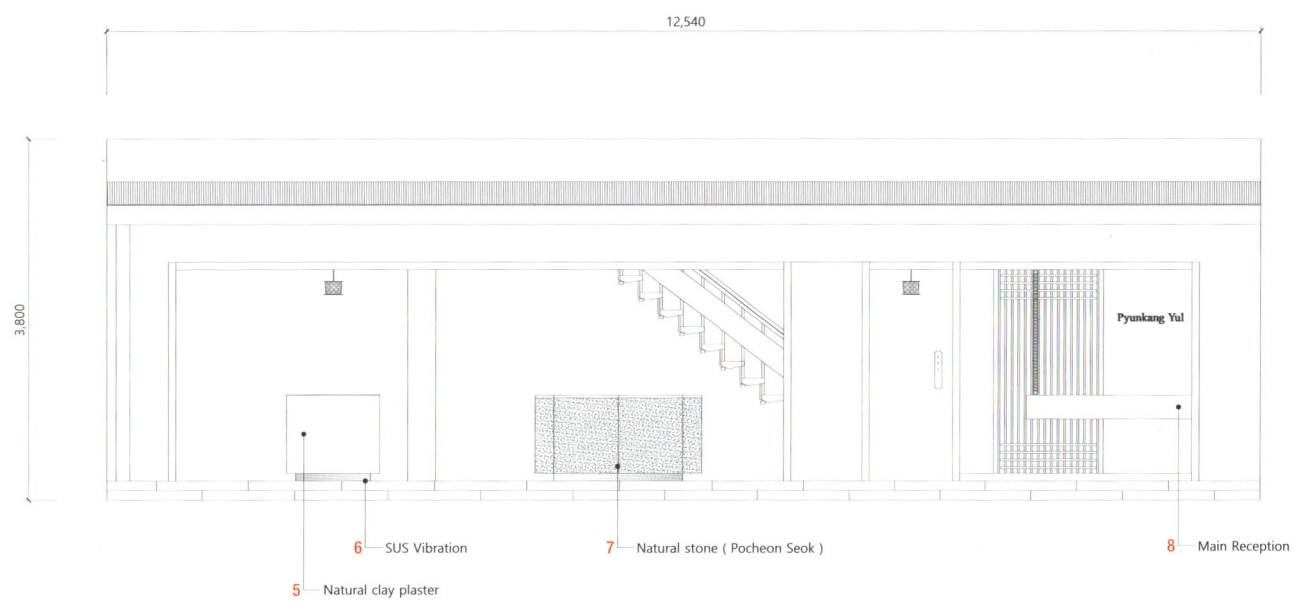

1층 홀 입면 B / 1F hall elevation B

1 배수, 수전용 Ø38 타공   **2** 스테인리스 스틸 걸레받이 1,200 pi

1 Drain and faucet perforation Ø38   **2** Stainless steel baseboard finish 1,200 pi

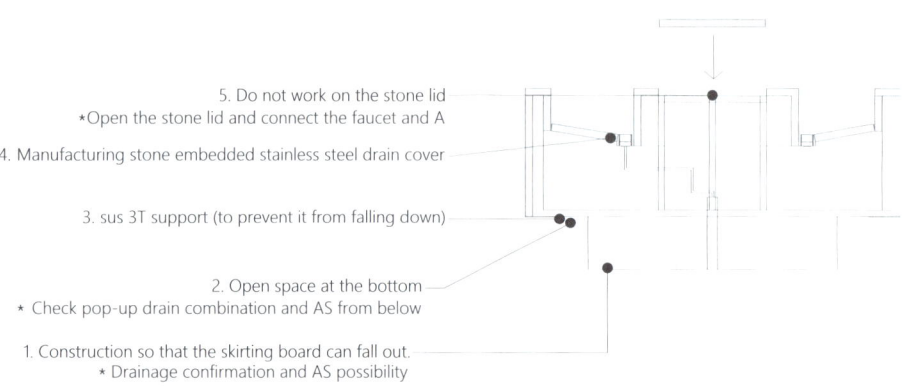

5. Do not work on the stone lid
* Open the stone lid and connect the faucet and A

4. Manufacturing stone embedded stainless steel drain cover

3. sus 3T support (to prevent it from falling down)

2. Open space at the bottom
* Check pop-up drain combination and AS from below

1. Construction so that the skirting board can fall out.
* Drainage confirmation and AS possibility

세면대 상세 / wash sink detail

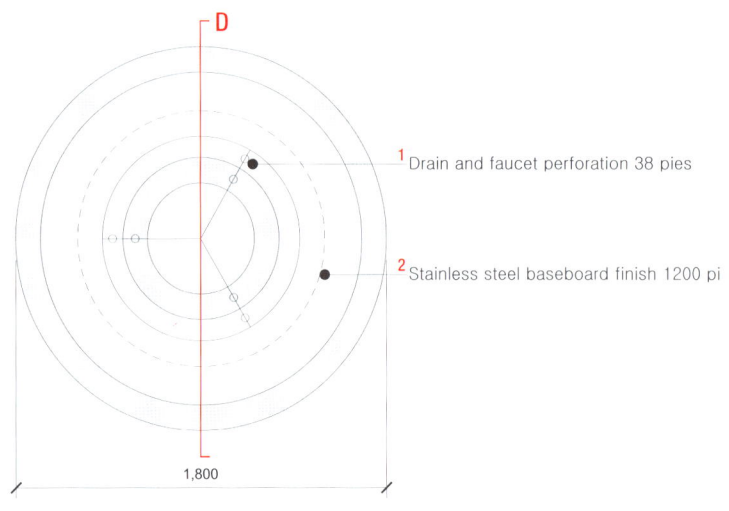

1 Drain and faucet perforation 38 pies

2 Stainless steel baseboard finish 1200 pi

세면대 평면 C / wash sink top view C

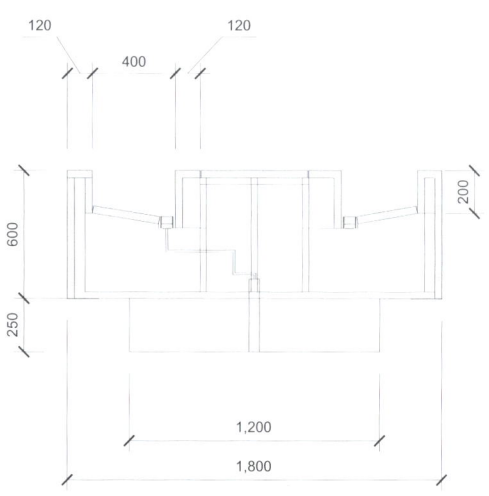

세면대 단면 D / wash sink section D

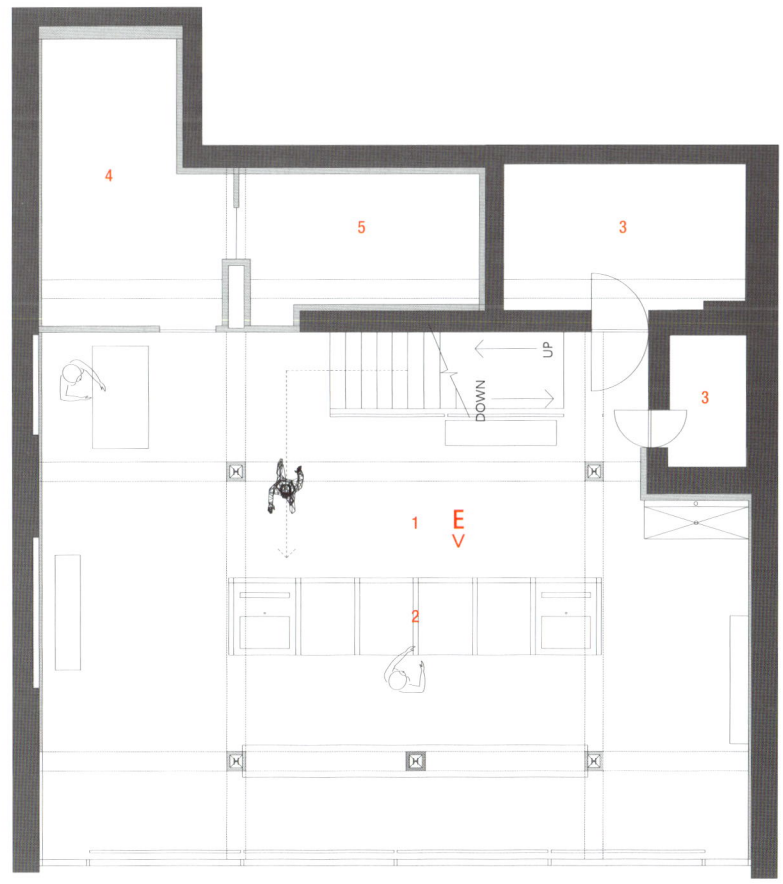

2층 평면도 / 2nd floor plan

1 홀 2 제품 체험 존 3 화장실 4 전시 공간 5 창고

1 Hall 2 Product testing zone 3 Toilet 4 Exhibition space 5 Storage

1 체리우드   2 스테인리스 스틸 폴리싱   3 포천석   4 에어컨

1 Cherry wood   2 SUS polishing   3 Pocheonseok   4 Air conditioner

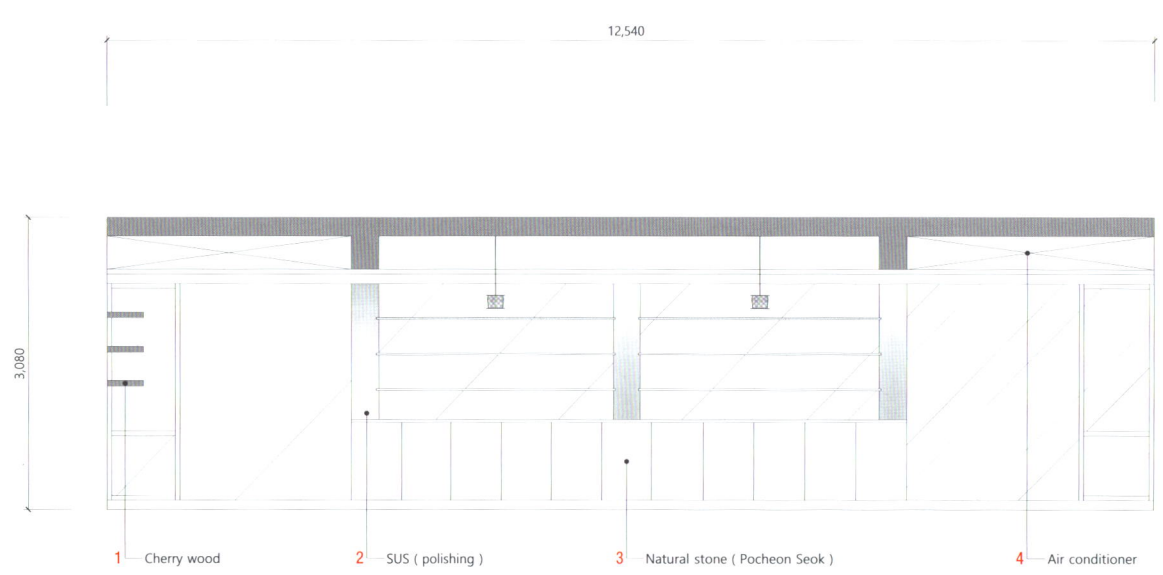

1 Cherry wood   2 SUS (polishing)   3 Natural stone (Pocheon Seok)   4 Air conditioner

2층 홀 입면 E / 2F hall elevation E

MEDICUBE FLAGSHIP STORE HONGDAE

FLYMINGO | Junho Kim, Jiyeon Hwang

지난 5월 홍대에 메디큐브의 첫 플래그십 스토어가 오픈했다. 첫 디자인 미팅에서 메디큐브측은 안티에이징에 효과가 큰 기능성 기초 제품 라인 출시에 맞춰 매장에서도 이러한 제품력이 한눈에 보였으면 좋겠다는 요청을 했다. 디자이너는 이와 관련된 수많은 키워드들을 도출해내던 도중, 광채가 나며 탄력 있는 피부와 닮아 있는 젤리'라는 키워드를 공간의 콘셉트로 내세우며 안티에이징 기능성을 강조한 매장을 구성하게 되었다. 파사드를 구성하는 둥근 모서리의 큐브들은 모듈화되어 플라스틱 성형으로 제작 후 핑크빛의 유광 페인트를 도장하여 만지면 말랑하고 탱글할 것 같은 젤리 질감의 모습을 구현해냈다. 공간은 전체적으로 화이트와 핑크 톤으로 구성되어 클린하고 통통 튀는 생기 있는 분위기를 자아낸다. 내부 공간의 집기 또한 모서리를 둥글게 굴리는 디테일과 유광 도장 마감을 보여주며 공간이 전체적으로 콘셉트에 충실하도록 연출했다. 그리고 제품을 자유롭게 테스트해볼 수 있도록 원형의 테이블을 중앙에 배치하였다. 또한 중앙 테이블과 어우러지도록 젤리 곰 캐릭터 조형물과 큐브를 함께 배치하여 제품이 자연스럽게 노출되는 포토존을 구성했다.

Medicube opened its first flagship store in Hongdae last May. During the initial design meeting, the client expressed their plans to launch functional skincare products with strong anti-aging effects and wanted the new store to showcase these products in a way that would catch customers' attention at first glance. After researching numerous related keywords, the designer chose 'jelly' as the space's key concept, drawing parallels between jelly and glowing, elastic skin, and designed a space that emphasizes anti-aging functionality. The façade's cubes with rounded corners are designed in modular system. They are plastic-molded and finished with a glossy pink coating to create a jelly-like texture that appears soft and bouncy to the touch. Overall, the space features white and pink tones, creating a neat and vibrant atmosphere. The interior fixtures also incorporate rounded corners and glossy finishes to maintain organic connection with the design concept throughout the space. A circular table is installed in the center to allow customers to freely test products. Additionally, gummy bear character sculptures and cubes are placed together with the central table, forming a photo zone that naturally exposes the products.

디자인 김준호, 황지연 / 플라이밍고
위치 서울특별시 마포구 홍익로6길 27
용도 플래그십 스토어
면적 76.7m²
마감 바닥 – 카펫 / 벽 – 페인트 / 천장 – 페인트
완공 2024. 5
디자인팀 고수민, 박서연, 강민주, 이해니
사진 최용준

Location 27, Hongik-ro 6-gil, Mapo-gu, Seoul
Use Flagship store
Area 76.7m²
Finishing Floor - Carpet / Wall - Paint / Ceiling - Paint
Completion 2024. 5
Photographer Yongjoon Choi

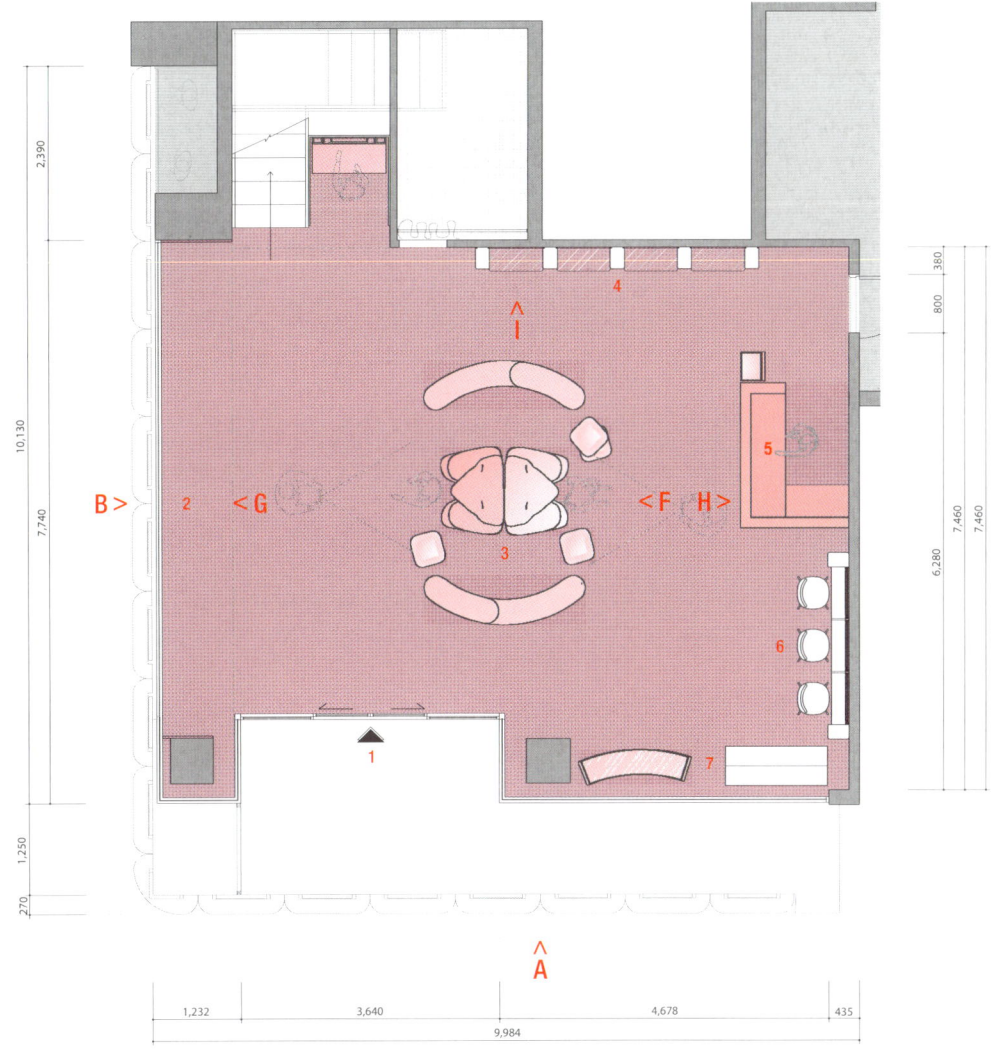

평면도 / floor plan

1 입구 2 쇼윈도 존 3 포토 존 4 스킨케어 존 5 카운터 6 테스터 존 7 프로모션 쇼윈도 존

1 Entrance 2 Show window zone 3 Photo zone 4 Skincare zone 5 Counter 6 Tester zone 7 Promotion show window zone

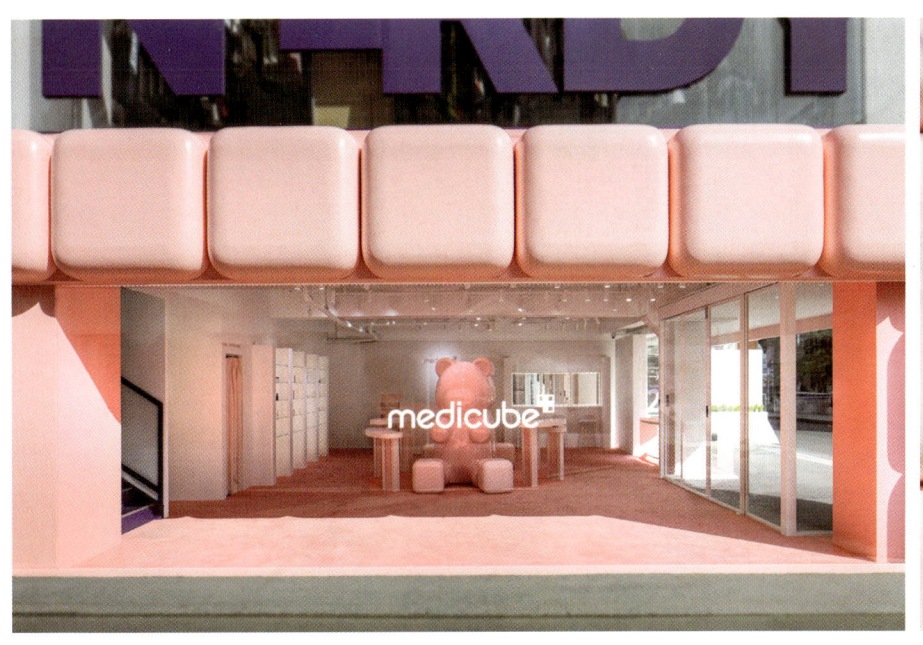

1 지정 분홍색 도장 2 조명형 수지 채널 5,000K 3 95인치 모니터 / 지정 흰색 도장 4 로고 흰색 시트 커팅 5 □50X50 파이프 6 지정 연분홍색 도장 7 지정 연분홍색 도장 / 지정 분홍색 도장 8 T8 투명 유리 9 T5 지정 프로스트 아크릴 / 지정 연분홍색 도장

1 App. pink painting 2 Illuminated resin channel 5,000K 3 95" monitor / App. white painting 4 Logo white sheet cutting 5 □50X50 pipe 6 App. light pink painting 7 App. light pink painting / App. pink painting 8 T8 clear glass 9 T5 app. frost acrylic / App. light pink painting

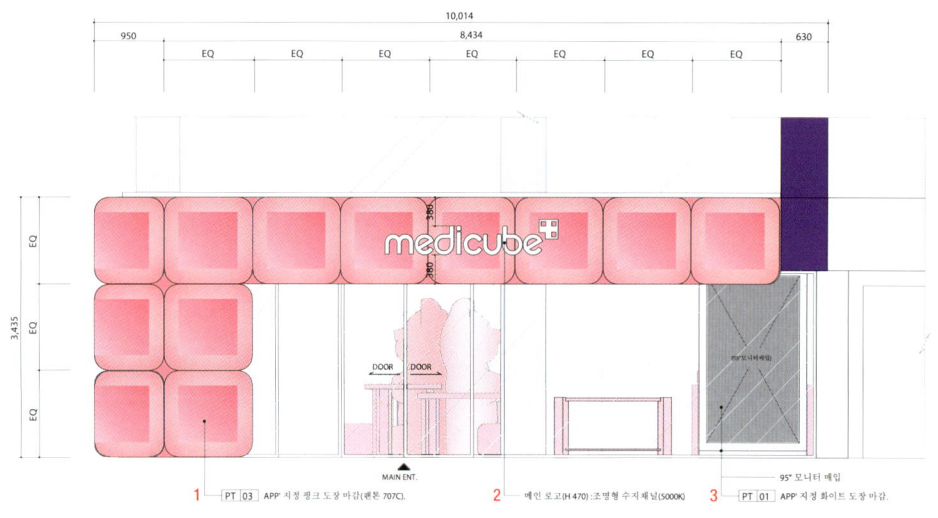

파사드 A / facade A

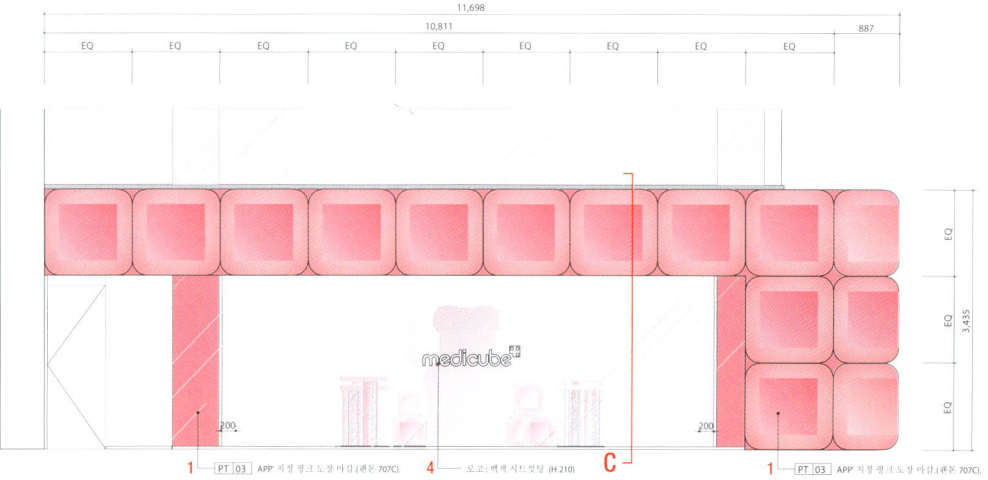

파사드 B / facade B

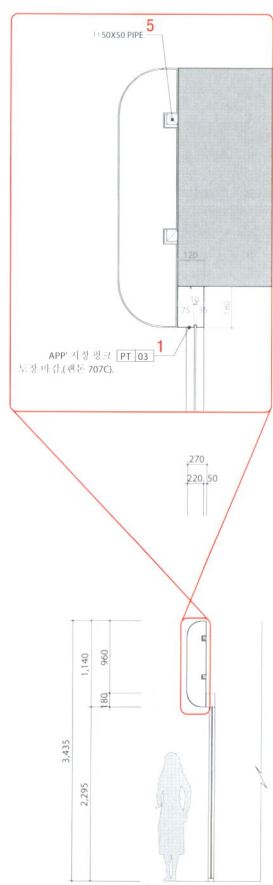

파사드 단면 C / facade section C

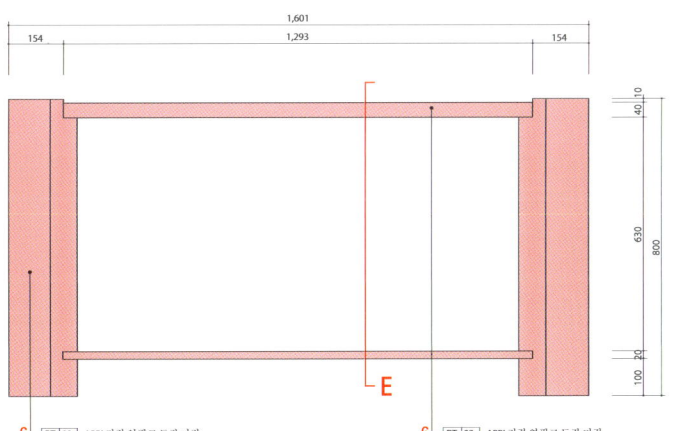

테이블 정면 D / table front view D

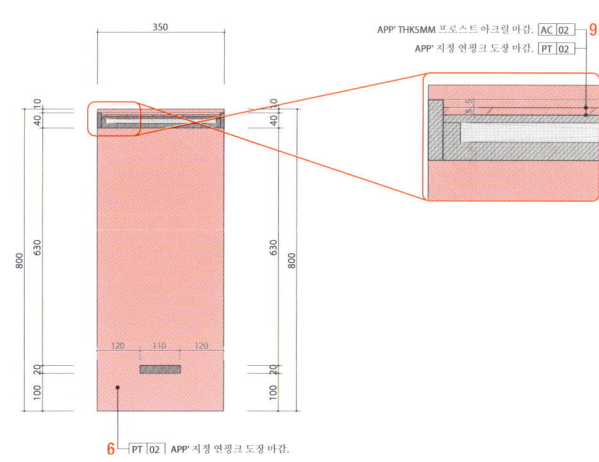

테이블 단면 E / table section E

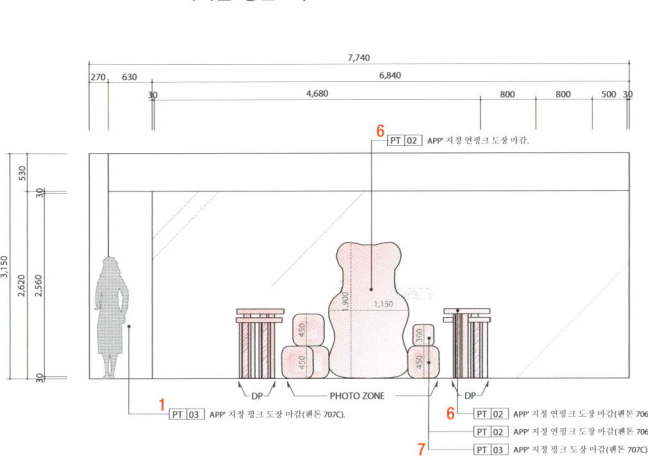

홀 입면 F / hall elevation F

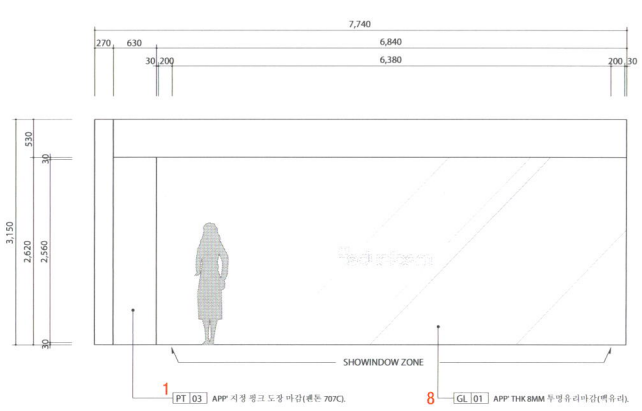

홀 입면 G / hall elevation G

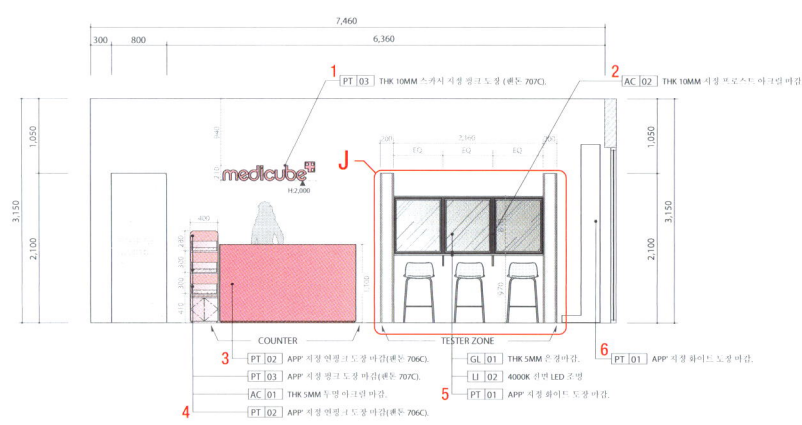

홀 입면 H / hall elevation H

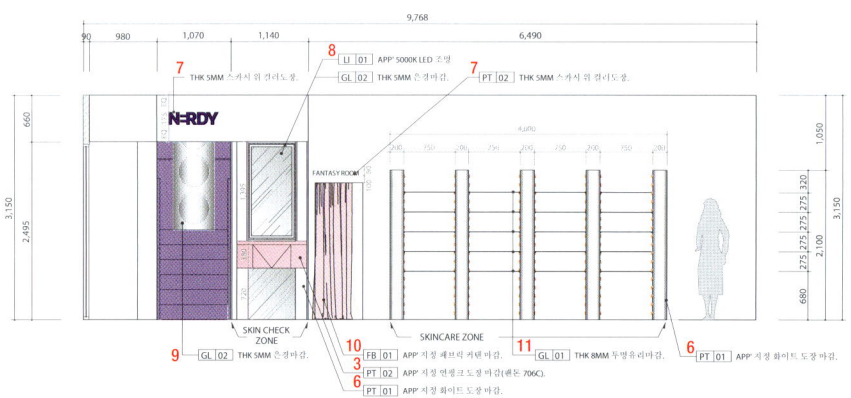

홀 입면 I / hall elevation I

1 T10 문자 커팅 위 지정 분홍색 도장 2 T10 지정 프로스트 아크릴 3 지정 연분홍색 도장 4 지정 분홍색 도장 / T5 투명 아크릴 / 지정 연분홍색 도장 5 T5 은경 / 4,000K 전면 LED 조명 / 지정 흰색 도장 6 지정 흰색 도장 7 T5 문자 커팅 위 지정색 도장 8 지정 5,000K LED 조명 / T5 은경 9 T5 은경 10 지정 패브릭 커튼 11 T8 투명 유리 12 지정 흰색 우레탄 도장 13 T5 애칭 미러 14 T3 확산 PC / T5 애칭 미러 15 지정 4,000K LED 조명 16 T2 갈바륨 / T5 애칭 미러 17 자석

1 App. pink painting on T10 letter cutting 2 T10 app. frost acrylic 3 App. light pink painting 4 App. pink painting / T5 clear acrylic / App. light pink painting 5 T5 mirror / 4,000K front iluminated LED light / App. white painting 6 App. white painting 7 App. color painting on T5 letter cutting 8 App. 5,000K LED light / T5 mirror 9 T5 mirror 10 App. fabric curtain 11 T8 clear glass 12 App. white urethane painting 13 T5 etched mirror 14 T3 PC diffusion plate / T5 etched mirror 15 App. 4,000K LED light 16 T2 galvalume / T5 etched mirror 17 Magnet

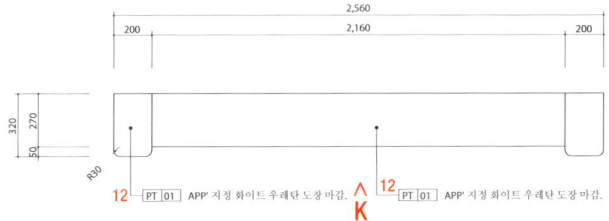

화장대 평면 J / makeup table top view J

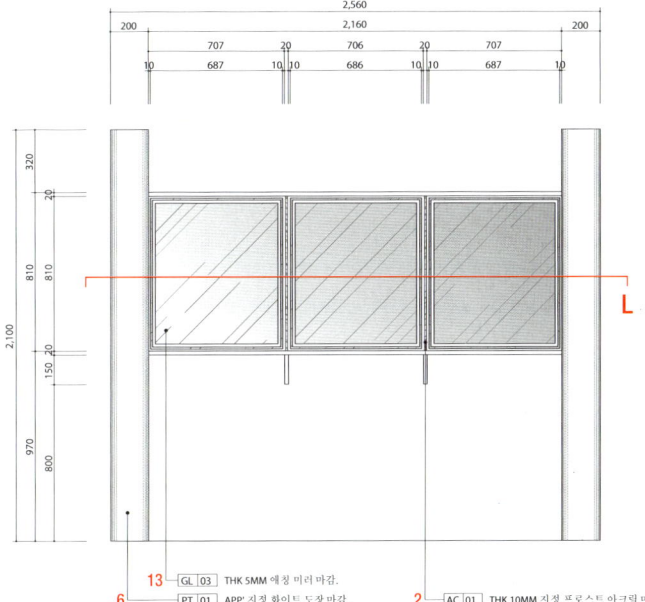

화장대 정면 K / makeup table front view K

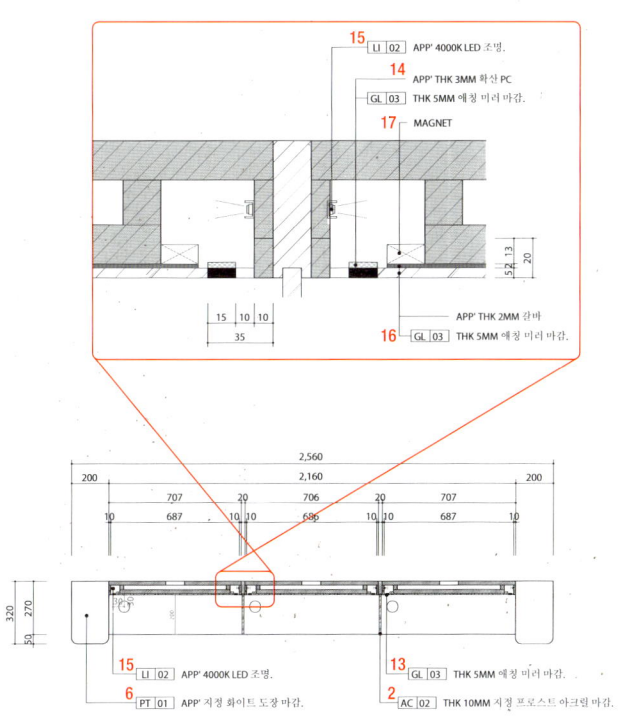

화장대 단면 L / makeup table section L

JAVIN DE SEOUL

MY NAME IS JOHN | Woonam Lee, Dongwook Shin

자빈드서울의 첫 플래그십 스토어는 미니멀하고 뉴트럴한 무드를 중심으로 브랜드의 철학과 가치를 공간 디자인에 최대한 구현하기 위해 세심하게 설계되었다. 자빈드서울은 뷰티 코스메틱 브랜드로서, 제품의 품질뿐만 아니라 그 제품이 전시되는 공간의 감각적 경험을 중요시한다. 따라서 디자이너는 공간의 조도와 조명 컬러에 특별히 신경을 기울였고, 공간의 구조는 직선과 사선을 활용하여 깔끔하고 세련된 느낌을 강조했다. 높은 천장고의 장점을 극대화하기 위해 천장으로부터 커튼을 행잉하여 시각적으로 공간을 확장시키고, 자연스러운 흐름을 만들어냈다. 이러한 설계는 공간에 들어서는 순간 방문자가 자빈드서울이 지향하는 미니멀리즘과 뉴트럴한 감성을 즉각적으로 느낄 수 있도록 돕는다. 특히 주목할 만한 부분은 자빈드서울의 인기 제품군인 '윙크 시리즈'에서 영감을 받아 디자인된 초승달 모양의 가구인데, 이들은 브랜드의 상징성을 담아내며 공간의 독특한 포인트로 작용한다. 블랙과 화이트를 기본으로 한 컬러 팔레트 역시 브랜드의 미니멀한 감성을 반영한다. 블랙은 흑단 마루를 통해 깊이감과 고급스러움을 더했고, 화이트는 가구, 패브릭, 거울 등을 통해 밝고 깨끗한 느낌을 전달한다. 공간의 또 다른 매력 포인트는 렌티큘러 기법을 활용해 표현한 자빈드서울의 시그니처 '윙크' 작품이다. 이 팝아트적 요소는 공간에 활기를 불어넣으며, 방문자들에게 감각적 즐거움을 선사한다. 이곳에서 방문자들은 공간, 가구, 제품을 자유롭게 즐기고, 각자의 감각과 취향에 따라 다양한 경험을 누릴 수 있다.

Javin De Seoul's first flagship store is carefully designed to express the brand's philosophy and values through its space design, centered on a minimal and neutral tone and manner. As a beauty cosmetics brand, Javin De Seoul values not only product quality but also the sensory experience of the space where products are displayed. Accordingly, the designer paid special attention to lighting levels and colors, while primarily using straight and diagonal lines in the spatial structure to convey a neat and sophisticated feel. Curtains are suspended from the ceiling to maximize the advantage of high ceilings. They create a visual expansion of space while ensuring natural spatial flow. This design element helps visitors instantly experience Javin De Seoul's minimalist and neutral sensibility the moment they enter the space. The most notable feature is the crescent moon-shaped furniture inspired by 'Wink Series', which is one of the popular products by Javin De Seoul. It serves as a unique spatial element that embodies the symbolism of the brand. The black-and-white color palette also reflects the brand's minimalist aesthetic. Black ebony flooring adds depth and luxury, while white elements in furniture, fabrics, and mirrors create a bright and pure atmosphere. Another highlight of the space is 'Wink', a signature piece by Javin De Seoul, rendered using lenticular techniques. This pop art object enlivens the space and offers visitors a sensory delight. Here, visitors can freely explore the space, furniture, and products and enjoy a variety of experiences tailored to their individual senses and tastes.

디자인 마이네임이즈존 / 이우남, 신동욱
위치 서울특별시 용산구 이태원로55가길 21, 4층
용도 플래그십 스토어
면적 100m²
마감 바닥 – 원목마루 / 벽, 천장 – 노출콘크리트
완공 2024. 3
디자인팀 김영현, 최창수, Manon Ary
사진 이해란

Location 4F, 21, Itaewon-ro 55ga-gil, Yongsan-gu, Seoul
Use Flagship store
Area 100m²
Finishing Floor - Wood flooring / Wall, Ceiling - Exposed concrete
Completion 2024. 3
Photographer Haeran Lee

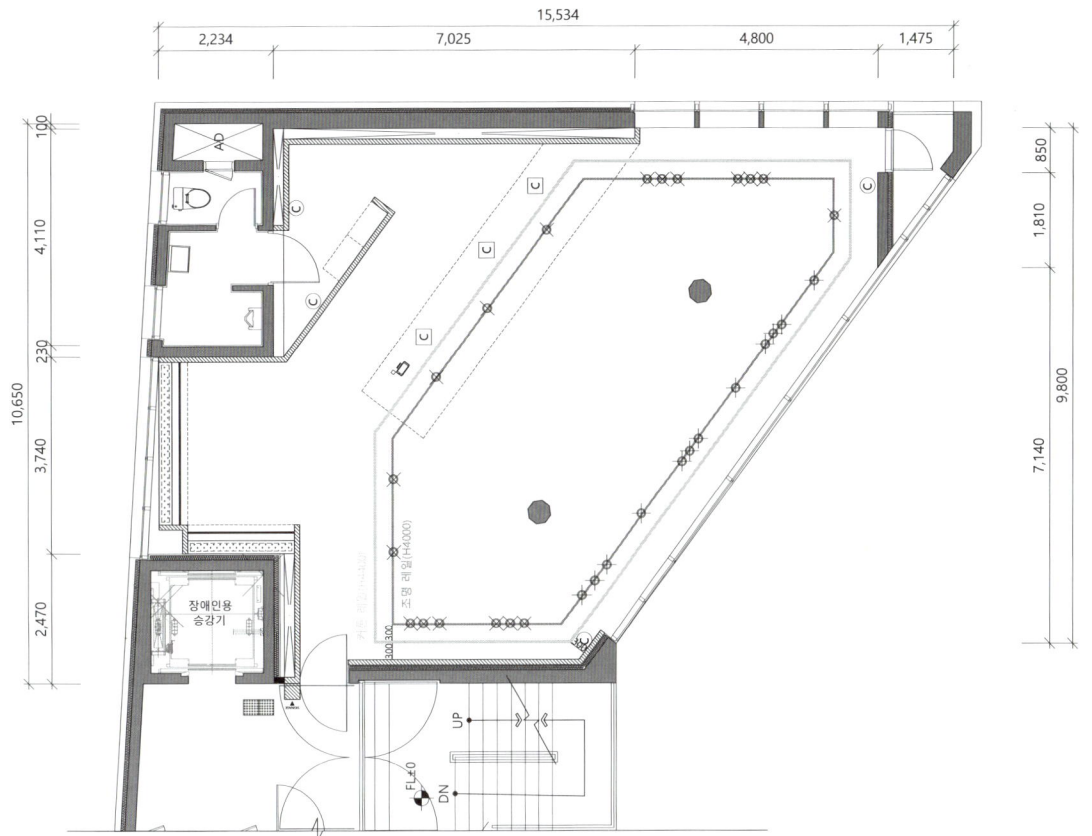

천장도 / ceiling plan

평면도 / floor plan

1 입구 **2** 홀 **3** 카운터 **4** 창고 **5** 쇼룸 **6** 피난 발코니 **7** 테스터 존 **8** 화장실

1 Entrance **2** Hall **3** Counter **4** Storage **5** Showroom **6** Fire escape balcony **7** Tester zone **8** Restroom

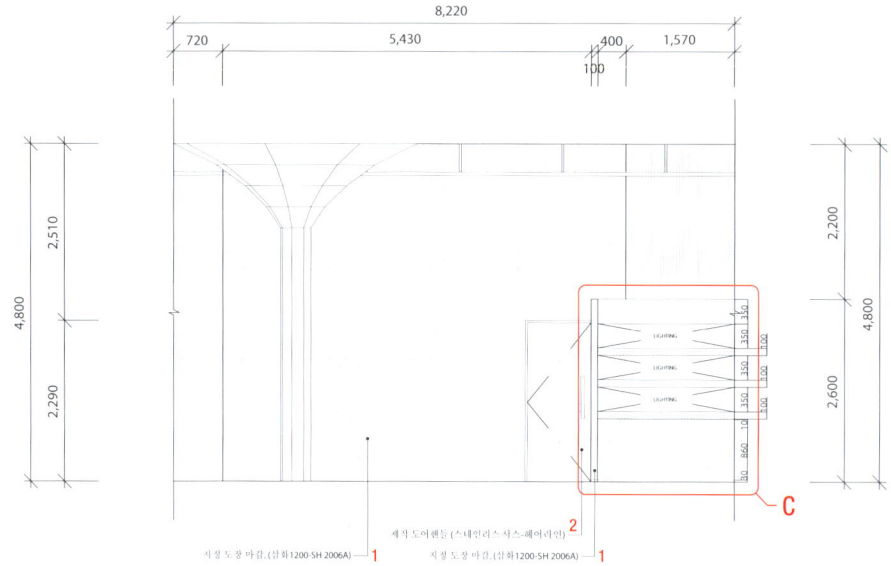

홀 입면 A / hall elevation A

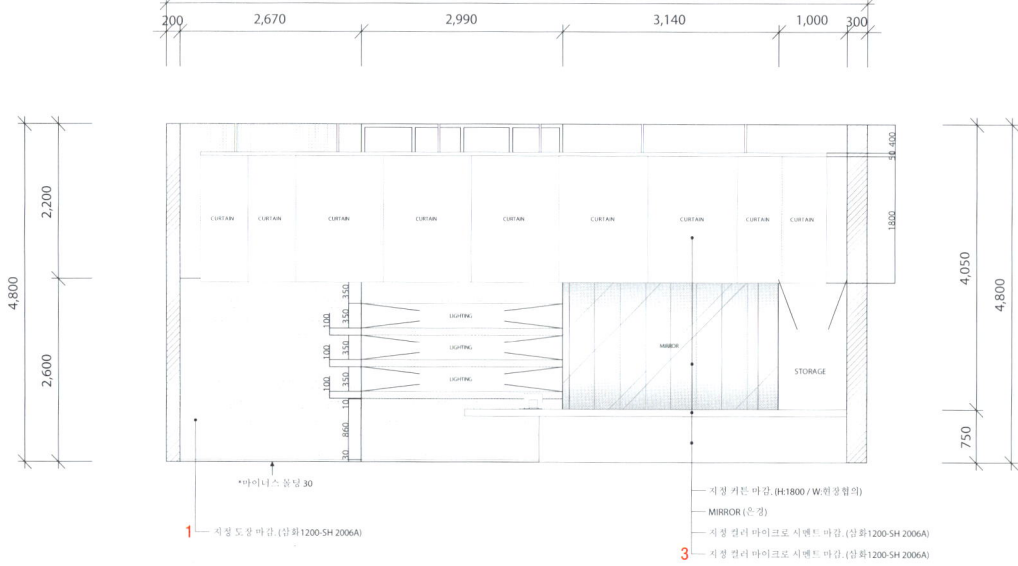

홀 입면 B / hall elevation B

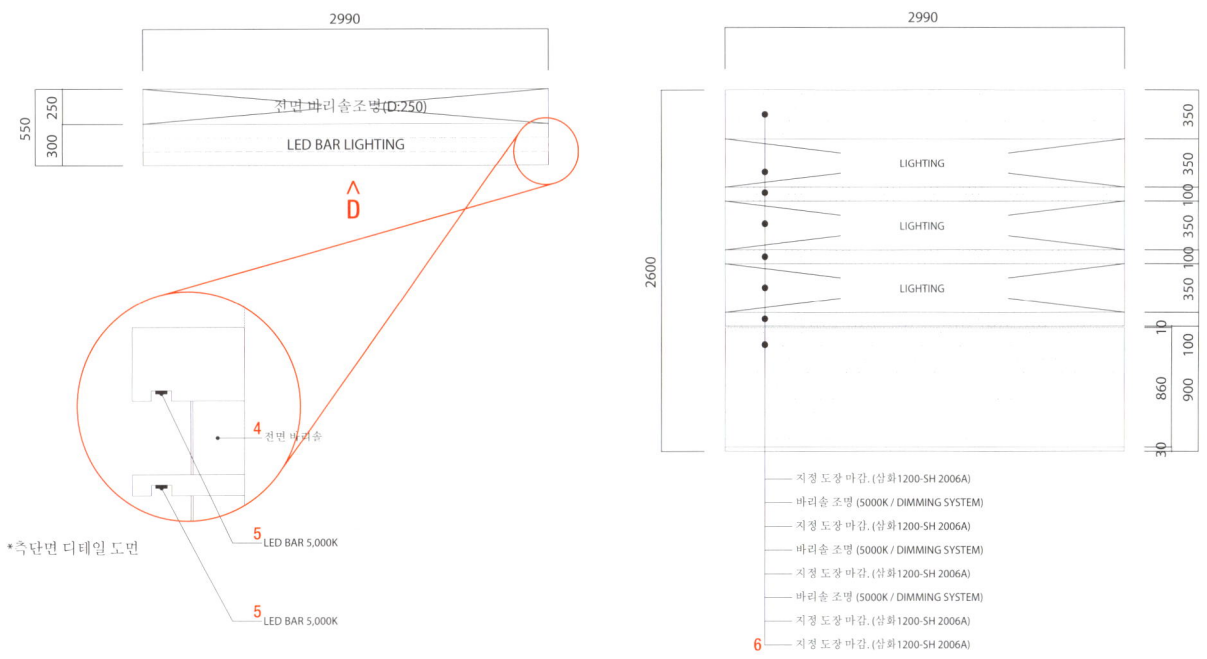

1 지정 도장 2 제작 손잡이 3 지정 커튼 / 거울 / 지정색 마이크로 시멘트 / 지정색 마이크로 시멘트 4 바리솔 5 LED바 5,000K 6 지정 도장 / 바리솔 조명 / 지정 도장 / 바리솔 조명 / 지정 도장 / 바리솔 조명 / 지정 도장 / 지정 도장

1 App. painting 2 Custom-made door handle 3 App. curtain / Mirror / App. color microcement / App. color microcement 4 Barrisol 5 LED bar 5,000K 6 App. painting / Barrisol / App. painting / Barrisol / App. painting / Barrisol / App. painting / App. painting

가구 평면 C / furniture top view C

가구 정면 D / furniture front view D

277

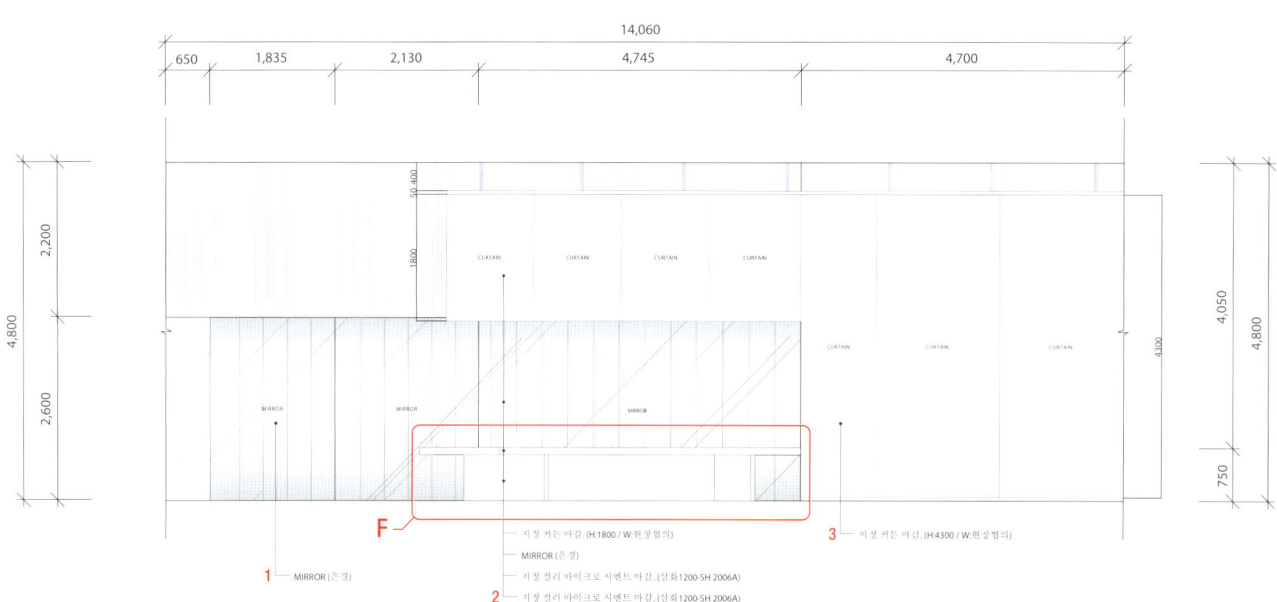

홀 입면 E / hall elevation E

1 거울 2 지정 커튼 / 거울 / 지정색 마이크로 시멘트 / 지정색 마이크로 시멘트 3 지정 커튼 4 지정색 마이크로 시멘트 / 지정색 마이크로 시멘트 5 지정 파티클 보드 가구 위 필름 6 지정색 마이크로 시멘트

1 Mirror 2 App. curtain / Mirror / App. color microcement / App. color microcement 3 App. curtain 4 App. color microcement / App. color microcement 5 Film on particle board furniture 6 App. color microcement

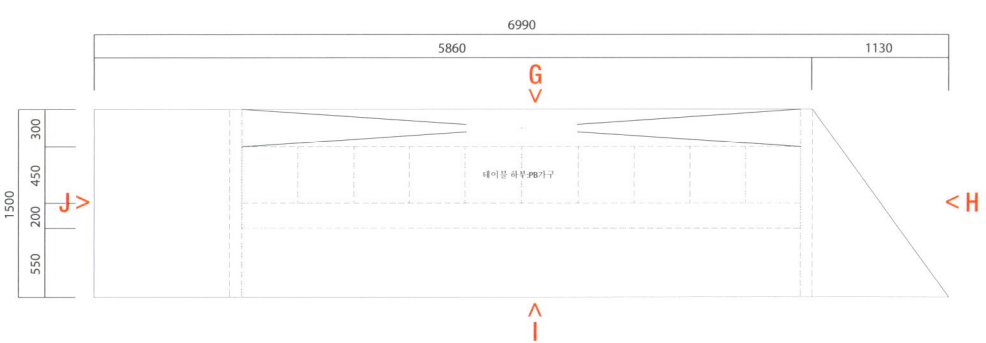

가구 평면 F / furniture top view F

가구 후면 G / furniture rear view G

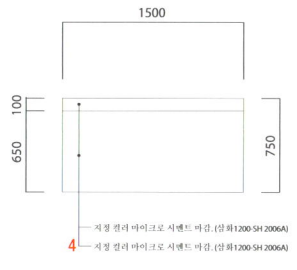

가구 측면 H / furniture side view H

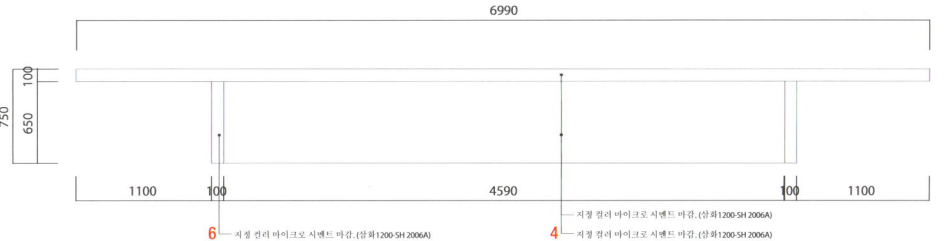

가구 정면 I / furniture front view I

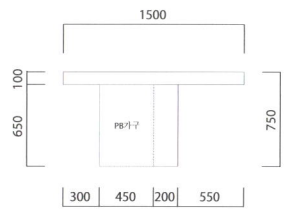

가구 측면 J / furniture side view J

INFORMATION

레스토랑

테스트 키친 | 헤드어반스튜디오 | 서울특별시 성동구 성수일로8길 5 서울숲 SK V1타워 A동 510호 | 02-3443-6300

이목 스모크 다이닝 | 프로젝트 마크 | 서울특별시 마포구 성지길 51, 3층 | 02-6261-1007

오발탄 삼성점 | 인투익스 | 서울특별시 성동구 성수일로 12길 26, 5층 | 02-543-3881

치즈웨이브 | 바이석비석 | 서울특별시 원효로 89길 13-9, 2층 | 02-792-0107

스팅키 베이컨 트럭 | 프로젝트 마크 | 서울특별시 마포구 성지길 51, 3층 | 02-6261-1007

덕분 한식당 | 플라이밍고 | 서울특별시 성동구 서울숲길 51, 8층 | 02-3409-2601

레인보우 브릿지 광교점 | 인투익스 | 서울특별시 성동구 성수일로 12길 26, 5층 | 02-543-3881

해남천일관 | 라보토리 | 서울특별시 용산구 대사관로31길 7-2, 5층 | 02-798-3778

비킹후스 | 디자인바이팔삼 | 부산광역시 수영구 남천바다로 9번길 18, 2층 | 051-802-8283

바리에가타 그로타 | 프로젝트 마크 | 서울특별시 마포구 성지길 51, 3층 | 02-6261-1007

카페

세트 | 디자인스튜디오 마움 | 서울특별시 마포구 토정로 24-9 | 02-523-3013

노티드 김포 | 서브텍스트 | 서울특별시 성동구 동일로 307, 2층 | 02-516-3639

인크커피 | 헤드어반스튜디오 | 서울특별시 성동구 성수일로8길 5 서울숲 SK V1타워 A동 510호 | 02-3443-6300

세컨드원 레이크 아산 | 다나함 어소시에이트 | 서울특별시 강남구 언주로174길 13 | 02-3445-2030

오베뉴 한남 | 셰르파 스튜디오 | work.in.sherpa@gmail.com | 010-2542-5118

레망도레 | 스토프 | 서울특별시 용산구 유엔빌리지길 254, 4층 | 02-322-3433

아이소 사운드 | 스토프 | 서울특별시 용산구 유엔빌리지길 254, 4층 | 02-322-3433

호롱 | 디자인스튜디오 마움 | 서울특별시 마포구 토정로 24-9 | 02-523-3013

필메이트 | 라보토리 | 서울특별시 용산구 대사관로31길 7-2, 5층 | 02-798-3778

수수커피 | 프로젝트 마크 | 서울특별시 마포구 성지길 51, 3층 | 02-6261-1007

병원 · 종교

오가나셀 피부과 의원 잠실점 | 인투익스 | 서울특별시 성동구 성수일로 12길 26, 5층 | 02-543-3881

셀린의원 홍대점 | 인투익스 | 서울특별시 성동구 성수일로 12길 26, 5층 | 02-543-3881

제일소망교회 | 다나함 어소시에이트 | 서울특별시 강남구 언주로174길 13 | 02-3445-2030

플래그십 스토어

인사일런스 성수 | 디자인바이팔삼 | 부산광역시 수영구 남천바다로 9번길 18 2층 | 051-802-8283

편강 율 플래그십 스토어 | 셰르파 스튜디오 | work.in.sherpa@gmail.com | 010-2542-5118

메디큐브 플래그십 스토어 홍대 | 플라이밍고 | 서울특별시 성동구 서울숲길 51, 8층 | 02-3409-2601

자빈드서울 | 마이네임이즈존 | 서울특별시 용산구 유엔빌리지길 1, 헤드빌딩 3층 | 02-2282-1056